EDUCATING THE SONS OF SUGAR

EDUCATING THE SONS OF SUGAR

Jefferson College and the Creole Planter Class of South Louisiana

R. Eric Platt

The University of Alabama Press
Tuscaloosa

The University of Alabama Press
Tuscaloosa, Alabama 35487-0380
uapress.ua.edu

Hardcover edition published 2017.
Paperback edition published 2019.
eBook edition published 2017.

Typeface: Caslon and Scala Sans

Cover image: (*Above*) Jefferson College student group, c. 1896–1897; (*Below*) Jefferson College Aerial, c. 1870. General Photograph Collection, Marist Houses and Communities, Jefferson College, Cabinet 189, Drawer 1. Images courtesy of the Archives of the Society of Mary, Atlanta, GA.
Cover design: Todd Lape / Lape Designs

Paperback ISBN: 978-0-8173-5951-5

A previous edition of this book has been catalogued by the Library of Congress as follows:
Library of Congress Cataloging-in-Publication Data
Names: Platt, R. Eric, 1981– author.
Title: Educating the sons of sugar : Jefferson College and the Creole planter class of South Louisiana / R. Eric Platt.
Description: Tuscaloosa : The University of Alabama Press, [2017] | Includes bibliographical references and index.
Identifiers: LCCN 2017006654| ISBN 9780817319663 (cloth) | ISBN 9780817391515 (e book)
Subjects: LCSH: Jefferson College (Convent, La.)—History. | Education, Higher—Louisiana—History. | Sugar growing—Louisiana—History—19th century. | Creoles—Louisiana—History—19th century. | Plantation owners—Louisiana—History—19th century. | Elite (Social sciences)—Louisiana—History—19th century.
Classification: LCC LD2601.J42 P53 2017 | DDC 378.763/31—dc23
LC record available at https://lccn.loc.gov/2017006654

For my parents, Robert and DeWanda

AMDG

How much better to get wisdom than gold,
to choose understanding rather than silver!
—Proverbs 16:16

JEFFERSON COLLEGE

Where runs the Mississippi with a song,
And winds the road along the country side,
'Tis here a college stood in strength sublime;
And still sheds its glory o'er the ceaseless tide.
Where ancient trees still lisp forgotten lore,
And scatter shade upon the silent way,
Louisiana's youth found wisdom's store;
And many of her sons its name display.
Long has it lived, e'en through the battle's roar;
While glories of the years have come and fled;
Yet lifts its head in grandeur as of yore,
And blesses those of hers who died and bled.
Yes, Jefferson, like patriot of old,
Withstands the gnawing tooth and touch of time;
Its precious gifts it still will here unfold;
It lifts its hands on high with faith sublime.
Then, bless these college walls known through this land,
And may our sons find welcome in its gate;
May heaven guide it on in teachings grand—
To future ages all its joys relate!

LaPlace, La., Sept. 15, 1923.
—Rixford J. Lincoln

Contents

List of Figures ix

Acknowledgments xi

Introduction: A Creole College in St. James Parish 1

1. Aristocracy, Education, and the Ancienne Population 14

2. The Rise and Fall of Jefferson College 39

3. The Forgotten Louis Dufau 68

4. Return of the Sugar Barons 94

5. Marists and Americanization 135

Conclusion: Class and College 168

Epilogue: Manresa, the Fifth Life of Jefferson College 179

Appendix A: Institutional Presidents at the Jefferson College Site 185

Appendix B: Acts to Incorporate and Support Jefferson College 187

Appendix C: Sale of Jefferson College to Louis Dufau 191

Appendix D: Property Deed: Valcour Aime to the Jefferson College Board of Directors 195

Appendix E: Acts of Transfer: The Jefferson College Board of Directors and Valcour Aime to the Society of Mary 197

Appendix F: An Act to Incorporate the Society of "The Fathers of the Society of Mary" 201

Appendix G: Sale of St. Mary's Jefferson College to the Society of Jesus 205

Notes 209

Bibliography 265

Index 285

Figures

Figure 1. Jefferson College, c. 1870 126

Figure 2. "Governor A. B. Roman Portrait," c. 1850s 127

Figure 3. "Louis Dufau to Joseph Copes," August 23, 1856 128

Figure 4. "Portrait of Francois Gabriel (Valcour) Aime," c. 1838 129

Figure 5. "Sacred Heart of Jesus Chapel," c. 1924 130

Figure 6. "Rev. Henri Bellanger, S.M.," undated 131

Figure 7. "St. Mary's Hall," c. 1905 131

Figure 8. "Rev. James Blenk, S.M., and Marist Faculty," c. 1890s 132

Figure 9. "Jefferson College Brass Band," c. 1905–1906 133

Figure 10. "Jefferson College Football Team," c. 1920 133

Figure 11. "Blenk Science Hall and Oak Ave, Jefferson College, Convent, La.," c. 1910s 134

Acknowledgments

The writing of this book could never have come to pass without the help of several outstanding individuals, all of whom deserve heartfelt thanks. First and foremost, I would like to offer profound gratitude to my devoted parents, Robert and DeWanda Platt. Often my parents have queried about my obsession with all things New Orleans and south Louisiana. My response has always been that Louisiana is a world entirely separate from the American South. It lives. It breathes. It has a life and allure all its own. For their love, forbearance, unending support, and ability to find value in every task, great or small, I offer the love and devotion that only a son can give. With immense joy, I dedicate this book to them. To my sister, Robin, her husband, Kent, and their children, Holton, Hayes, and Rowyn, thank you for being a part of my life and an extension of my happiness. I also owe sincere gratefulness to my grandparents, living and deceased, for countless hugs, weekend visits, and faithful support. My family has been and remains a steadfast foundation for all of life's endeavors.

I would also like to thank several colleagues who stood behind this project. Dana Hart, at Louisiana State University, and I often engaged in detailed conversations about nineteenth-century Louisiana, the region's complex social makeup, and its early forms of higher education. Our chats over coffee and tea made for a pleasant addition to my many Louisiana escapades. A debt of thanks goes to Ray Brown at Westminster College for his collection of information on closed colleges and universities. Ray's database, *Colleges that have Closed, Merged, or changed their Names*, has yielded several leads to many defunct educational institutions that impart a wealth of insight into the history of American higher education. Gerald McKevitt, SJ, at Santa Clara University, proved instrumental in motivating the production of this text. He offered helpful suggestions and provided a much-appreciated connection to the Catholic Church and Ignatian spirituality. Sadly, Father McKevitt passed away on September 18, 2015. His legacy of scholarship remains, as do cherished memories of his brilliance, humility, and encouragement. Thanks also to Tim Murphy, director of the Manresa House of Retreats in Convent, Louisiana, for providing important supplementary material related to the restoration and preservation of former Jefferson College. Owing to such architectural safeguarding, Manresa remains one of the most picturesque religious sites on the Mississippi River.

As any researcher of closed colleges will tell you, relevant primary material is not always easy to come by. Should a college close, the fate of its archives (if such a repository even existed) is tied to the institution, and material contained therein is dispersed. Records, letters, academic catalogues, and related documents are transferred to other college or university campuses, private collections, state repositories, and local and regional libraries, or they are lost to time. Learning that the Society of Jesus purchased Jefferson College in 1931, I first consulted the Archives of the New Orleans Province of the Society of Jesus at Loyola University. There, I found the genesis documents that led to several years of data collection. Upon discovering the Manresa House of Retreats, however, I was engaged in another writing project altogether and shelved this historical undertaking. For years, I squirreled away information in notebooks referencing the locations of documents connected to the present-day spiritual retreat/former Creole college. If allowed, I took pictures or transcribed information and placed copies in binders. Years later, I opened those binders and began to assemble the unique history of a college and associated social class long since deceased.

In addition to the New Orleans Jesuit archives, I frequented several other collections and made the acquaintance of many friendly archivists. Before I parade their names, I would like to thank Joan Gaulene at Loyola University. Joan is not only an archivist; she is a dear friend. Her support largely made it possible for the publication of my first book and has substantially contributed to the completion of this text. Words alone cannot express the depth of my appreciation for her years of service to the Society of Jesus and countless scholars—of which I am delighted to be one. Susan Illis at the Provincial Archives for the Society of Mary in Atlanta, Georgia, was incredibly helpful concerning the St. Mary's Jefferson College portion of this book. The Marist archives house a variety of interesting documents, and in one place I was able to trace the educational work of the Society of

Mary in Louisiana. My thanks go to Susan, the Marist Provincial, and the Marist fathers and brothers who have spent innumerable hours preserving their history in the United States. Contributing to the annals of the Louisiana Marists, particularly Archbishop James Hubert Blenk, was Dorenda Dupont at the New Orleans Archdiocesan Office of Archives. Dorenda retrieved valuable information about Blenk, his life in Louisiana, and his rise to become a prominent regional religious leader.

Sean Benjamin and Ann Case at Tulane University's Louisiana Research Collection, Howard-Tilton Memorial Library, made available significant details connected to Louis Dufau and his educational exploits. Dufau may have had a disappointing career, but the information Sean and Ann provided was anything but. Barry Cowan at Louisiana State University's Special Collections, Hill Memorial Library, helped locate valuable curricular catalogues related to the four academic regimes of Jefferson College. Thanks to Barry I was able to more accurately recount the intellectual intricacies of an institution that shaped the minds of many Louisiana leaders. Working with Joan, Susan, Dorenda, Sean, Ann, and Barry, I constructed the core of this text. Following my investigations at these vital archives, I proceeded to locate and examine additional information housed at the University of Louisiana at Lafayette Special Collections; University of New Orleans, Earl K. Long Library; Nicholls State University, Ellander Memorial Library; Southeastern Louisiana University, Center for Southeast Louisiana Studies; Louisiana State Museum Collections Historical Center in New Orleans; the Historic New Orleans Collection, Williams Research Center; and the Archives of the University of Notre Dame, Illinois. To the archivists and staff at these well-organized collections, thank you for your time, expertise, and service. Similarly, two private "beat cops" of south Louisiana history, Glenn Falgoust and Dee Dee DiBenedetto, contributed contextual insights into the region's social environs during the antebellum and postbellum eras. Glenn and Dee Dee helped shore up the accounts of several persons described in this book, including Valcour Aime and André Bienvenu Roman.

I offer special thanks to Melandie McGee, Shamekia Woods, and Rebecca Holland at the University of Southern Mississippi for critiquing the manuscript that evolved into this text. I am just as indebted to the careful editorial scrutiny of my longtime friend Rhonda Lott at the University of Tennessee. Over the years, Rhonda has been kind enough to accompany me on various trips to New Orleans and south Louisiana for research and regional exploration. It is a blessing to have friends and colleagues with which to share one's work. For their diligent efforts and fastidious review, I wish each of them the very best for continued success in work and life. I must also pay homage to Claire Dehon at Kansas State University for scanning French terms, phrases, and interpretations for accuracy. Linguists like Claire are invaluable for the accurate translation of supporting documents. Likewise, Diane Powell was immensely helpful regarding the indexing of this volume. Diane, her husband, Joseph, at the University of the West Indies, and their daughter, Nisa, have been a welcome source of fellowship. My gratefulness also extends to Dan Waterman, Jon Berry, and the staff of the University of Alabama Press for their tireless efforts. Dan and Jon have provided a hearty education pertaining to the world of book publication and gave generously of their time to ensure this book's publication.

Penultimate thanks goes to the Honor Society of Phi Kappa Phi for funds granted through the Love of Learning Award. This endowment financed the reprinting of images found in the pages that follow. Finally, singular recognition is owed to my pup, Jolie. For years, Jolie has sat by my side as I toiled away at this book and other research ventures. She has been a loyal companion and a constant source of joy. Jolie, you are an irreplaceable addition to my life of research, travel, and adventure. Thanks for being you!

Introduction

A Creole College in St. James Parish

> Jefferson College stands today . . . calmly overlooking the mighty [Mississippi] river . . . and which would, if it could, tell . . . the story of the changing fortunes of men.
>
> —Dagmar Renshaw LeBreton

Regarding Americans' fascination with the Old South, Louisiana in particular, the historian Craig A. Bauer put it plainly: "Whether from the 'moonlight-and-magnolias' romanticized vision of plantation life as has been presented in the fictionalized media or from a study of the historical tomes written about the antebellum period, there remains a strong interest among the general public in the details of this . . . society and culture of long ago."[1] Branching out from Bauer's observation, I would add that there is a lingering curiosity surrounding the evolution of south Louisiana's River Road region, otherwise known as the "gold coast," in reference to the prosperity of antebellum sugar planters. The parishes between Baton Rouge and New Orleans were historically significant due to a litany of social and agricultural circumstances that shaped affluence, custom, and plantation life; however, planters throughout the American South differed from those in south Louisiana due to the latter's predominant French Creole background.[2] In response to such cultural consequence, authors have chronicled Louisiana's wealthiest sugar barons and described their palatial plantation homes set amid lush green fields of their cash crop, sugarcane.[3] Fronting the Mississippi River, these plantations and their inhabitants witnessed the mechanization of sugar production; the transformation of river traffic from sailing vessels to steam ships; the cruelty of slave labor and sharecropping; social class rivalries between Anglo-Americans, Creoles, and French immigrants; and the demise of French Creole aristocracy in the wake of Reconstruction and the advent of the twentieth century. Still, a seminal sociopolitical and, at times, religiously affiliated phenomenon has been left largely unexplored in the cannon of Creole history: education—particularly college administration and instruction.

While exploring Louisiana's sugar parishes by way of books, journal articles, and road trips, I came across an architectural marvel situated in the parish of St. James: a college built by wealthy Creole sugar planters. During the eighteenth century, St. James Parish was a portion of the County of Acadia, named for Canadian migrants known as Acadians. In 1807, Acadia was divided, giving birth to both St. James and Ascension Parishes.[4] On the east bank of the Mississippi River, in the small town of Convent, stands the imposing pillared edifice of the Manresa House of Retreats, formerly Jefferson College. Arguably one of the largest antebellum structures still standing on the River Road, the whitewashed main building comprises three stories of brick, stucco, and cypress fronted with twenty-two massive Doric columns (each measuring approximately nine feet in circumference), two embedded columns, and, at the very top, a roof of slate tiles. The retreat's main building, including its matching porters' lodges, president's home and detached kitchen, surrounding brick wall, brick workmen's shed (formerly the campus powerhouse), and gothic Sacred Heart of Jesus Chapel, encompasses the remnants of a once larger nineteenth- and early twentieth-century college campus. Behind the main building lies an impressive, early twentieth-century science hall, today used as a dining facility for those visiting Manresa in pursuit of spiritual reflection and renewal. The remaining structures, though designed to resemble the aged edifice, are late twentieth, early twenty-first-century additions. These buildings include (but are not limited to) the Manresa director's office, a library (St. Peter Canisius Hall), a spiritual counseling center (Our Lady of Good Counsel Hall), and Loyola Hall, which houses the Blessed Peter Faber Conference Room. Modern additions aside, the sheer size of the main building, St. Mary's Hall, dwarfs that of the antebellum plantation homes that dot the Louisiana sugar parish landscape, while the historic president's home itself resembles a miniature plantation house.

Adjacent to the brick and black iron gated fence is a historic marker with gold gilt letters that reads as follows:

MANRESA HOUSE
OF RETREATS

Since 1931 The Society of Jesus (Jesuits) have operated a retreat house here for the spiritual development of the laity. The grounds and several

> of the buildings were the site and home of Jefferson College founded for the education of the youth of St. James Parish.
>
> The main building dates from 1842; the President's house and Gate Houses from 1836.
>
> After Jefferson College failed in 1848, Louis Dufau of New Orleans operated the Louisiana College here; this college failed in 1856. In 1859, Valcour Aime purchased the site and erected a chapel in memory of his children. From 1862–1864, the Federal Troops occupied the buildings during the Civil War.
>
> In 1864, Mr. Aime donated the properties to The Society of Mary (Marists) who established St. Mary's Jefferson College, which operated until 1927, when it closed.
>
> The Jesuits purchased the properties in 1931 and since that time have conducted retreats based on the Spiritual Exercises of St. Ignatius Loyola, the founder of the Jesuits. Each year over 5,000 laypersons, religious and priests make retreats here.
>
> For the Greater Glory of God[5]

The above description is decidedly epigrammatic and, like many pieces of literature pertaining to the history of Jefferson College, not entirely forthcoming or correct as concerns the lineage of events connected to the former educational institution. Nevertheless, the marker does depict an important transition: the change in governance from completely secular to entirely religious.

In 1831, Jefferson College was chartered and promoted by Louisiana governor André Bienvenu Roman. Built by wealthy, Catholic (though poorly practicing) Creole sugar planters (known as Louisiana's *ancienne population* or, more infrequently, *ancien régime*) to provide an intellectual foundation for their sons and heirs, the institution reflected aristocratic planter ethos through architecture, gentlemanly instruction, and secular liberal arts curricula resonant of late eighteenth- and early nineteenth-century Paris.[6] Dedicated to Thomas Jefferson, the third president of the United States, the college was originally established to impart a local education for ancienne population progeny, oppose the educational regime of encroaching Anglo-Americans, and assist in the preservation of Creole heritage via academic study. Principally, as Rodney Cline summarizes, "it became a matter of interest to rich planters to have a college in their midst for the proper education of their sons."[7]

Despite French Creole efforts to establish a premier educational institution, the agricultural economic collapse of 1837 and a devastating fire mired the college's first era. Although Jefferson College was rebuilt on an extravagant scale that mirrored a shift in architectural design, support waned, and the college closed in 1848.[8] In 1853, Louis Dufau, a French immigrant and refugee from 1848 revolutionary France, purchased the campus and reopened it as Louisiana College. However, Dufau's administration was laden with personal frustrations, poor social support, and ultimate failure. Louisiana College ceased to operate in 1856.[9] Then, in 1859, former Louisiana governor A. B. Roman approached Valcour Aime of St. James Parish to inquire if the Creole sugar baron would purchase the defunct college. Aime, who was known as the "Louis XIV of Louisiana" on account of his immense wealth and plantation home, "Le Petite Versailles," bought the institution and allocated its administration to his four sons-in-law. The college closed for a third time due to federal military occupation during the second year of the American Civil War.[10] Before the war ended, Aime donated the college to the Society of Mary, which oversaw the institution's reopening in 1864 as St. Mary's Jefferson College, the first Marist college in the United States.[11] Now, under the management of a religious order instead of a singular individual or group of Creole planters, the college shed its secular origins and slowly adapted to encapsulate "American" regulatory, curricular, and extracurricular trends in higher education.

Subject to Marist control, the newly donned "Catholic college" contributed to the revival of Catholicism in the sugar parishes while reflecting the modernization of Louisiana higher education through athletics, scientific instruction, student organizations, and relaxed discipline. Students, largely silent before and after the Civil War as a result of intense course loads and harsh daily schedules, were encouraged by their clerical guardians to write letters (albeit censored) and produce print material that recounted lived experiences and activities at the St. James Parish campus. Unfortunately, as the nineteenth century gave way to the twentieth, St. Mary's Jefferson College suffered academic intransigence, decreased enrollments, and financial woes. As Lillian C. Bourgeois shrewdly elucidates, "Jefferson [College] was born with a gold spoon in its mouth, although through the near-century of its life, it often ate from one of tin."[12] The ailing college remained open until the Mississippi River flood of 1927, which severely crippled south Louisiana's human and capital resources. In 1931, the campus was purchased by the Society of Jesus as a silent religious retreat for clergy and laymen, and it remains so to this day.[13]

This telling of the multiple lives of Jefferson College offers a brief glimpse into the detailed structure and complex progression of an educational institution built by Creole planters to instill their sons with a cultured, aristocratic heritage while keeping them close to home and exposed to plantation management practices. In essence, this "Creole college" was a social thermometer for sugar planter culture, Creole resistance to and eventual compliance with Anglo-American custom, the revival of Louisiana Catholicism, and the "Americanization" of regional educational practices. Although fraught with difficulties, Jefferson College served as the intellectual foundation for several people who played key roles in Louisiana education, law, politics, and even the Catholic Church. Some of these former students include Paul Octave Hébert, the fourteenth governor of Louisiana; James Hubert Blenk, the seventh archbishop of New Orleans; and, for a brief period, Sydney Story, the alderman who put forth legislation to establish the New Orleans prostitution district known as "Storyville."[14] Jefferson College not only produced a bevy of leaders, it also epitomized changing pedagogical and organizational ideals as its faculty and administrators addressed or ignored evolving didactic standards.

Prior to the Civil War, Jefferson College championed period-specific definitions of higher education with curricula centered on liberal arts instruction. It was a college, like so many, reserved for a select few who could afford the high cost of tuition. Students entered the college with rudimentary knowledge acquired through private tutors. These highborn males were taught to contemplate philosophy and engage in linguistic instruction that would create genteel polyglots prepared to inherit their fathers' fortunes.[15] Hindered by disaster, war, political reconstruction, and the advent of emancipation, Jefferson College, like various southern academies, was slow to adopt modern science and technology curricula.[16] After Reconstruction both state-supported and private higher education embraced curricular arrangements largely centered on practical, scientific reasoning. Jefferson College, on the other hand, lagged behind and, after having been converted to a Catholic institution, begrudgingly clung to a bygone, European instructional regime with only the slightest traces of practical, job-centric instruction. As higher education in the American South shook off preparatory departments and sought academic accreditation through the Southern Association of Colleges and Schools (SACS), Jefferson College entered the twentieth century acquiescing to intellectual legitimacy standards regarding scientific instruction, though much too late. At the end of the nineteenth century, state-supported colleges and universities had evolved into competitive, industrialized centers of modern thought that enrolled young men of approximately eighteen years of age who had previously earned high school diplomas.[17] Jefferson College, with its dwindling student body, lingering preparatory division, and inadequate instructional routine, failed to rival universities with progressive academic units. As such, the college closed partly due to the 1927 flood, but mostly as a result of stale pedagogical practices, inadequate course offerings, and poor financial decisions. Even so, Jefferson College's detailed history illuminates relevant details pertaining to the changing nature of nineteenth- and early twentieth-century Louisiana social and intellectual life via the actions of its instructors, students, and institutional leaders.

As Darryl Peterkin explains, research relevant to the history of any educational institution reveals its underlying "human condition" where buildings, administrations, and instruction are shaped by motives often removed from purely altruistic purposes.[18] Linking Peterkin's observation to this book, a study of the vibrant yet tumultuous years of Jefferson College allows for a deeper understanding of French Creole culture and social class decline as it relates to the creation and multiple closures of an insular educational institution where Greek and Latin were taught within and sugar production was taught without. This book also recounts the religious reawakening of an ethnic group characterized by antebellum secularism, as well as the reshaping of Jefferson College's purpose from one of prestige to one of necessity. Such a narrative not only depicts the grudging curricular changes of rural Louisiana higher education set within a cultural context of shifting social regimes, it does much, in the words of Dagmar Renshaw LeBreton, to "tell . . . the story of the changing fortunes of men."[19]

In recounting the history of south Louisiana, scholars have illustrated antebellum class struggles between French Creoles and Anglo-Americans for political authority while detailing the tenuous, unequal alliances between wealthy Creoles and French immigrants to topple American dominance. Still, other researchers have explored the racially bifurcated yet genetically linked white Creoles and Creoles of Color (otherwise known as the *gens de couleur* or *gens de couleur libres*—a title that identified the latter as "free" persons of color, born outside of slavery) in order to untangle a complicated class system that went beyond ethnic origins to include people of mixed-race parentage, the offspring of Louisiana's European and African populations.[20] Historians have also examined the waxing and waning relationship between Louisiana sugar planters and the Catholic Church before and after the Civil War.[21] Scant research, however, investigates sugar planter educational ideology as it pertains to social class preservation and the only college built to serve members of the ancienne population. Creole planters, proud of their French heritage, deplored notions that their sons could be transformed into "American" engineers and businessmen. In order to safeguard forthcoming generations of Creole aristocrats, liberal arts education was key. What there was to learn about plantation management and sugar production was taught at home, father to son. Prior to Jefferson College's founding, well-to-do Creoles shipped their heirs to European institutions, select Catholic academies, or to the impermanent Collège d'Orléans in the port city of New Orleans.[22] With the closing of the Collège d'Orléans in 1825, where all courses were taught in French, and the establishment of the American, "English only" College of Louisiana in East Feliciana Parish, Creole sugar planters pressed for an institution that would instill their heirs with a gentlemanly education while keeping them familiar with their birthright and social standing.[23]

Cut off from metropolitan society, south Louisiana planters gleaned farming knowledge from parents, peers, and profit. Antebellum sugar barons were largely indifferent to agricultural and mechanical education in spite of the instructional efforts of the Louisiana Agriculturalist and Mechanics Association or relevant course offerings at the University of Louisiana (present-day Tulane University).[24] If formal education was sought, Creoles expected it to enhance upper-class values by means of linguistics, philosophy, and rhetoric. When Jefferson College was constructed alongside plantation homes and sugar refineries, curricula were principally liberal and only slightly scientific. As the college progressed through the first half of the nineteenth century, its physical structure dramatically changed to better match surrounding architectural partialities. Sailing around a bend in the Mississippi River known by riverboat pilots as "College Point," "the east bank . . . bursts forth with a parade of white pillars as though the entire parish of sugar planters had grouped their mansions on a single lot" to form the Creole institution.[25] Jefferson College not only matched the regional Greek-revival architectural style, its geographic position was dictated by the all-important waterway, similar to neighboring plantations. As planters relied on the river to transport crops, people, and supplies, plantations were symmetrically positioned facing the riverbank to enhance access.[26]

Nineteenth-century institutions of higher education, like Jefferson College, were regarded as points of civic pride, but geographic location became an issue for differing ethnic classes as communities vied for colleges that paralleled their own social mores. Donald G. Tewksbury explains, "These colleges were affiliated with the various political, racial, and religious interests that were struggling for a position of dominance."[27] As a result, disease, fire, natural disaster, internal factions, and/or external social unrest was, at times, enough to trigger closure. In Louisiana, this certainly was the case as colleges were created to settle squabbles between ethnic groups. Jefferson College experienced all of the aforementioned, particularly a fire in 1842 that allowed for the campus's structural redesign to reflect changes in plantation architecture. Similarly, John R. Thelin argues, "Transformations in campus architecture are symptomatic of changes—and problems—in American higher education."[28] This being the case, campus architecture serves as a relevant conveyance of symbols and motifs that represent unique societal ideals.[29]

After the Civil War and throughout Reconstruction, sugar planters worked to rebuild their once powerful economic industry. By instituting sharecropping and selling land to farming families and private companies, plantations persisted. By the end of Reconstruction in 1877, sugar planters had sold much of their property, making the former Creole elite powerless to dominate sugar capital.[30] Even though Jefferson College had been converted into a Catholic institution managed by the Society of Mary, its fate remained intertwined with its surrounding social milieu. As a religious institution reliant on local, likeminded people, the renamed St. Mary's Jefferson College was highly dependent on faithful Catholics. "The decisive factor determining the survival [of a Catholic college]," as Raymond H. Schmandt elucidates, "was always the growth or the stagnation of the local Catholic population."[31] As sugar planters, Catholic or otherwise, depleted their personal wealth and property, they moved out of the sugar parishes to join urban kin in New Orleans or elsewhere. According to Cline, "By the second decade of the twentieth century, the plantation aristocracy of the River Parishes, for whose sons the college was largely intended, was rapidly disappearing."[32] Due to corporate ascendancy and decreased private fortunes, St. James Parish was, at the start of the twenty-first century, described as a poor parish.[33] With the disappearance of planter support, the once-celebrated Creole college faded from Louisiana's higher education scene.

While there is expansive literature dedicated to Louisiana economics, regional social hierarchies, agriculture, and the ills of human chattel, few histories describe the educational mindset of Creole sugar planters or the only institution built to educate and preserve the plantation elite. Those that do are partial narratives fraught with inaccuracies. In the majority of texts, sugar planter education and educational thought has been relegated to a footnote, a handful of sentences, or a concise set of paragraphs. Regarding those relevant pieces of extant literature, it is essential to clarify their contributions to the formation of this study.

One of the earliest published accounts of education in Louisiana, with a short section dedicated to Jefferson College, is Alcée Fortier's *Louisiana Studies: Literature, Customs and Dialects, History and Education* (1894).[34] Fortier, a professor of Romance languages at Tulane University, later printed additional snippets pertaining to the St. James Parish institution in the first of his three-volume 1914 series titled *Louisiana: Comprising Sketches of Parishes, Towns, Events, Institutions, and Persons, Arranged in Cyclopedic Form.*[35]

Relying on Fortier's 1894 book, Edwin Whitfield Fay, professor of Latin at Washington and Lee College, published *The History of Education in Louisiana* in 1898. Fay's text, crafted as a formal report to the Federal Board of Education, describes the Creole college's founding and genesis years, but no more. Thereafter, little was written concerning Louisiana higher education until Otto S. Varnado, an early twentieth-century history student at Louisiana State University, completed his 1927 thesis, "A History of the Early Institutions of Higher Learning in Louisiana." Varnado is perhaps the first to compile a descriptive account of Louisiana's nineteenth-century colleges as separate entities from state-supported primary schools. Varnado's description of Jefferson College, however, is limited and focuses on the institution's final years.[36]

Twenty-two years after Varnado's study, Vernon Elmer Kappel, in his Tulane University master's thesis, "Louis Dufau's Louisiana College," attempted to piece together the transformation of Jefferson College into Louisiana College. As explained in the September 5, 1949, issue of the *Times Picayune*, "old letters, ledgers and accounts which Kappel has found in attics and dug out between walls of buildings in St. James Parish led the Tulane student to believe that Louisiana College was founded in 1851, by Louis Dufau, a political exile from France."[37] Kappel was the first to compose a biography of Dufau, a man who has been neglected in the larger historical narrative of Louisiana higher education. Although Kappel's work adds a substantial dimension to the patchwork history of Jefferson College, his thesis omits several telling conflicts between Dufau, local clergymen, fellow academics, and south Louisiana Creoles—all of which are pertinent examples of the social tumults experienced by French immigrants in the New World.

Other published and unpublished accounts of Jefferson College include the 1954 Tulane University thesis of Marist priest Earl Francis Niehaus, "Jefferson College in St. James Parish, Louisiana: 1830–1875," and his 1955 *Louisiana Historical Quarterly* article "Jefferson College: The Early Years." Niehaus does an exemplary job of presenting the origins of the college, but his work provides only partial insight concerning the political context and administrational modifications related to the institution's adolescent years. A disappointing follow-up to Niehaus is Edward John Power's 1958 text, *A History of Catholic Higher Education in the United States*. Power includes no more than three sentences on the subject of Jefferson College and inaccurately credits it as being the oldest college in Louisiana.[38] Clearly, Power failed to consult the works of Fay, Kappel, Niehaus, and Varnado. Despite Power's lack of information, he does make a thought-provoking statement: "Little is known of the history of this college."[39]

Surpassing Power in 1963, Phillip H. Dagneau (a Marist like Niehaus) published a series of missionary narratives pertaining to the Society of Mary in the United States. Dagneau's book, *Memoirs of Seventy-five Years for the Centenary of the Marist Fathers in America*, recounts the transfer of Jefferson College to the Marist order and biographical information related to the institution's first clerical president, Henri Bellanger. The book also contains vivid imagery depicting French priests struggling to contend with stifling humidity, mosquitoes, muddy paths, and lively pupils eager to embrace a more "student friendly" administration in contrast to the strict secular governance of times past.[40] Dagneau, a former instructor at St. Mary's Jefferson College, candidly describes daily life at the Creole college, including the presence and educational leadership of New Orleans archbishop, James Hubert Blenk. The archbishop's relationship to St. Mary's Jefferson College is best detailed in Mary Bernardine Hill's 1968 dissertation, "The Influence of James Hubert Blenk on Catholic Education in the Archdiocese of New Orleans, 1885–1917." Hill describes Blenk's formative education, admission to the Society of Mary, college teaching, institutional presidency, and rise to become a predominant religious leader.[41] Two years after Hill defended her dissertation and received the Doctorate of Education Degree from Louisiana State University, the noted architect and historian Samuel Wilson Jr. published "Manresa—Jefferson College, Convent, Louisiana" in *Louisiana's Architectural Heritage*. Wilson speculatively depicts construction efforts surrounding the college before and after the ruinous fire of 1842. However, he provides no citations, and sufficient primary evidence to support his claims has yet to surface.

Judge Oliver P. Carriere, former member and past president of the Manresa Board of Directors, authored a fifty-six-page pamphlet titled *A Sketch of the History of Jefferson College and Manresa House of Retreats, Convent, Louisiana: 1830–1973* in 1974. This brochure, written as a chronology with biographic vignettes, supplies the most accurate description of events at Jefferson College prior to and succeeding the 1931 campus sale to the Society of Jesus. Carriere's work, though predominantly concerned with Manresa, includes a dense, carefully organized ten-page college timeline. Not only is Carriere's text relatively free of inaccuracies, it is the first to include Louis Dufau as a notable figure related to the institution's overarching history.[42] Following Carriere's publication, an additional twenty-two years passed before another researcher attempted a history of Manresa. In 1996, William D. Reeves published *Manresa on the Mississippi: For the Greater Glory of God*. This illustrated volume focuses on the lives of priests and laymen as they served at or experienced spiritual direction at the former college campus. Although Reeves's book contains only two pages devoted to the annals of

Jefferson College, he compiles fascinating information with clear images, which establishes the lasting significance of a Jesuit house of retreats in south Louisiana.[43]

In addition to the handful of authors who have recounted period-specific portions of Jefferson College's history, other scholars have mentioned people or issues connected to the Creole college in relation to the grand narrative of Louisiana. Certainly, other manuscripts with relevant information exist and are critiqued in this text. These books, articles, and unpublished studies serve as points of reference to an assortment of archived documents that fueled this investigation. Emerging from the slew of interesting yet often disagreeing texts, extensive primary source analysis allowed for an extrapolation of misaligned dates, omitted names, and individuals credited for deeds performed by others. In hopes of making sense of the litany of inconsistent literature, I decided to visit the site of the former Jefferson College and find, if possible, connections between conflicting accounts. After perusing Manresa's website, I contacted the retreat's director and was granted a tour of the campus. Though Manresa is "off limits" to sightseers due to weekly religious retreats, its external beauty captivates the mind and hints at the existence of a deeper, multifaceted account that challenges traditional assumptions of Creole sugar planter educational thought and undertakings.

Approaching Manresa, visitors are met by a long white fence that transforms into a brick wall topped with wrought iron railing, which runs parallel to the River Road, an imposing metal gate, the St. Ignatius House of Prayer and Chapel of the North American Jesuit Martyrs (the college's original president's house and antebellum detached kitchen, reserved for long-staying retreatants), and a set of porters' lodges, repurposed as shrines. One is dedicated to the Virgin Mary, the other to her husband, St. Joseph. Standing tall among its peer facilities, the considerable main building, St. Mary's Hall, presides over the well-manicured grounds, reflecting the grandeur of a bygone era of agricultural prosperity. Manresa's interior is equally impressive, however, not in physical adornment but in quiet simplicity. While the external façade is embellished with rich structural detail, oak-lined paths, and strategically placed religious statues, the interior is sparsely decorated. Walls are painted white with exposed cypress beams; framed images of the original college and its students run the length of the main corridor; and rooms for visitors on retreat are meekly but comfortably furnished. All rooms are equipped with a single bed, nightstand and reading lamp, desk, rocking chair, crucifix, and icon of the Holy Family. This simple configuration permits little distraction while those attending Manresa gaze inward, focusing on their spiritual life and the presence of "God in all things." Large oaks draped with Spanish moss shadow walkways leading to the former science hall, now cafeteria. The covered walkway to the retreat's gothic chapel is lined with metal Stations of the Cross, which were, in reputation at least, cast in France at a cost of 4,000 francs. The Stations, crafted in 1867, were originally intended for the cemetery behind St. Joseph's Church in Thibodaux, Louisiana. As a result of the church's destruction by fire in 1916, they lay dormant until Harvey Peltier, a Louisiana attorney, purchased and donated the set to Manresa in 1946.[44] Leaving the covered walkway and entering the chapel, guests are struck with awe-inspiring detail. While the main building's interior is provocative in its simplicity compared to its outer complexity, the chapel is an internally and externally ornate affair resonate of the Creole planter who paid for its construction in 1859, Valcour Aime.

The chapel's gothic architecture is notably different from the Greek-revival design of nearby campus buildings. Those who enter experience a stimulating ambience imbued with religious reverence. The interior walls are built of exposed brick made from Mississippi River clay. The scent of incense fills the air as candlelight conjoins with daylight streaming through modern stained glass windows. Contained in the sanctuary is a wooden altar with a gold tabernacle in front of a painted wall of gold nails grouped in threes against a burgundy background. These colors, gold and burgundy, represent the sixteenth-century Loyola family, while the nails symbolize the vows of religious life: poverty, chastity, and obedience. Also present is an impressive crucifix with the phrase "Jesus Nazarenus Rex Judaeorum" (Jesus of Nazareth, King of the Jews) writ large over the corpus. Statues of Saints Joseph and Mary flank the nave while an icon of St. Ignatius Loyola hangs nearby. Printed above the chapel's main entrance is the Jesuit seal accompanied by the order's motto, *Ad Majorem Dei Glorium*, translated as "For the greater glory of God"—a fitting slogan for this formerly secular college turned sacred haven dedicated to thoughtful introspection.[45]

Exiting the chapel, my tour came to a close. Over the course of several hours, I was permitted to view remaining structures dating from the 1830s campus, the 1840s rebuilt college structure, the chapel commissioned by Aime in the late 1850s, the science hall from the 1910s, and additional buildings constructed after the Marists sold St. Mary's Jefferson College to the Jesuits. This deliberate type of physical exploration coupled with archival and literary investigation reverberates in the writing of Gary McCulloch and William Richardson: "A further considerable gain for both the author and the reader of historical studies of education lies in the opportunity they offer to ponder the nature of continuity and change in education and the societies with which it interacts."[46] By presenting the intricate, contextualized history of Jefferson College, inclusive of its administrational shifts, response to external pressures, and religious conversion, we can better understand how French Creole sugar planters endeavored to preserve an ethnic aristocracy, but ultimately conformed to and merged with Anglo-Americans to form a unified Caucasian populace in response to racialized white Creole/Creole of Color etymological confusion and diminished financial stability.

Rarely does literature focus on the closure of colleges, the reopening of educational institutions, or their reformation into organizations with altered purposes, goals, identities, and missions. Granted, numerous colleges and universities that failed to survive have been physically demolished—their corporeal footprint forcibly removed from history. This alone makes the study of defunct higher education difficult. In a similar vein, there is negligible literature centered on the relationship between social class fruition and the progress or decline of higher education appertaining to individual planter classes (cotton, indigo, rice, sugar, tobacco). Beyond this limitation, research does not convey connections between Louisiana's white French Creole inhabitants and the existence of a college meant to bolster the ancienne population. As such, this study not only attempts to illustrate Creole planter class maintenance and evolution via higher education, it also provides a sharpened view of social class decline and religious revivalism as observed through a century of college life, leadership, and instruction.

1

Aristocracy, Education, and the Ancienne Population

What is a Creole? . . . there seems to be no more serviceable definition of the Creoles of Louisiana than this: that they are the French-speaking, native portion of the ruling class.

—George Washington Cable

Louisiana's antebellum sugar parishes were, as Rodney Cline details, "characterized by the wealth and prestige of the plantation aristocracy."[1] In Richard Follett's *The Sugar Masters: Planters and Slaves in Louisiana's Cane World, 1820–1860*, he goes beyond Cline's description to paint a vivid picture of expansive sugarcane fields with large, columned plantation homes flanking the Mississippi River. Follett portrays French Creole planters overseeing the sugar extraction process including cane harvesting, mashing, boiling, and crystallization. In the same vein, he carefully dissects sugar planters' brutal reliance on slave labor to produce Louisiana's "white gold" as well as the vast amounts of money spent on "land, labor, and machinery . . . to maximize . . . harvest yields."[2] Dagmar Renshaw LeBreton, biographer of the Creole planter Valcour Aime, artfully illustrated the south Louisiana plantation landscape (inclusive of family names and slave quarters) as follows: "What a magnificent sight it must have been in that particular part of the 'Coast' known as 'Rich Bend' where here situated the Aime, the Caprice, the Roman, the Chopin, and the Delogny plantations . . . for there was no cheap shabbiness about them . . . luxuriant cane fields back of them, a white line of negro cottages on the one side of them, on the other the impressive group of sugar house buildings, the region d'atre [*sic*] of it all."[3] Indeed, the sugar parishes were positively littered with handsome plantation vistas, but these scenes were set within a context of harsh manual labor and solitary leadership.

Sugar barons, as lone masters of vast plantations, were islands unto themselves. Surrounded by slaves and lower-class laborers, wealthy French Creoles reached out to their contemporaries to form communal and financial partnerships to fortify their social class status.[4] Although French Creoles believed that alliances with peer planters would facilitate regional control, these elite Creoles, as well as Anglo-Americans, understood that whoever controlled education controlled the future of Louisiana. For this reason, both ethnic groups built colleges for their descendants.[5] As social class struggles were heightened by a perceived need to keep offspring familiar with Louisiana, plantation economy, and cultural heritage, arguments for the development of educational institutions were compounded by a firm agreement that sons sent away for instruction returned home alien and averse to Deep South mores. As stated in the proceedings of Louisiana's 1844 Constitutional Convention, "When our children return from the north, after having received an education there, they have to be re-acclimated. . . . All this can be remedied by entering upon the work [of education] ourselves, with a determination to accomplish it. A good education furnished to the rising generation, will afford a better guarantee of protection than fleets and armies."[6] Essentially, it became a matter of some urgency that Southern sons be reared *and* educated in the South. But, before the emergence of Louisiana's class-dominated higher education scene can be explored, the etymological legacy and cultural ideologies of French Creoles must be detailed. In short, what characterized wealthy French Creoles, and what exactly does Creole mean?

The word Creole originates from "the Portuguese and Spanish . . . *criar*, meaning 'to raise; (as in, to raise a child).'"[7] In time, *criar* was colloquially modified to become *criollo*, then *crioulo*, and finally Creole. In spite of its Spanish etymology, this label, propagated as a somewhat ambiguous Franco-Louisiana categorization, caused centuries of ethnic strife and misunderstandings based on blood and breeding.[8] Joseph G. Tregle Jr. explains that early Louisiana officials such as the founder of New Orleans, Jean-Baptiste Le Moyne, Sieur de Bienville, used the term Creole in reference to "new world" offspring of families claiming French or Spanish descent. Similarly, French religious officials and Catholic priests listed ethnic-specific native Louisiana parishioners as "Creole" on birth, marriage, and death certificates.[9]

When France ceded Louisiana (otherwise known as the Territory of Orleans) to Spain in 1762, elements of Spanish culture invaded colonial New Orleans. Architecture, in particular, was influenced after two devastating fires (1788 and 1794) destroyed large portions of the French Quarter. The

Vieux Carré was rebuilt, but Spanish architectural features replaced French building designs.[10] Though originally displeased with Spain's presence in Louisiana, French Creoles eventually accepted the addition of a second group of white Europeans as an opportunity to swell the numerical size of their social class. In *The Story of French New Orleans: History of a Creole City*, Dianne Guenin-Lelle asserts that, with the sale of Louisiana to the United States in 1803, "Spanish heritage was explicitly linked to the French."[11] Negotiated marriages and business agreements ensured that Spaniards merged with affluent Creole families, thus presenting a unified front to counter the noticeably present Anglo-American population. Indeed, French Creoles, along with their Spanish accompaniments, used the title of Creole to distance themselves from Anglo-Americans, a growing social class following Louisiana's eighteenth-century sugar industry boom. "White Creoles" as Emily Epstein Landau points out, "were held distinct from Americans with different heritages and immigration histories, just as Creoles of Color understood themselves as different from enslaved, formerly enslaved, and even freeborn African Americans."[12]

In the eighteenth century, Creole was recognized as a class demarcation that held little racial connotation. This understanding remained steadfast until the term became bifurcated as two Creole classes emerged, separated by skin color. By the time Louisiana became an American territory, Creole was sometimes used to refer to persons of French, Spanish, or African ancestry.[13] Nevertheless, an etymological distinction between class and race lay in genetic associations. Traditionally, Creole referred to the ancestrally "white" population (predominantly French), while the phrase "Creole of Color" or *gens de couleur* demarked those of racially mixed parentage.[14] Miscegenation was not uncommon as the custom of *placage*, liaisons "between white men and free women of color in antebellum New Orleans," resulted in the birth of mixed-race individuals who claimed the term Creole as part of their biological heritage.[15] Historian Lawrence N. Powell, in his book *Accidental City: Improvising New Orleans*, explains how Creole acquired severe racial overtones during Reconstruction as members of the *gens de couleur* drew upon the label to separate themselves from former slaves.[16] Creoles of Color were not only racially different from enslaved or free African Americans; many were educated and possessed substantial financial means but were never considered equal to white Creoles.[17]

Regardless of their similarities in business acumen, elitism, and white supremacy, Anglo-Americans criticized French Creoles for their illicit *placage* prodigy. Many Creoles of Color carried their white father's surname, consequently making the identification of paternal heredity across the color line relatively simple.[18] Even though several Anglo-Americans engaged in their own *placage*, the American social class as a whole was far more obsessed with race than class and heavily propagandized the "torrid" interracial love affairs of French Creoles. This variety of mudslinging polarized the region's stance on "pure" and "mixed" blood.[19]

Politicizing Creole with racial associations did little to assuage members of the ancienne population from proclaiming their regional authority. Indeed, the notion that Creole demarked members of the white French Creole aristocracy became so firmly linked with south Louisiana's history that the nineteenth-century novelist and historian George Washington Cable, in his 1884 book *The Creoles of Louisiana*, described the term Creole as one reserved for "the French-speaking, native portion of the ruling class."[20] Cable's description is concise, but it does not explain the racially divided nature of this significant title. Scholars such as Virginia R. Dominguez have expanded and modernized Cable's definition by demystifying contextual factors of race and geographic locale as concerns the historic perception of white Creoles and Creoles of Color. More than any other historian, Dominguez provides a blunt, hierarchical definition of white Creoles as it relates to the century of class dominance, struggle, and downfall examined in this study:

> The long history of African slavery in the United States and of white ownership of African slaves left in Louisiana, as in other parts of the United States, a traditional association of whites with upper status and blacks with lower status. To white Creoles today the mere suggestion of possible African ancestry invokes a lowering of social class and economic status for the people in question. To colored or black Creoles, on the other hand, the claim of at least partial European ancestry accords the group in question a status (or an expectation of status) higher than that accorded to "pure" blacks. . . . What many of us normally assume to be likely *connotations* of membership in a particular group are, in the case of southern Louisiana, often, if not always, the crucial variables that individual New Orleanians manipulate in making themselves members of a group, or in identifying others as members of a group.[21]

In order to reinforce portrayals of social class and racial supremacy, white French Creoles obsessed over genealogy and paid for the construction of vast, questionable family trees replete with titled rankings and European nobility that demonstrated the "purity" of their pedigree.[22] As Creole became racially confused by Anglo-Americans and parties external to the cultural haughtiness of south Louisiana, Creoles pressed their European heritage into the public sphere via publications and speeches centered on the "truth" behind their title.[23]

Creole Class, Creole Mindset

For all intents and purposes, white Creoles considered themselves to be purely French; they were not, however. The "pure" French were immigrants who had fled to Louisiana as a result of eighteenth- and nineteenth-century transatlantic revolutions. These refugees, referred to as *française étrangers* (foreign French) were, for the most part, highly educated and quickly rose in Louisiana's cultural and political spheres, but they were barred, more often than not, from the lofty social strata of white Creoles. Generally, Creoles resented French immigrants but relied on them to oppose Anglo-Americans through conjoined political numbers.[24] Content to count French immigrants as voting "party members," Creoles asserted their regional dominance and crowned themselves with the title ancienne population ("old inhabitants"). This designation further distinguished French Creoles from foreign French as well as the growing number of *gens de couleur*.[25]

Despite a change in title, French Creoles, throughout the nineteenth and early twentieth centuries, contended that the term Creole was reserved for those of pure European ancestry and should under no circumstance be used to refer to individuals of mixed Afro-European parentage.[26] After the 1803 Louisiana Purchase, Creoles, more often than before, referred to themselves as the ancienne population, which in their minds reflected a "time-honored" position as Louisiana's "first" aristocracy.[27] Albert E. Fossier elucidates,

> This "ancienne population" of white, wealthy Creoles were described as . . . high minded, irascible, social and generous; much addicted to the sport of the turf and the gaming table. They are fond of hunting and keeping large packs of dogs. Having overseers for the most part over their lands and are too apt to become dissipated. There is a rising spirit of literature and a disposition to read among them, which will employ many of the hours that used to be spent around the gambling table. The people generally are adverse to care, deep thinking, and profound impressions; and are volatile, gay, benevolent, easily excited to joy or sorrow; and a common maxim of a sickly climate where life is precious, is "a short life and a merry one."[28]

The self-styled moniker, however, was cumbersome and in time the ancienne population again referred to themselves as Creole, though not exclusively.[29] This terminology shift goes far to illustrate that "creole whites [were], of all persons, the most anxious to be deemed of pure white blood."[30] In addition to titles, the ancienne population promoted a caste system much more intricate than the traditionally held tripartite of whites, free persons of color, and slaves. In their scheme, white French Creoles maintained the upper crust, seconded by French immigrants and, when it proved profitable through familial or political alliance, Anglo-Americans. Beneath Americans were poorer Creole families, Creoles of Color, Acadians, non-French immigrants (Germans, Irish, Italians), and, finally, individuals of pure African descent (slaves).[31]

Not only were Creoles concerned with their zenith in south Louisiana's pecking order, they also desperately wished to establish a simulacrum of Parisian gentry. In an 1848 essay, Henry Augustine Washington wrote about Virginia's planters as "lords of soil and masters of slaves" who harkened to hierarchies of English nobility.[32] If antebellum Virginians mirrored England's social "chain of command," wealthy white French Creoles drew hegemonic inspiration from France, the mother of Louisiana.[33] Carl A. Brasseaux comments that "Creole planters, descendants of European settlers in Louisiana, generally sought to recreate in Louisiana a romanticized vision of feudalistic France, with themselves as the New World aristocracy. Inspired by the French nobility's belief that a person was only as rich or powerful as he appeared to be, ambitious Creoles mimicked the Old World aristocracy and sustained their social pretensions by building grand homes and purchasing carriages, fine furnishings, and domestics."[34] As such, French Creoles fabricated a conceptualization of pre-revolutionary French provincial aristocracy—a tiered system that encompassed nobility, peasants, and servants.[35] Therein lay a considerable problem for Creole planters and politicians. While Creoles focused on their personal plantation kingdoms, regional class hierarchy maintenance, and governmental control within the limited confines of south Louisiana, Anglo-Americans concerned themselves with national politics and expanded westward across the United States. Anglo-Americans, as Nathan A. Buman makes clear, "understood their mission in a national framework while the Creoles saw the world from a Franco-centric view," a critical difference that ultimately allowed Anglo-Americans to numerically and politically overpower the ruling Creole class.[36]

Even though wealthy Creoles could, at a moment's notice, produce family lineages that pointed to Europe, they were, in actuality, "self-made men." Creoles garnered fortunes by trading, purchasing, and clearing/cultivating land, and/or marrying into wealthier Creole families.[37] The most affluent married strictly with other Creoles of equal financial and political prowess.[38]

Likewise, inbreeding kept family names safe from extinction: upper-class antebellum Creoles forbade marriages between their children and lower-class Americans, Irishmen, Germans, or Acadians, but cousin marriage was an acceptable option. This practice kept the ancienne population small but powerful. Although they took precautions to protect themselves from "marrying down" the social ladder, French Creoles were disturbed by the presence and growth of lower-class Acadian families. During the eighteenth and nineteenth centuries, waves of Acadians settled throughout the sugar parishes. For the most part, these exiles from British-controlled Canada were small farmers and herdsman, Catholic, and fundamentally reliant on family, faith, and personal labor to subsist.[39] While the ancienne population needed allies in order to outnumber Anglo-Americans, both Creoles and French immigrants shunned Acadians.[40] To Creoles, the française étrangers were at least educated freethinkers who emulated the secular atmosphere of their beloved France.[41] As a result, Creoles were more than willing to accept them as seconds in their tiered class system. Acadians, however, were relegated to the lower ranks of the social order.[42]

As John L. Wakelyn explains, south Louisiana's antebellum period was an era of Creole authority where the ancienne population "formed their own political blocks [and] eventually gained much political power."[43] Although French Creoles relied on their privileged status, they were subject to an inter-class system based entirely on economics and property. Just as there were wealthy Creoles, there were also poor Creoles who owned small farms and an expansive middle class of Creoles with plantations of 500 or fewer acres. Poor and middle-class Creoles were conscious of their station below their wealthy kinsmen.[44] Ethnicity alone was not enough to garner social respect and acclaim. Land, capital, material property, and slave ownership gave Creole planters prestige in the eyes of their peers.[45] Cultural clashes marked by property ownership enhanced already staunch societal delineations that, according to Shannon Lee Dawdy, included "large slaveholders and merchants, small slaveholders, [and] non-slaveholders."[46]

By the mid-eighteenth century, the acquisition of wealth crystalized with a small percentage possessing the vast majority of Louisiana's capital. At the end of the eighteenth century, the best way to enter the upper class was through marriage.[47] Admission into the topmost Creole echelon hardened as the gap between social classes widened. By the 1850s, quality land was owned by a select few, which created a considerable improbability for newcomers to merge with the plantation elite.[48] The richest Creole sugar barons grew wealthier with relative ease and maintained plantation homes that physically evolved from merely functional to stylistically magnificent. As the price of land and slaves increased, poorer farmers became ever more unlikely to enter the middle or upper classes.[49]

Visible status, in and of itself, was the rod by which Louisianans measured themselves. Wealthy Creoles preserved their social position by spending, not saving fortunes; material possessions of expensive clothing, fine art, opulent adornments, and rare books were just as valuable as money in the bank. This ruse of external wealth went far to hide the fact that many Creole planters maintained heavy mortgages. Upon their husbands' deaths, more than a few Creole widows found themselves with debts to pay instead of an inheritance to enjoy.[50] All the same, as Jennifer Green explicates, "no amount of money could give an African American equal status to whites."[51] Even in debt, the white planter aristocracy was well preserved. Status was more than financial resources; it was social capital reliant on contextual conventions that excluded those of African descent.[52]

Concerning the conditions of poor Creoles and Acadians who lived in the shadows of wealthy sugar planters, Roger W. Shugg, in his 1939 book *Origins of Class Struggle in Louisiana: A Social History of White Farmers and Laborers during Slavery and After, 1840–1875*, states, "For more than a generation, from 1840 to 1875, they endured a succession of disastrous circumstances, the competitive disabilities and oligarchic rule of a wealthy slaveholding society, the holocaust of a war that destroyed chattel slavery but little else, and the racial exploitation and misgovernment of reconstruction. Economic depression spread with the passing years as Louisiana failed to recover from the losses of war and plunged into a protracted period of internal strife. The people who were poor and white had changed neither their color nor their condition in any appreciable sense."[53] Impoverished Creoles and Acadians were hired for plantation work; Irish were employed when necessary and only to carry out the most dangerous of jobs such as clear-cutting bayous, digging ditches, draining disease infested swamps, and repairing levees. As plantation owners saw it, slaves were too expensive to waste on hazardous tasks when expendable Irishmen could be hired for minimal fees.[54]

Shugg makes clear that, although middle- and lower-class Creoles regularly complained of their status, they were well aware of their advantage over black men, women, and children, either free or enslaved. "A white man might be as poor as a slave, but at least he remained free and did not have to work beyond the elementary needs of subsistence. From this privileged position, he could look down upon one lower than himself, the enslaved black, with the contempt which might otherwise have been directed against those who were his equal or superior. Race prejudice, in other words, filled the void of class hatred."[55] In spite of their diminished circumstances, lower class Creoles and Acadians saw fit to politically align themselves with Creole Whigs.[56] In 1833, the Whig Party sprang forth from the National Republican Party in support of the US sugar tariff. Though Creole Whigs opposed the predominantly Anglo-American Democratic Party, both groups supported the tariff, as it mandated a national reliance on Louisiana sugar, in lieu of imported sugar from Latin America and the Caribbean islands.[57] Lower-class Creoles voted on the Whig ticket for two reasons: one, the regional sugar crop provided a steady source of income, and two, despite their lowered status, Creoles supported their upper-class kinsmen due to communal loyalties. Until 1860, French Creole Whigs led parishes such as Lafourche, St. James, St. Landry, St. Martin, and Terrebonne.[58] Acadians and poorer Creoles might have been used as a cheap form of supplementary labor, but they were applauded for politically supporting their aristocratic betters.

The Wealth of Sugar

In addition to politics, French Creoles had, of course, capitalized on agriculture. When it came to the sugar enterprise, Creoles proved clever yet highly autonomous, and always willing to purchase additional land, slaves, and, at times, innovations tested by their peers to enhance production. Early attempts to cultivate sugarcane used cane transplanted from Saint Domingue (present-day Haiti). During the 1750s, Jesuits residing on their plantation outside the fortified colonial city of New Orleans cultivated minor crops of sugarcane with a small army of forty slaves.[59] While Spain governed the Territory of Orleans from 1763 until 1802, indigo was the preferred crop. The Spanish crown discouraged efforts to grow sugarcane as the country owned profitable sugar plantations elsewhere, particularly in the Caribbean. Indigo, in time, proved a loss, as the tropical plant failed to flourish in Louisiana's sub-tropical climate. Spanish-controlled indigo plantations in South America outstripped Louisiana production, and plant disease in the late eighteenth century decimated territorial harvests. As a result, Louisiana planters lost considerable capital on indigo.[60] If Louisiana was to prosper, residents desperately needed some stable means of accumulating financial capital. One French Creole, Jean Étienne de Boré, disregarded the Spanish government's colonial crop preference, gambled on an alternative method of sugar production, and paved the way for generations of ancienne population sugar planters.

De Boré, born in 1740 to a New Orleans family, lived and studied in France for twenty-eight years before returning to Louisiana. Upon his return, de Boré invested in sugarcane and a risky granulation venture that could have cost him his family's fortune. Ignoring the risk, he triumphed and received credit for discovering the first successful method for granulating sugar.[61] Before de Boré's success, planters, unable to discern a lucrative granulation method, had boiled sugarcane juice to produce molasses. But, as molasses was difficult to preserve, it turned a poor profit. By virtue of de Boré's granulation discovery, sugar production in Louisiana exploded. De Boré was promoted by period authors as the "savior of Louisiana," grew considerably rich, and served as the first mayor of New Orleans.[62] Descriptions of his anxiety-provoking granulation experiment have been preserved accordingly:

> Boré's attempt had not been without exciting the keenest interest; many had frequently visited him during the year to witness his preparations; gloomy predictions had been set afloat, and on the day when the grinding of the cane was to begin, a large number of the most respectable inhabitants had gathered in and about the sugar house to be present at the failure or success of the experiment. Would the syrup granulate? Would it be converted into sugar? The crowd waited with eager impatience for the moment when the man who watches the [concoction] of the juice determines when it is ready to granulate. When the moment arrived, the stillness of death came among them, each one holding his breath and feeling that it was a matter of ruin or prosperity for them all. Suddenly the sugar maker cried out with exultation, "It granulates!" and the crowd repeated, "It granulates!" . . . Each one of the bystanders pressed on to ascertain the fact on the evidence of his own sensed, and when it could no longer be doubted, there came a shout of joy and all flocked around Étienne De Boré, overwhelming him with congratulations, and almost hugging him, and whom they call their savior—the savior of Louisiana.[63]

By the time of his death in 1820, de Boré had amassed a fortune estimated at $300,000. Consequently, Creole planters converted their crops en masse to sugarcane and erected large redbrick processing mills to house requisite granulation machinery.[64]

Patterning their methods after de Boré, Creoles relied on peer-approved techniques to advance sugar production. Wary of new granulation methods, intuitive Creoles honored evaporation procedures tested and deemed beneficial through their own experimentation or that of fellow planters. A mistake due to poor boiling tactics could yield low quality, unsellable sugar.[65] By slowly implementing successful granulation methods, Louisiana planters constructed a reliable process that resulted in high sugar yields and sizable profits. Concordantly, when slave revolt in 1791 ruined the highly competitive Saint Domingue plantation system, Louisiana was well positioned to address US and European sugar demands and production increased. As new technologies emerged, the antebellum sugar industry morphed from horsepower and kettle boiling to steam engines and vacuum panning. Before the Civil War, the vast majority of Louisiana sugar mills had incorporated steam pressure to enhance output. Large engines actuated conveyer belts and heavy iron rollers. Freshly cut cane was delivered to mills via carts, placed on belts, and crushed. Mashed cane exited the mill by way of a second series of belts, while juice was piped into a reservoir. Multiple-effect vacuum panning, patented by Norbert Rilleux, a Creole of Color, in 1843, enhanced already available vacuum panning methods and further decreased the risk of scorched sugar. Though costly to implement, sugar barons switched from kettle production to vacuum panning. Employing the fruits of industrialization, antebellum Creoles outperformed Caribbean sugar planters.[66]

Not only were mills powered by steam engines, planters purchased locomotives and laid iron rail to increase cane transportation from fields to mechanized mills. In order to match the new engines' production speed, slaves worked harder and longer during the grinding season. As sugar planters "raced the clock" to beat the devastating frosts of November and December, slaves often worked eighteen- or twenty-hour days.[67] Follett describes the nature of back-breaking labor systematically enhanced by industrialized automation as follows: "Marching to the methodical beat of pistons and flywheels, slaves and masters alike faced an arduous schedule in which labor continued at a frenzied pace until the grinding season drew to a close."[68]

In order to produce quality sugar, planters depended on a careful calibration of personal wealth, land ownership, vetted refining processes, and grueling slave labor. John C. Rodrigue sums up this social, racial, and capitalistic production equilibrium as "an individual process . . . that . . . could only be conducted with large concentrations of land, labor, and capital. . . . Yet if custom shaped planters' attitudes, so too did racism and the refusal to allow black people to exercise independent initiative. Thus the traditions to which sugar and slavery gave rise fused with structural impediments to produce an inertia that proved almost impossible to overcome."[69] In short, slavery underpinned all aspects of antebellum plantation wealth. Slaves, tired of sweating in fields or spending long hours in noisy, dangerous sugar mills, fled to the swamplands in pursuit of freedom. For many, however, the journey proved impossibly arduous, so they returned to the plantation or were captured by patrols. Either way, they were punished through such cruelties as beating, branding, and imprisonment.[70]

Indifferent to the injustices of slavery, the antebellum plantation belt between Baton Rouge and New Orleans was heralded as a successful merger of art and industry. Per Henry Marie Brackenridge, mechanized plantations were powerful examples of an industrialized civilization.[71] That civilization, nonetheless, was "a society based on planter control, minimal governmental influence, and the tenets of white racial supremacy" where human chattel turned "white gold" into actual gold.[72] During the 1840s, south Louisiana's slave population was estimated at 42,000. By 1860, that number had more than doubled.[73] In St. James Parish alone, enslaved people comprised over seventy percent of the total population.[74] This large number mattered little to sugar barons, as "sugar remained their first interest. It was the basis of a livelihood that included comfortable homes, interesting readings, and stimulating travel" tantamount to a feudal lordship where the plantation master was king, his wife queen, and his children heirs to the throne.[75]

Regardless of their lordly lifestyles, Creole planters often felt removed from the busy, captivating urban center of New Orleans and, despite vocalized opposition, could do little to halt the transfer of the Territory of Orleans to the United States and its American government in 1803. As Michael O'Brien surmises, "A Creole could stand before the New Orleans city hall on 30 November to observe the Spanish flag being displaced by the French tricolor and then return three weeks later to see the French flag replaced with the Stars and Stripes. Seeing three imperial flags hoisted and lowered, he knew that his preference had not been consulted."[76] Their removed geographic location, however, did little to keep wealthy Creoles away from the southern metropolis. Sugar planters attended the French opera, slave markets, and met with well-to-do New Orleans factory owners, bankers, merchants, lawyers, and urban Creole kin.[77] Sugar barons used the Mississippi River to maintain a presence in New Orleans. As a result, Creole planters were closely tied to white, urban Creoles and enjoyed city entertainment, albeit far more infrequently than their New Orleans–based counterparts.[78] To facilitate stays in the port city, it was not uncommon for wealthy planters to own houses or rent lodgings in the French Quarter.[79]

Even though Creoles were the majority of sugar planters in the late eighteenth and early nineteenth centuries, a host of Anglo-Americans originating from Alabama, the Carolinas, Kentucky, Mississippi, Tennessee, and Virginia descended on Louisiana after the 1803 purchase to take up the mantle of "sugar planter." By the mid-1830s, Anglo-Americans had become a decisive presence in Louisiana's plantation communities. However, most settled in central and northern Louisiana, leaving the gold coast, for the time being, to Creoles.[80] Members of the ancienne population predominantly owned plantation properties above and below New Orleans. Conversely, Anglo-Americans purchased lands closer to Baton Rouge or near rivulets that branched from the Mississippi River and seeped into marshes such as Bayou Lafourche or Bayou Teche. In addition to proximity, Creoles noticed with some concern that Anglo-Americans reproduced faster than Acadians or French immigrants. As Protestants, Americans questioned south Louisiana's Catholic traditions, preferred English common law to French legal custom, were understood to possess boisterously outspoken personalities, and hungered for political power.[81] As the nineteenth century progressed, Americans and immigrants continued to invade south Louisiana, and populations soared. At the start of the nineteenth century, approximately 43,000 people occupied the gold coast. By 1810, sugar parish residents were estimated at 76,000.[82] In the face of increasing Anglo-American numbers, antebellum French Creole culture remained ubiquitous.[83]

Anglo-Americans not only desired political power, they had received formal educations far superior to those of their Creole contemporaries and quickly adapted to Louisiana's cultural climate. Carlyle Sitterson writes, "The Creoles, with their narrow experience and limited education, could not compete with the aggressive Anglo-Americans. Averse to change and complacently content, they built a social barrier of bitter resentment between themselves and the Anglo-American planters."[84] Although upper-class Creoles supported Whig principles, their political networks were meek compared to fierce Anglo-American unity. As American numbers grew, ethno-classist factions fueled staunch, heritage-based political lines—the Creole ancienne population versus the ever-increasing Anglo-Americans.[85] Ultimately, political alliances with immigrant populations decided which group held command.

Political Strife and the Burr Conspiracy

As German and Irish immigrants assimilated into American communities, French Creoles were outnumbered. Creoles disliked and distrusted Irishmen and Germans for proclaiming American allegiance and "waged a long war of cultural and social exclusivism to preserve their Creole culture and to brace up their slipping political strength in lower Louisiana."[86] Eventually, Creoles allowed their sons to marry American daughters in a futile attempt to squelch rivalries via breeding. Creole surnames, it was believed, would pass on to the next generation while American surnames disappeared. This practice, to any measurable degree, had little effect.[87] Resulting from a fear of Anglo-American ascendancy, the Creole elite formed alliances with the française étrangers.[88] Toward the end of the eighteenth century and during the first half of the nineteenth, most Southern port cities had experienced a substantial influx of Francophone migrants.[89] These immigrants were exiles from insurrections including the 1791 Saint Domingue slave revolt, the antireligious French Revolution (1789–1799), the *coup d'état* of Napoleon Bonaparte (1799), and the later French revolutions of 1830 and 1848.[90] By 1860, the foreign French made up approximately one-third of south Louisiana's French-speaking population.[91]

As hoped, political coalitions between Creoles and French immigrants successfully offset Anglo-American supremacy.[92] In addition to political support, French immigrants were notoriously well educated and more than a few had served as instructors at Parisian academies and universities. Prior to the founding of formal educational institutions in Louisiana, instruction was often in the hands of French immigrants. Members of the française étrangers served as private tutors and educated young men at their urban townhouses or rural plantation homes.[93] In accordance with Creole desires, lessons were geared toward culture, languages, and polite behavior instead of math and science. Wealthy Creoles, on occasion, forged business alliances with rural French laborers but never allowed these liaisons to "go beyond dollars and cents."[94] As Paul F. Lachance explains, "Once in Louisiana, immigrants struggled for position in a new society. Most immigrants had little more than reputation on which to stake a claim, while arriving Frenchmen of the petty nobility and members of the aspiring merchant-bureaucrat class . . . had their own aspirations."[95] The establishment and maintenance of schools and colleges was one such ambition; diplomatic position was another. Due to the fact that Louisiana's antebellum businessmen and politicians comprised Creoles, Anglo-Americans, and French immigrants, political feuds were inevitable.[96] With three weighty groups vying for power, the ethnic background of future governors was of immense concern, but in territorial Louisiana, governors were selected by distant figureheads, not elected.[97]

Therefore, when the twenty-eight-year-old, Virginia-born William Charles Cole Claiborne was appointed by US president Thomas Jefferson to serve as governor of the Territory of Orleans in 1803, Creoles immediately distrusted him. In return, Claiborne was suspicious of the self-proclaimed ancienne population and wrote disparagingly about their willingness to join the Union as members of a "state" rather than a "territory." Claiborne's comments only exacerbated Creole dissatisfaction.[98] In sum, "The inhabitants of Louisiana," as John Charles Dawson concludes, "were opposed to [United

States] territorial status with a governor and other officials to be sent from Washington to be over them, officials who knew nothing of the people, their manners and customs, and their language."[99] Determined that Louisiana become wholly American, Claiborne encouraged the progressive influx of US citizens, as he believed their presence would foster an overt sense of nationalism. To Claiborne's surprise, a conspiracy that spanned the eastern United States harmed American loyalty in New Orleans. This alleged plot, involving former US vice president Aaron Burr, gave Claiborne reason to doubt his Anglo-American brethren and forced the young governor to reconsider his ethno-political alliances.[100]

After his infamous 1804 duel with former US secretary of the treasury Alexander Hamilton, Burr finalized his vice presidency and journeyed south in March of 1805. Burr planned to reconnect with an old acquaintance in Louisiana, General James Wilkinson, and rebuild his shattered political career. Wilkinson, ex-pensioner for the Territory of Orleans's Spanish governors, oversaw upper Louisiana and commanded the United States Army in the Mississippi Valley. Aware of Burr's precarious post-duel state of affairs, Wilkinson was less than thrilled to receive news of the erstwhile vice president's impending arrival. Gossip circulated among politicians and Southerners alike that Burr schemed to incite discord between Spain and Louisiana in an attempt to divide western territories from the Union. Such a split, it was rumored, would induce Mexico to form a new western territorial government with Burr at the helm. The authenticity of these reports remains debatable. Not willing to risk the recently purchased Louisiana territory, Jefferson alerted southern officials in November of 1806 that a "friendly power" (Burr) might instigate war. The president urged all faithful Americans to apprehend duplicitous parties that provoke disloyal actions. When Burr's ship arrived in New Orleans, Wilkinson had to make a choice: help Burr or support Jefferson, the man who had raised him to his current political rank. Keen to ensure his station, Wilkinson broke ties with Burr.[101]

Before the vice president's duel, Wilkinson had supported Burr as a bureaucratic ally. Afterward, Wilkinson hedged away from this potentially treacherous connection, informed Jefferson of Burr's whereabouts, and began arresting Anglo-Americans who supported the former vice president. Learning of Wilkinson's betrayal, Burr fled but was captured in Alabama and extradited to Virginia to stand trial for treason. Tried and acquitted in 1807, Burr set sail for Europe to live in exile.[102] The conspiracy was settled, but Wilkinson's reign of arrests had created intense discord in New Orleans. Claiborne, though not fond of Wilkinson, saw promise in the general's anti-Burr activities. Several of Wilkinson's detainees were Anglo-American members of the territorial government and among Claiborne's fiercest detractors. No doubt Claiborne felt at ease with the removal of so many dissenters. Still, the governor had to contend with their supporters. A defender of the incarcerated Americans, Daniel Clark Jr. challenged Claiborne to a duel. The "gentlemanly" pistol fight resulted in agony when Claiborne was shot in the thigh and had to spend three months recuperating. As Powell hypothesizes, "From a political perspective, the wound was probably a blessing in disguise."[103] French Creoles saluted Claiborne for his courage and the elimination of Anglo-American administrators. In turn, Claiborne found a novel respect for the zealously sympathetic Creoles and appointed more than a few to offices once held by Americans. The governor's purge went so far as to cancel contracts with American newspapers and forge new printing agreements with Creole publishers.[104]

Wilkinson's arrests, though daring and highly criticized, served Claiborne well. "The thrust of the terror was," according to George Dargo, "against the English-speaking community. Not a single member of the French community stood accused of complicity." By the end of the Burr fiasco, Claiborne, a former supporter of common law, changed his legal tone in favor of the Creole preference, French Civil Code.[105] Recovering from the Clark duel, Claiborne signed into law the Civil Digest of 1808.[106] Creoles throughout south Louisiana applauded Claiborne for his approval of the new legal code as well as the appointment of their peers to political offices. When the American governor welcomed French immigrants from Saint Domingue and Napoleonic era France, Creole political support increased, and appreciation for the American governor swelled. In 1811, Claiborne firmly secured sugar planter loyalties by suppressing the largest slave revolt on American soil. Similar to the Saint Domingue insurrection, Louisiana slaves meant to abolish the institution of slavery and win their freedom. Planters feared that, if not dealt with promptly, the revolt would spawn the downfall of Louisiana's plantation economy. In response, Claiborne permitted swift yet atrocious actions that resulted in the deaths of ninety-five slaves. Creoles, pleased with their sympathetic, territorial governor, elected him to become Louisiana's first state governor in 1812. Claiborne ultimately fused with the Creole elite by marrying into an ancienne population family, a symbolic attachment that counted the governor as a kindred compatriot.[107] Fully immersed in Creole politics, Claiborne turned his attention to reproducing an ambitious endeavor from his Mississippi Territory administration (1801 to 1803): education.

The Failed Collège d'Orléans

As Lee Morgan points out, "Claiborne saw education as a way of uniting disparate elements of the population—ethnic, religious, and political."[108] The American governor's personal education, first at Richmond Academy in Virginia and later at the College of William and Mary, where curricula included basic sciences, languages, literature, and philosophy, no doubt influenced his personal ideologies.[109] Thomas Jefferson also inspired the young Virginian. In fact, Morgan contends that Claiborne was Jefferson's protégé. At the age of sixteen, Claiborne met Jefferson in New York and the two became comrades. Claiborne later moved to Tennessee and served as a congressional representative. All the while, he supported the president. Jefferson rewarded Claiborne's dedication by "naming him governor of the Mississippi Territory, then co-commissioner for the Louisiana Purchase, and finally governor of the Territory of Orleans."[110] Since the president had not traveled to Louisiana, he relied on Claiborne for regional news and leadership. Though Jefferson never visited Louisiana, "no one had more influence over the territory."[111]

While serving as governor of the Mississippi Territory, Claiborne authorized a bill to establish Jefferson Military College, an institution dedicated to his mentor. Impressed by Jefferson's educational ideals, the Mississippi governor believed that a centrally located college would "dispel the mental darkness" that plagued the territory. On May 13, 1802, the Mississippi government approved Claiborne's college proposal and formed an administrative board composed of residents from the communities of Natchez and St. Stephens. The new Mississippi institution had the right to confer both bachelor's and master's degrees, and, according to the governor's plan, would head a system of regional feeder schools. Claiborne even served as the governing board's first president from January of 1803 until May of 1805, albeit from afar in New Orleans.[112] Although Claiborne was involved in the college's early development and insisted that the institution be centrally located, board members squabbled over its exact location. Finally, it was settled that the college would be constructed in Washington, Mississippi, approximately six and a half miles east of Natchez. On January 7, 1811, Jefferson Military College, consisting of two redbrick buildings, opened with fifteen students in attendance.[113] Though named after the secular US president, a theologian, Reverend A. Stephens, was one of the first leaders to preside over the institution.[114] The Mississippi college was renowned for superior cadet training and remained in existence until the latter half of the twentieth century. Owing to low enrollments and financial insecurity, the college closed in 1967.[115] Restoration efforts to preserve the historic site began in 1977 and were completed in 1984. Today, the Jefferson Military College Foundation (organized in 1980), in conjunction with the Mississippi Department of Archives and History, preserves Mississippi's first formal institution of higher education.[116]

Similar to his Mississippi plan, Claiborne, in 1805, proposed a system of

Louisiana schools and libraries surmounted by a college to be located in New Orleans.[117] Until this point, Louisiana lacked government-supported education. Certainly, boys and young men had profited from informal New Orleans-based apprenticeships and private tutelage. Women in particular experienced quality education as the Ursuline Nuns and Sisters of the Sacred Heart established long-lasting academies—a notable achievement given the instability of both secular and religious education in eighteenth- and early nineteenth-century Louisiana.[118] In 1725, the Capuchin Friars opened a short-lived French Quarter boy's school on St. Ann Street. This venture, however, failed six years later due to insufficient capital and damaging arguments between the colony's Capuchins and Jesuits.[119] Instead of proselytizing apprenticeships or the sectarian school, French Creoles hired tutors to educate their sons. Overall, eighteen-century planters lacked interest in creating and supporting formal schools for either their own children or those born to less privileged parents.[120] Shugg adds, "To the richer slaveholders, who could provide their children with tutors and private schooling, the cost of educating the poor seemed prohibitive."[121] Personal libraries, in addition to private instruction, exposed Creole offspring to an array of literature and intellectual thought.[122] Novels purchased from Paris further reinforced quasi-Parisian lifestyles by instilling an appreciation for French literature in the home.[123]

As boys matured into Creole youth, they were sent to Europe to compound primary studies with personal experiences. This "grand tour" of the eastern continent heaped prestige on the ancienne population as they, unlike lower social classes, could afford costly educational ventures.[124] Sons of Anglo-Americans, on the other hand, were sent to northern institutions such as Harvard College in Cambridge, Massachusetts, or Yale College in New Haven, Connecticut. Creoles who wished to keep their sons on the continent enrolled them in Georgetown University in Washington, DC, St. Mary's College in Baltimore, Maryland, St. Joseph's College in Bardstown, Kentucky, or Transylvania College in Lexington, Kentucky.[125] While historians like Shugg and Sitterson reported a low level of concern among Louisiana's populace regarding formal education, others, including Shannon Dawdy and Sarah L. Hyde, indicate that a significant portion of early nineteenth-century Louisianans wished to craft an educational system that would provide superior public instruction.[126] To be sure, many Southerners disliked patronizing distant institutions of higher education, so they petitioned their state governments to develop southern colleges for their southern sons.[127] Under Claiborne's direction, the Louisiana legislature appropriated $15,000 to establish the Collège d'Orléans. This New Orleans institution, like its Mississippi predecessor, was meant to draw enrollment from a system of schools conjoined with supportive libraries, but as funding was insufficient, the college alone was constructed.[128]

French Creole and Anglo-American legislatures may have agreed to build a college, but they disagreed on the language of instruction. As New Orleans and south Louisiana were historically French, Creoles insisted that the college employ a purely French form of instruction that emulated ideals of nobility, gentry, and enhanced social graces sans moral and religious instruction. Conversely, Anglo-Americans believed that Louisiana, now a US territory, needed a college vested in the English language. To the latter's dismay, the ancienne population prevailed, and French became the pedagogical language at the New Orleans college.[129] Regarding ethnic predilections for language and learning, the early twentieth-century historian William Hamilton Nelson states, "There is no use in trying to disguise the fact that intense rivalry existed between the English-speaking people of Louisiana and the French. They were compelled to work together at times, but they realized that on the question of education as well as other questions there was a wide gulf."[130]

Designed by architects Claude Gurlie and Joseph Guillot and erected along St. Claude Street, the two-story Collège d'Orléans was fronted with slender columns and a double gallery.[131] The institution was situated on former Morand Plantation, which had been purchased by real estate developer Claude Tremé. City officials hoped that the college would attract residents to an otherwise poorly populated suburb, known locally as a *faubourg*. Most New Orleans residents, according to Henry C. Castellanos, did not approve of the muddy, swamp-like sector and "were loath to locate their domiciles in this noisome neighborhood."[132] As a result, the college languished until 1811.[133] When it finally opened on November 4, Jules d'Avezac, a migrant lawyer from Saint Domingue, was nominated to serve as its first president.[134] D'Avezac was not the only refugee to teach at the college. Other immigrant professors included Pierre Lambert, Paul Antoine Lanusse, Louis Moreau-Liset, and James Pitot.[135]

The college's first cadre of students consisted of seventy boys, but before long, enrollment swelled to one hundred. Despite a brief break in classes due to the 1815 Louisiana invasion of British troops in the Battle of New Orleans (the last major skirmish of the War of 1812), the new institution had overcome its suburban confines and experienced a healthy beginning. However, with d'Avezac's resignation and return to the legal field, Professor Carr, the institution's English instructor, assumed the presidency, and the situation turned dire. Carr was implicated in a financial scandal associated with the college's budget and was discharged.[136] Soon thereafter, a Saint Domingue poet, M. M. Nicolas Rochefort, took over as president. The refugee poet, like his predecessor, was a piteous leader. While versed in Latin, ancient history, and prose, Rochefort was, as a former pupil tells, averse "to mathematics, and could not bear any allusion to that science. If such a subject was introduced, he became nervous and fretful. It was like presenting water to a mad dog."[137] Even among his peers, Rochefort was negatively perceived. The college's professor of mathematics, Teinturier, particularly disliked the poet and his dedication to the written page. "Mr. Rochefort! Ha! Ha! A mere coiner of rhymes! A manufacturer of jingling sentences. A fine authority truly! A man who could not go through one of the simplest operations of the multiplication table!" In like manner, Rochefort abhorred Teinturier. "Good God! that some of my best pupils should be exposed to be spoiled by that man, their imagination chilled, and their poetic fire extinguished just as it began to expand!"[138] To be sure, early accounts depict the Collège d'Orléans's faculty as a motley crew of eccentrics who rarely saw eye to eye.[139] Under Rochefort's direction and without any intervention from Claiborne (the American governor had died in 1817), conditions worsened, and student enrollment declined to a mere twelve boys by 1821. Heavily in debt, the college's enrollment further decreased to seven pupils, and Claiborne's original idea of a system of primary schools surmounted by a stable form of higher education was abandoned. With the college on the brink of collapse, city officials began to wonder if a "gentlemanly" education was what Louisianans really needed.[140]

When Rochefort resigned, arguments between French Creole and Anglo-American legislatures erupted over who should administer the college.[141] On May 30, 1822, the institution's governing board decided that Joseph Lakanal, a French immigrant living outside of Louisiana, would serve as president. Lakanal's administration, however, sealed the Collège d'Orléans's fate.[142] Prior to his career as a professor of rhetoric in France, Lakanal studied theology at the Seminary of Saint Magloire of the Doctrinaires in Paris and was ordained a priest. Religious life must not have appealed to him, as he departed the vocation. (Other accounts posit that Lakanal was defrocked.) In 1782, Lakanal received a doctorate of arts from the University of Angers and taught philosophy at Bourges and later at Moulins until the outbreak of the French Revolution. The former priest, now philosophy professor, became a public spokesman for the rebellion. In 1792, Lakanal rose to the position of Deputy for the National Convention, and, as Simone de la Souchére Deléry states, "voted death for Louis XVI."[143] Lakanal was ultimately appointed director of the Paris Public Education Committee, helped establish the École Normale Supérieure, and opened a string of public schools throughout Paris. His reputation as an educator, his association with the French Academy, and status as a co-founder of the École Impériale Polytechnique followed Lakanal to the United States.[144] In America, as in Paris, he was noted for his outward dislike of French nobility and royal hierarchies in general. Lakanal's opposition to the French monarchy and death vote for Louis XVI marked him as a regicide, a title that, along with his status as an allegedly defrocked priest, clouded his social and educational reputation in Louisiana.[145]

Subsequent to Napoleon Bonaparte's first abdication in 1814, Lakanal grew leery of the returning Bourbon monarchy. Indeed, following Bonaparte's renunciation, the deceased Louis XVI's brother, Louis Stanislas Xavier, the Count of Provence (later known as Louis XVIII), announced his succession to the throne as his nephew, Louis-Charles, Duke of Normandy and Royal Prince of France, had died in prison after his parents' (Louis XVI and Marie Antoinette) arrests and beheadings. He also declared his intent to punish all regicides. The count's announcement was accompanied by the publication *Procès ou Assassinat juridique de Louis XVI, Roi de France et de Navarre* (Trial or Legal Murder of Louis XVI, King of France and of Navarre), in which Lakanal was listed among others who supported the former monarch's beheading. Before the text could be widely circulated, Lakanal liquidated his estate and crossed the Atlantic Ocean with letters of introduction from a handful of prominent French individuals, including Marie-Joseph Paul Yves Roch Gilbert du Motier, the Marquis de LaFayette, a name esteemed by US citizens for his involvement in the American Revolutionary War (1775–1783).[146] Leaving France in 1806, Lakanal arrived in New York, and after one month he traveled to Philadelphia.[147]

Lakanal was not a champion of royal hegemony, but he did enjoy a comfortable lifestyle, which he was denied in the New World. His sparse living conditions reminded him that if Napoleon had remained in power, the former priest would have increased in wealth and prominence as a loyal Bonapartist.[148] Perhaps for this reason, while residing in Philadelphia, Lakanal commiserated in a plan to rescue Napoleon from his exile on the island of St. Hélène in the South Atlantic.[149] American officials, in 1817, intercepted a package addressed to Joseph Bonaparte, Napoleon's nephew, which contained six letters detailing arrangements to smuggle the deposed emperor to New Orleans. All were secured by a wax seal that read, "Lakanal, Deputy to the National Convention." However, lacking sufficient evidence to indict Lakanal, the US government refused to take action.[150]

Relieved to have escaped incarceration, Lakanal quieted his political activities and moved to Kentucky.[151] Though the plan to liberate Napoleon never transpired, a testament to the aborted scheme exists to this day. In the New Orleans's French Quarter at 500 Charters Street stands the former home of mayor Nicholas Girod. Now a bar and restaurant known as the Napoleon House, Girod purportedly offered the building to shelter Napoleon should he be rescued and brought to the American South.[152] Despite the fact that Napoleon never landed on American soil, wounds from the Burr conspiracy were relatively fresh and the plot to rescue the ex-emperor was, at least to Louisiana's Creoles, disturbingly similar to the Burr debacle.[153]

While in Kentucky, Lakanal received a letter from the Collège d'Orléans governing board encouraging him to relocate and take up the vacant college presidency. Members of the board were no doubt aware of his educational reputation, but, perhaps, were not fully aware of his supposed involvement in the failed plot to rescue Napoleon. Pierre Guillot, fellow regicide, mathematics professor, and friend of Lakanal, informed the board of his colleague's experience and presence in Kentucky. Admitting that the Collège d'Orléans was in decline, the board believed that adding Lakanal's pedigree to the faculty might improve wealthy Louisianans' perception and patronage of the institution.[154] When he arrived in New Orleans, Lakanal acted quickly to reform the ailing college. He expanded curricula to include botany, engineering, furniture construction, Louisiana geography, and woodcarving. Lakanal believed that enhancing students' practical knowledge of levees might aid in diverting disastrous Mississippi River floods. Therefore, he drew literature from countries that dealt with similar flooding issues, specifically the Netherlands. By the end of 1822, Lakanal had increased enrollment from seven to eighty students.[155] In spite of his good works, however, pride overtook the new college president.[156]

Lakanal had succeeded in reforming the Collège d'Orléans, but his ego wore on the public's nerves. Boasting of his meritorious administration, Lakanal declared himself to be the college's rescuer and savior of New Orleans's youth. New Orleanians quickly tired of the vociferous foreigner and wanted him removed from the college presidency, New Orleans, and Louisiana as soon as possible. Deléry explains: "Lakanal, pompous, pedantic, sarcastic, and domineering, never succeeded in being accepted by local citizens."[157] New Orleans's Creoles sought any reason to vilify the French immigrant. Soon, they discovered, although it was no great secret, that Lakanal had sought refuge in America after the conclusion of the French Revolution because he supported the insurrection, acted as a member of the National Assembly of France, voted for the decapitation of a French monarch, and, finally, played a role in the plot to rescue Napoleon.[158] In the end, as Alfred N. Hunt makes clear, "New Orleans residents did not approve of having their sons educated by a regicide."[159] As a direct result, Lakanal's presidency ended on July 22, 1823, after the college's board voiced their disfavor.[160]

College authorities removed Lakanal not only due to his political background but because he rarely included them in institutional decisions. For the most part, New Orleanians were suspicious of Lakanal. Those loyal to France's Bourbon monarchy denounced him not only as a regicide, but an ex-priest, a deist, and even an "anti-Christ." His defenders implored the public to judge Lakanal on his successful college administration alone, but New Orleans's citizens declined, preferring to close the college rather than allow the vilified educator to remain in the city.[161] Not long after, he relocated to Mobile, Alabama, and remained there until 1837. That year, Lakanal, tired of the New World, returned to France. Arriving in Bordeaux, the aged educator traveled to Paris alone (his wife had died and was buried in Mobile). By the end of 1844, it was reported that Lakanal suffered from a cold or possibly pneumonia, which "compelled him to take to his bed." His condition worsened, and, in 1845, the former president died. His body was laid to rest in the Père Lachaise Cemetery.[162]

Regardless of Lakanal's termination from the Collège d'Orléans, his opinion still held some sway, as his son-in-law, Lucien Charvet, was appointed acting president following his father-in-law's recommendation. The college's governing board soon discovered that Charvet, like Lakanal, proved a poor choice. Under his administration, the institution once again struggled, and in 1825 it closed.[163] In the years that followed, the surrounding marshes were drained, and the *faubourg* became heavily populated. On August 12, 1826, the city of New Orleans sold the college property to Félix Pinson and Evariste Blanc, who then divided the campus into smaller lots for private sale.[164] With the suburb's expansion, today known as Faubourg Tremé, Ursuline Street was extended beyond the French Quarter and a portion of the campus was demolished. Thereafter, the first New Orleans Catholic church devoted to African Americans, St. Augustine's Church, was built over a section of the college's foundation. When Castellanos published his 1895 book, *New Orleans As It Was: Episodes of Louisiana Life*, two wings of the old college building still stood on St. Claude Avenue (present day Henriette Delille Street).[165] The remaining structures were torn down years later, and today, a grassy patch exists where antebellum students once studied.[166]

Notwithstanding the Collège d'Orléans's closure, it produced several educated Creoles who became prominent New Orleans citizens. For example, the noted Louisiana politician, judge, historian, and grandson of de Boré, Charles Étienne Arthur Gayarré, studied at the college.[167] Gayarré remembered his academic mentors, particularly Rochefort, well. While on his deathbed, Rochefort summoned his pupil to impart a lasting reminder. Upon entering his professor's domicile, the Saint Domingue native saluted Gayarré in his best Latin, "*Moriturus te salutat, O pater conscriptus!*" ("He who is soon to die, salutes thee, O conscript Father"). Watching his instructor slip into oblivion, Gayarré listened as Rochefort affirmed, "You are my work, boy; you are my work—never forget it!"[168] In *Imagining the Creole City: The Rise of Literary Culture in Nineteenth-Century New Orleans*, Rien Fertel candidly points out the permanent results of Gayarré's early education: "Rochefort and his colleagues, all refugees of the Haitian Revolution, impressed on Gayarré the conception of white Francophone exceptionality across the Atlantic World, an idea that informed his future histories."[169]

Thereafter, the closure of the Collège d'Orléans spawned years of debate between ancienne population and Anglo-American politicians, which gave rise to a feudalistic pair of colleges, one American and one Creole.[170] Without hesitation, Anglo-American legislatures led by Governor Henry Johnson secured funds formerly reserved for the Collège d'Orléans and reallocated them to create the Anglo-American College of Louisiana.[171] One can imagine French Creoles' fury when the "American" college was chartered. Creole anger only intensified when, in 1825, the "legal language" of New Orleans was converted to English.[172] In defense of their linguistic legacy, Creoles pronounced French to be a language far superior to English.[173] These schismatic attacks increased regional sectionalism concerning education and, more specifically, the geographic location of colleges. Indeed, as John P. Dyer explains, "nothing seemed to provoke a rash of [sectionalism] more than the location of institutions of higher learning."[174] Tempers mounted when Anglo-American legislatures decided to locate the College of Louisiana in East Feliciana Parish—predominantly populated by Americans.[175] A governing board, principally consisting of Anglo-Americans, not only administered the new college, they rooted it in Protestantism. Antebellum Creoles may not have been devout Catholics, but an American Protestant college was too much to bear. Infuriated by yet another Anglo-American triumph, Creoles pushed for the election of a French Creole governor and the establishment of a Creole college.[176] Despite their efforts, a French immigrant, Pierre Augustin Charles Bourguignon Derbigny, won the 1828 gubernatorial election.[177]

Amid Creole grumbling, Derbigny, a member of the National Republican Party, not the Creole-preferred Whig Party, took office with little intent to appease the ancienne population. French Creoles did not have to wait long before a second chance at government control asserted itself. On October 6, 1829, Governor Derbigny "died of injuries sustained a few days earlier when bolting horses . . . flung him from his carriage."[178] With Derbigny's death, a discovery surfaced that the Louisiana Constitution of 1812 included no provision for a lieutenant governor. Instead, the senate president would become governor until his annual replacement. State control then fell to Armand Julie Beauvais. When the senate reconvened in January of 1830, Jacques Dupré became the acting governor. Halting the senate-governor tumult, Louisiana legislatures, as John M. Sacher details, "realiz[ed] the absurdity of this gubernatorial merry-go-round . . . [and] scheduled an election for July 1830, despite the absence of any constitutional sanction for such an event."[179] Eleven politicians including representatives from the American and Creole camps vied for Louisiana's top administrative position. Before the election ended, several candidates resigned in support of an experienced, well-connected, and highly persuasive French Creole candidate, André Bienvenu Roman. A ten-year legislative veteran and prosperous sugar planter, Roman surpassed Anglo-American contenders, William S. Hamilton and David Randall, to become Louisiana's ninth governor on January 31, 1831. He is to this day remembered "as one of Louisiana's most distinguished Creoles in antebellum days."[180]

The Creole governor was not only a successful state leader and plantation owner; he was also an unyielding supporter of education.[181] In the years succeeding Roman's election, an estimated one and a half million dollars were spent on education.[182] With a Creole sugar planter leading Louisiana, members of the ancienne population were poised to retake political control and design a college in their own image that would rival and eventually overtake the College of Louisiana. Although this Creole college experienced years of economic strife, war, rebuilding, and reform, it lasted well into the twentieth century. Dedicated to secular aristocratic ideals embodied by Thomas Jefferson and emboldened by Parisian academic ideology, the new institution was christened Jefferson College. Replicating the sectionalist choice to locate the College of Louisiana in East Feliciana Parish, Jefferson College was constructed in St. James Parish, the heart of French Creole planter aristocracy. Apart from numerous administrational shifts, multiple closures, name changes, and a foundational conversion from secularism to Catholicism, this early form of Louisiana higher education beheld the changing social strata of Louisiana's sugar empire and echoed ethnic ideals synonymous with privileged plantation society. Above all, the naissance of Jefferson College, endorsed by the newly elected Governor Roman, reinforced Creoles' claim to political and educational ascendancy.

2

The Rise and Fall of Jefferson College

> Ten years, Jefferson shone in the heaven of the humanities and of the sciences; ten years, it lived brilliantly, carefree and skeptical. . . . But, one day, God grew tired . . .
>
> —René de Sennegy

Thomas Jefferson, one of the most influential minds apropos early American schools, colleges, and universities, insisted that boys and young men sent to Europe for their education returned foreigners.[1] Although Louisiana sugar barons respected Jefferson as a fellow plantation owner and defender of freehold farming, they dismissed the president's home-based instructional ideals and continued to send their sons across the Atlantic for studies in law, liberal arts, and medicine.[2] This practice persisted until the ancienne population found sufficient cause to establish a college in south Louisiana. When the need arose to craft their own educational institution, Creoles hearkened to Jefferson's reasoning for local higher education, but took his assertion a step further and concluded that northern institutions had the same alienating effect as European academies. For example, Jean-Bernard Xavier Philippe de Marigny, the New Orleans Creole plantation owner whose lands later became Faubourg Marigny, sent his son north for higher education. After studying a short while, Marigny's son returned chewing tobacco and swearing, possessed with a general boredom for all things Louisiana.[3] Profanities, however, were the least of Southerners' worries. They were chiefly concerned that abolitionist-minded Northerners might influence their "impressionable" plantation legatees.[4] French Creoles found "American" and "anti-slavery" ideals noxious and would not tolerate such characteristics in their offspring.[5] Indeed, the reasons to create a local college were myriad: proximity and the necessity to maintain a sense of regional familiarity were key justifications. To be sure, Jefferson

College was crafted to be a true "Jeffersonian" institution—secular, strict, and intellectually meticulous—that would help preserve Louisiana's Creole elite by instilling all that was customarily correct per ancienne population thinking. However, there was another, more immediate purpose for the Creole college's founding: to oppose Anglo-Americans and their College of Louisiana.

To settle enmities between Creoles and Americans, the Louisiana legislature appropriated funds to maintain both institutions.[6] Edwin Whitfield Fay surmised that Jefferson College "owes its origin without doubt to the divided sentiments of the French and English populations of the state. The Collège d'Orléans had been under the former influence, but the College of Louisiana with its American traits, had supplanted it. Now Jefferson College was set up in opposition to the latter, and the efforts of both were doubtlessly paralyzed by class rivalry."[7] French Creoles, tired of meddlesome American politicians, viewed the College of Louisiana as an affront to their gentrified culture, regional heritage, and yet another attempt to usurp authority. If the College of Louisiana was meant to be a "stronghold" for Anglo-American students, then Creoles required a college to safeguard Creole youth.[8] Mary Bernardine Hill put it bluntly: "Rich French plantation owners . . . desired an education for their sons befitting their state in life."[9] Certainly, Jefferson College was created to oppose American education, but the long-lasting purpose was to bolster the ancienne population by way of a centralized institution devoted to teaching and retaining Creole youth.

The College of Louisiana, though intended to be secular, absorbed Protestant traits with the inauguration of its first president, Reverend Jeremiah Chamberlain. A Presbyterian minister, Chamberlain acted as president of Centre College in Danville, Kentucky, before serving at the College of Louisiana.[10] Creoles watched in contempt as the American college grew under its Protestant administration. Before Pierre Derbigny's gubernatorial administration came to a cataclysmic end, Creoles realized that in order to regain political prowess and promote their own educational enterprise, they needed a Creole leader, preferably a plantation owner, as governor. With great fanfare, they installed André Bienvenu Roman. His first term as governor lasted from 1831 until 1835. Later, in 1839, he was elected a second time and remained governor until his retirement in 1843.[11]

Born March 5, 1795, to Jacques Étienne Roman and Marie Louis Patin, A. B. Roman could, like most French Creoles, trace his family lineage directly to France.[12] As a young man, he received his education first at St. Joseph's College in Bardstown, Kentucky, and later at St. Mary's College in Baltimore, Maryland. After returning to Louisiana in 1815, he took up the sugar planter trade in St. James Parish. There, Roman married Aimee Francoise Parent and fathered eight children. In 1818, he was elected to the state legislature. Roman later served as speaker of the lower house for four years, St. James Parish judge for two, and in 1829 he was promoted to Speaker of the Louisiana house.[13] Not only was Roman respected in the sugar parishes, he was regarded in New Orleans as a prominent and influential Creole. In addition to his St. James Parish plantation home, Cabanocey, Roman occupied rooms in a French Quarter townhouse at 611 Royal Street. From this location, he entertained Creole peers, politicians, and visiting celebrities.[14] Before leaving the state speakership, Roman served as one of eight Creoles on the predominantly Anglo-American College of Louisiana board of directors (twenty members in all). This first-hand experience reviewing the American college served him well when he propositioned the state to found and support Jefferson College.[15]

In the wake of Roman's gubernatorial installation, he and his contemporaries put forth legislation to incorporate the College of Louisiana's academic equal.[16] With a throng of Creole supporters, Roman was well positioned to oppose American legislatures. Twenty-six years after his death in 1866, the governor was described as "the most remarkable creole that fine race has thus far produced. . . . Roman founded Jefferson College, in the parish of St. James, and there built up, for the state of Louisiana, an educational institution on par with Princeton and Yale," an epitaph that would have made any Creole brim with pride.[17] Although Jefferson College never matched Princeton or Yale, education clearly interested the governor. Roman was particularly keen to promote the establishment of an educational institution in his home parish of St. James. He described the proposed college as a cultural haven where the sons of Louisiana's original population could be educated without leaving their native land.[18] Before Jefferson College's charter was approved, Roman led a fundraising campaign in September of 1830 that yielded $50,822.[19] Wealthy Louisianans, including but not limited to Dominique Francois Burthe, attorney general Étienne Mazureau, and James H. Shepherd, provided most of the aforementioned funds. With their contributions, the college's destiny was wed to the regional aristocracy.[20]

Founding, Construction, and Early Curricula

After garnering enough money to launch Jefferson College's construction, Roman petitioned Robert Carter Nicholas, St. James Parish's legislative representative, to introduce a bill that incorporated the new institution on February 28, 1831.[21] While housed in the New Orleans Masonic Library, the original Jefferson College charter, signed by James Shepherd, A. B. Roman, Étienne Mazureau, Dominique Burthe, and Valcour Aime (Governor Roman's brother-in-law), was reportedly lost in a fire.[22] Nevertheless, other charters were attached to the institution as it passed through the nineteenth century.

As Roman's loyalties lay with south Louisiana's French Creoles, he championed Jefferson College and criticized the College of Louisiana. His gubernatorial predecessor, Jacques Dupré, propositioned legislators for the continued improvement of the American college, but Roman, with sufficient knowledge of the institution's performance from his service on the board of directors, argued against Dupré and insisted that the East Feliciana Parish college had fallen short of the state's expectations to provide valid instruction. The Creole governor proclaimed that Jefferson College would succeed where the College of Louisiana had failed.[23] The argument worked, and Jefferson College's charter was approved.

While the Creole college's governing board could have chosen New Orleans as the new institution's locale, they decided against the southern metropolis, claiming three reasons: the failure of the Collège d'Orléans, the city's high crime rate, and the regular occurrence of yellow fever.[24] New Orleans and other densely populated Louisiana cities were prone to the deadly disease, but the rural St. James Parish remained relatively free of epidemics. As well, the Creole-populated St. James Parish site significantly distanced Jefferson College from the College of Louisiana.[25] Parish residents had supported education in the past by enrolling their daughters in a female academy connected to the local Convent of the Sacred Heart. Creoles were so pleased with the convent and educational work of the Sisters of the Sacred Heart that when the surrounding township was recognized as a city, citizens chose to name it "Convent."[26] Later, in 1869, Convent became the parish seat of government.[27] There was yet another reason for the placement of Jefferson College in St. James Parish: river access. As Robert Mark Rudd explains, "The Mississippi River, the economic lifeline of the region . . . was the primary vehicle for cotton, sugar, and finished goods."[28] Riverboats not only transported agricultural cargo, they ferried students to and from Jefferson College announcing their arrival with songs trumpeted on steam calliopes.[29] This mode of transportation was typical for Creole planters, as members of the ancienne population used the red paddle-wheeled ships with their black smoke stacks to trek to and from New Orleans. It is not hard to imagine these all-important vessels transporting architects and building materials to Convent in the early 1830s to aid in the college's construction.

On July 2, 1831, $3,500 of the originally gifted $50,822 was used to purchase sixty-five acres of Jean Vavasseur's plantation property.[30] Continuing to purchase necessities in accordance with antebellum plantation life, the new college's governing board spent $5,000 on slaves for building purposes.[31] While some slaves molded and baked bricks of clay dredged from the Mississippi River, others planted oak trees to form adjoining allées.[32] By the mid-1830s, Jefferson College, as an independent corporation, owned sixty slaves.[33] The Creole college was not the first educational institution to participate in slavery. The practice of owning and selling human life had, in fact, been a part of the American higher education landscape since the colonial era. Professors and students at Harvard College (founded 1636) owned slaves, while southern colleges, including but not limited to the College of William and Mary (1693) and Emory University (1836), also engaged in chattel servitude.[34] Not only did slaves build antebellum Jefferson College, its architecture and landscaping reflected traditional, symmetrical plantation arrangements with the main building facing the river. According to Jessie Poesch and Barbara Sorelle Bacot's description of Louisiana sugar plantations, the campus layout appears identical.[35]

"According to the Roman family tradition," Oliver P. Carriere explains, "Joseph Pilié was the architect of the Jefferson College buildings. He was the father-in-law of Governor Roman's brother. The family used no other architect for any of its buildings."[36] Pilié, a refugee from the 1791 Saint Domingue slave revolt, served as New Orleans's city architect from 1818 until his death in 1842.[37] He also planned and oversaw the construction of Bon Séjour, a second Roman family plantation (known today as Oak Alley), between 1837 and 1839.[38] Architectural designs of the first college building have been lost; however, Samuel Wilson Jr. alleged that the central structure was three stories of exposed redbrick with white columns. Wilson hypothesized that the college's facade was meant to resemble Thomas Jefferson's University of Virginia but was remarkably similar to Jefferson Military College in Washington, Mississippi. He further posits that the major difference between Jefferson Military College and Jefferson College was that the latter had white fluted columns and a promenade. Pilié, the presumed architect of both Bon Séjour and Jefferson College, might have employed similar structural features at both sites, but Wilson's architectural theories are based on little primary documentation. Aside from newspaper advertisements requesting professional plasterers and furnishers, no other evidence has been uncovered to support his assertions.[39]

Historians Robert E. Hunter and Edwin L. Dooley Jr. assert that the original Jefferson College building was "300 feet long and 44 feet in depth, with Greek columns on all four sides."[40] Behind the main building was a second wing, one hundred feet deep by forty-two feet wide and constructed at a right angle to the main college building. Combined, the two buildings could accommodate approximately 300 students.[41] A high brick and cement slurry wall enclosed the main building, front lawn, and two porters' lodges (built in 1836).[42] Inside the surrounding wall, adjacent to the main college building, were two covered recreation rooms that provided protection for students during inclement weather. Five additional brick houses to accommodate the college's professors were planned, but only the president's residence was built, which for the vast majority of the college's existence functioned as a hotel. Like other Jefferson College buildings, the president's house reflected regional architecture. Patricia L. Duncan, while studying the Creole college's architecture, observed that the president's house, upon completion, resembled "a small version of the typical Louisiana plantation home."[43] The campus also possessed privies, a cesspool, and, to the dismay of those enrolled, a "place where the students undergo solitary confinement during the day time as a punishment . . . [which] may be called a prison. This is a small building in the rear of and detached from the main building, about forty yards distance, the upper story is divided by a narrow passage running the whole length of the building and on each side are situated cells, nine in number." The effectiveness of this punitive practice to curtail negative dispositions was not recorded, but by the mid-1840s the Louisiana legislature "recommend[ed] the immediate suppression of this mode of punishment."[44] Though the college was chartered in 1831, almost three years passed before the physical plant was complete and students could enroll.[45] In December of 1833, Louisiana newspapers advertised that classes would begin in January of 1834.[46] The college's governing board restricted admission to students between the ages of nine and fifteen who could read, write, and produce evidence of respectable conduct. Applicants over the age of fifteen were not accepted "without special vote of the Board of Directors."[47]

In total, the cost of construction (not including slaves or educational equipment) was $124,586.97. As well, additional land was purchased for $10,000.[48] The governing board also spent $8,710.15 on 7,000 library books, $600 on equipment for the physical science lab, and an additional $600 on tools and materials for the chemical laboratory. Instructor salaries for the first year were estimated at $28,120. As costs exceeded budgetary limits, the college generated considerable debt. Despite Roman's solicited $50,822, an additional $62,591.20 allocated by the Louisiana Legislature, and yearly tuition revenue that generated between $250 and $300 per student depending on curricular choices, the college owed $61,849.55 in unpaid bills for ancillary property expenditures, furnishings, mathematical equipment, and other unrecorded miscellanea.[49] If the buildings, along with their apparatuses, books, and furnishings, were destroyed, the college would have been cleared of debt, as the physical plant was insured for $62,000.[50]

As Creole planters were accustomed to immense financial liability, the college debt appeared not to have caused overt concern. Likewise, French Creoles registered no complaint when students were required to supply, upon entrance, a set of silver utensils and goblet. Creoles wanted their college to reflect Parisian aristocracy through material worth as well as secularism. Even though some Catholic and Protestant Louisianans viewed France's worldly government with suspicion, Michael O'Brien explains, "irreligion was part of the sensual thrill of French culture." Ergo, casual levels of indifference toward religion denoted a mark of social elevation, but, as time passed, religion slowly seeped into the daily social and educational lives of Jefferson College's wards.[51] Echoing Parisian religious separation, it is little wonder, then, that the college bore the moniker of Jefferson. Louisianans remembered Thomas Jefferson as the US president who provided the territory with civil and religious liberty.[52] To French Creoles, Jefferson not only represented exemplary plantation ideals, he expressed a deep appreciation for Parisian lifestyle, a sentiment that the ancienne population held dear.[53] Creoles considered Jefferson to be a model plantation owner who "embodie[d] a social type" and who "was born to and lived upon land and slaves, inhabited a place for a lifetime, and wrote about its meaning."[54] Albeit an American, Jefferson was an archetype of genteel nobility that could be recreated in a world set apart from Europe.

Creoles wanted their college to provide a liberal education reflective of Jeffersonian ideals and devoid of theology. As Charles P. Roland reasons, "In the lives of many [sugar] cane proprietors religion was not a compelling force for perhaps they were more nearly able to achieve a 'heaven on earth' than was the general run of mankind."[55] To reinforce irreligious philosophical instruction, the college employed French immigrants who embodied early nineteenth-century Parisian thought. As a result, Jefferson College became immensely popular among Creole planters.[56] Understandably, the college's nonsectarian nature caused conflicts with Catholic clergymen.[57] Priests, nuns, and deeply religious lay-persons disliked the secular culture of the gold coast and pointed to Jefferson College as a torch bearer for worldly teachings reinforced by Voltaire, Masonry, and pure ignorance.[58] In spite of clerical angst, "Jefferson saw its courses followed by the elite of Creole youth."[59]

Concerning the teachings of Masonry, Phillip H. Dagneau explains, "There was a Masonic tone in the teaching of these men of the early college faculty and administration."[60] To be sure, Freemasonry had found its way into south Louisiana long before the Louisiana Purchase of 1803. From the late eighteenth century forward, Masonic lodges had been established in New Orleans and replicated the Anglo-American/French Creole ethnic split, as some lodges adhered to the York Rite, while others paid homage to the French Rite.[61] As wealthy sugar planters traveled to New Orleans for business or pleasure, they also attended Masonic lodge meetings. Even though Masons were present in the overtly secular Jefferson College faculty and administration, Catholic students were not barred from observing their personal faith. Indeed, they were permitted to attend Mass at the nearby St. Michael's Church and study the Catechism with a parish priest if they or their parents so desired.[62]

Because of their parentage, social status, and educational achievement, French Creole youth attained prominent positions in law, medicine, and education.[63] By and by, students of varying ethnic backgrounds attended Jefferson College and, like their Creole peers, experienced varying levels of post-college success. The first bachelors of arts degrees conferred at Jefferson College were bestowed on Ignace Hebert, Godegrande LaBranche, and Similien LaBranche. Certificates of mercantilism were issued to Louis Hebert and Charles Robin Delogny.[64] Later graduates included Ernest Canonge, member of the Louisiana State House of Representatives; Michel Jean Fortier, private in the Louisiana Artillery, Fourth Battalion, St. Mary's Cannoneers; Samuel M. Robertson, Louisiana congressman; and Wilhelm J. Waguespack, New Orleans attorney and member of the Waguespack planter family.[65] Alfred Roman, son of Governor A. B. Roman, also attended the Creole college. Admitted to the Louisiana bar in 1845, Alfred later served in the Confederate Army. In 1880, he was appointed judge of the New Orleans criminal court and penned *The Military Operations of General Beauregard*.[66] George Eustis, Louisiana congressional representative, was also educated at Jefferson College. He went on to attend Harvard Law School, practiced the legal trade in New Orleans, and served as secretary to the Confederate States Legation in Paris. After the Civil War, Eustis remained in Europe as a US delegate during the Franco-Prussian War.[67]

While wealthy Louisianans attended Jefferson College, the state legislature charged the institution to provide tuition-free education for approximately twenty impoverished students.[68] Governor Roman was reputed to support gratis education and stated that "common schools, wholly free, . . . break down the odious distinction which exists in Louisiana between the children of the poor and those of the rich."[69] While Roman permitted free admission for a few poor students, he did nothing to reduce the price of tuition at Jefferson College. Apparently, aside from public schools, higher education was a privilege reserved for a select few. As Lillian C. Bourgeois explains, "Educational opportunities were not equally distributed and the Acadian children shared little with the wealthier Creoles."[70] Eventually, sons of Acadians attended Jefferson College despite their parents' perception of liberal education as an impractical and expensive luxury that did little to enhance an agrarian lifestyle. To an Acadian farming family, money exhausted on a child's education could have been better spent on the cultivation of land. Also, sending sons to college or preparatory school meant the loss of able field hands.[71] What college life was like for these pauper students is relatively unknown. At Jefferson College's predecessor, the Collège d'Orléans, young men from upper-class families hurled insults at "plebian" students to remind them of their diminished social status.[72] Disregarding possible taunting, enrollment increased during the 1830s. Academic catalogues containing enrollment data for the genesis years of Jefferson College were either never printed or have been lost; however, Robert M. Lusher, a nineteenth-century Louisiana superintendent of education and graduate of Jefferson College, approximated that 115 students were enrolled by 1835, a decent number given that most rural Louisianans were too poor to attend or generally disinterested in formal education.[73]

With a healthy student population, the Creole college maintained an active daily agenda of classes, recreation periods, regimented meals, and strict curfews—all without religious instruction.[74] This element, however, was not the only aspect missing. In particular, the college lacked a bathhouse. According to the *Statutes, Plan of Education, and Rules, Adopted in Jefferson College, Louisiana*: "Until a proper Bathing house shall be provided, the students bathe in the [Mississippi] river. . . . They are not, however, allowed to bathe without the permission in writing of their parents or guardians or their agents. . . . The students bathe under the supervision of at least one Prefect, and while they are bathing there is a boat in attendance with oarsmen to afford assistance as occasion may require."[75] Perhaps to compensate for the lack of facilities and the conceivability that bathers could be swept away by river currents, students were allowed personal servants. The original college statutes describe the presence of male servants (slaves) who slept near the dormitory always ready to "render the students such services as they may require."[76] Allowing slaves to follow their young masters to college was not unusual.[77] Michael David Cohen remarks that these practices "reproduced the hierarchy of the antebellum plantation" and kept planters' sons attuned to their position as commanders of human life.[78] When Creole planters sailed to New Orleans with their children in tow, they brought slaves to look after the young masters and mistresses, carry luggage, and transport packages.[79] While the specific duties of personal servants were not detailed in the Jefferson College statutes, the mere mention reinforced students' familiarity with the servant/master relationship: one dominant, one subordinate.

Although students were afforded the luxury of servants, the order of the day was militant. Even if Jefferson College was designed to resemble the University of Virginia, the new college's administration did everything possible to repel raucous, devil-may-care behaviors exhibited by rebellious students at Thomas Jefferson's experimental East Coast institution.[80] Jefferson College students, in contrast, rose early without talking, proceeded to a washroom, dressed, and formed two lines where roll was called. Prefects maintained morning silence and escorted students to their classes where professors again took attendance. Silence, it would seem, subjugated students' lives. "Those who make disturbance may, if the Professor think proper, be sent immediately to prison."[81] Recreation periods were permitted, albeit limited. All athletic activities were monitored to ensure propitious physical activity and general safety. Card games or gambling of any sort were strictly prohibited. To enforce these and other rules, a select group of older students were chosen as prefects. These always present, ever-watchful rule enforcers frequently doled out punishments to their younger charges. Recreation periods were brief, but meal times were shorter still. Twenty minutes were allotted for dinner, while only fifteen minutes were apportioned for breakfast and supper. (Students were not allowed to remove food from the dining hall.) At the close of each day, prefects led students in brief prayer before a final roll call and bed. If students chose to attend Mass at St. Michael's Church, prefects escorted them to and from services.[82] If anything, permitting nightly prayers and Mass attendance illustrates that Jefferson College administrators and Creole board members understood the presence of religious observations in daily life. Creoles may have imitated French secularism, but they were, after all, Catholic and well acquainted with the spiritual requirements imposed by their faith.

Much like the daily schedule, discipline was staunch and punishments resembled those used in French lycées (secondary schools). College leaders, like regional Creoles, looked to France for models of instruction as well as chastisement. Pamphlets describing the French model of regulation indicated that it was "believed to be much superior in efficiency to the methods generally employed in this country. For those who are duly impressed with the paramount importance of education, this distinction will not be regarded as a trifling advantage."[83] Punishments for rule-breaking were severe and increased in severity according to the crime: "I. Privation, wholly or in part, of recreation, with performance of an extraordinary task, for one or two days. II. Privation of the customary walk, with an extraordinary task. III. Arrest for one, or at most two days. The student [was] placed in a light and airy room, which [was] easily observed, and in which he perform[ed] an extraordinary task. IV. Bread and water only, for one or more meals. V. Imprisonment for one, or at most, two days. VI. Privation, either wholly or in part, of vacation. VII. Expulsion from the College."[84] What these "extraordinary" tasks were we do not know. Even so, students well understood the college's commandments, as the list of rules and associated punishments were read aloud each week. Conversely, students who distinguished themselves in the positive were rewarded with verbal honors from the college president. Strict discipline and firm instruction were not students' only restrictions. Those who attended Jefferson College were cut off from their families and only permitted to journey home on the last Saturday of each month. Students could write to and receive letters from their parents alone, rarely entertained guests, and were forbidden to possess a "gun, pistol, sword, dagger, sword-cane, or any other deadly weapon whatever." Finally, students were barred from imbibing alcohol or using tobacco.[85]

Not only was the daily schedule onerous and penitential in nature, so too were classes. Despite the desire to reproduce Thomas Jefferson's educational ideals, which included agricultural and mechanical sciences, privileged Creoles insisted that the college only expose their sons to pure liberal arts curricula.[86] Regardless of new technologies such as steam engines and vacuum panning, students were purposefully denied practical instruction.[87] Antebellum Creoles clung to aristocratic liberal arts mores that epitomized European colleges and universities.[88] Education, in short, was not meant to create mechanics or engineers, but lordly, cultured gentlemen able to converse in various tongues. Along with French, ancient languages were considered essential. Latin was useful for those interested in studying law or medicine while Greek was considered the sublime language reserved for Aristotelian deliberation.[89] Thus, the ability to read, write, and speak Greek or Latin was a sign of refinement and social grace. French Creoles, though displeased at the phenomenon, understood that English was becoming the language of choice in Louisiana. Therefore, it was necessary for Creole youth to master "American English" in order to converse with peers, despite ethnic differences, about politics and plantation management. Still, French was to be retained as the upper-class language. To venerate French custom while permitting English language learning, professors alternated between the two tongues for course instruction.[90]

Administrators and instructors, ever aware of Creole liberal arts curricular demands, also understood the changing nature of Louisiana regarding industrialization, river travel, and scientific thought. Accordingly, they modified instruction to inculcate elements of scientific reasoning to prepare particular students who, although attending a college built and funded by sugar barons, sought careers outside of the plantation gentry. In order to delineate curricula to best serve future vocations, Jefferson College faculty developed two branches of instruction: the classical and special. As dictated by the college's statutes, "it is left to the discretion of the parent to determine which of these the student shall follow. The Classical Course is designed more particularly for young men intended for the liberal professions and the sciences: the Special for those who progress to engage in occupations for which the learned languages are not indispensable."[91] While the classical branch was meant to mold and refine Louisiana's adolescent nobility, the special branch concentrated on subjects deemed essential for business and trade. Consequential to their parents' desires, students subjected to the classical branch could expect eight years of instruction. In the first year, classical students attended courses in English, French, and Latin, as well as mythology. Students in their second year were taught grammar and mythology but also partook in mathematics instruction, Louisiana geography, and Louisiana history. Third-year students continued in grammar with additional courses in geography, Greek, mathematics, and US history.[92]

Passing into their fourth year, classical students were exposed to additional courses in grammar but were also expected to study Latin prosody (Anacreon, Ovid, Virgil, and Xenophon). At the same time, they attended classes focused on algebra, geometry, English and French history, and European geography. In the fifth year, pupils continued grammar with the addition of advanced algebra, English and French prosody (Racine and Shakespeare), Roman history, and solid geometry. Sixth-year students, by this time considered competent to converse wholly in Latin, studied rhetoric (English, French, and Latin prose and verse, including Blair, Boileau, Demosthenes, Juvenal, Livy, Milton, Sallust, Tacitus, Terence, and Voltaire), ancient history, and advanced mathematics (algebra as applied to geometry, descriptive geography, and trigonometry). Progressing from the sixth to the seventh year, students engaged in advanced courses of ancient and modern English and French literature, mathematics of fluxions and hydrostatics, moral philosophy, natural philosophy and chemistry, and universal history. In their eighth and final year, if still enrolled, students were tutored in linguistic eloquence, attended a final French literature course (Bossuet, Cochin, Dagesseau, Flechier, Massillon, and Mirabeau), an English literature course (Burke, Erskine, Sheridan, and selections from American speeches), Greek with an emphasis on Demosthenes, Latin with a focus on Cicero and Livy, and finally a second dose of natural philosophy and chemistry.[93] If students found the classical course overwhelming, their complaints fell on deaf ears: as per the college's mandate, the parent, not the child, chose the course of study.

If, by chance, parents selected the special branch for their son, he could expect six years of study. The first year, much like the classical branch, consisted of English and French grammar, geography (use of globes in particular), mathematics, and mythology. In the second year, the student took a grammar course that centered on English and French translation, geography, mathematics (algebra and plane geometry), a second mythology course, and an abridged class in sacred and profane ancient history. In the third year, special branch students practiced composition and recitation and were required to take US geography, history of America and the United States, advanced algebra, and solid geography. Entering their fourth year, students studied the critical examination of written language, figures of speech, and verse; English and French geography; and US statistics. Aside from these classes, special branch students received additional instruction on English and French history, as well as continued mathematics (algebra applied to geometry, descriptive geometry, and plane and spherical trigonometry). In the fifth year, special branch pupils engaged in classes much like the classical branch with a continuation of grammar, rhetoric (including composition, declamation, the rules and exercises of written and oratory style), geography (statistics of England and France), ancient history with a particular emphasis on Rome, ancient and modern literature, mathematics encompassing fluxions and hydrostatics, natural philosophy, and chemistry. Lastly, in year six, students enrolled in comparative literature of English and French, mathematics (architecture, astronomy, and mechanics), moral philosophy, advanced natural philosophy and chemistry, and universal history.[94]

With the addition of the special branch, Jefferson College was gently modified to meet the region's educational expectations or at least prepare students, with an exhausting regime of coursework, to compete with their American counterparts. Although curricula were modified, Creole gentry adhered to the classical branch while lower-class students partook in the special branch. As gratis, lower-class pupils were commonplace among wealthier students, college administrators understood that coursework needed to provide a basis for gainful employment in addition to philosophy, linguistics, and the highborn preferences of the ancienne population. Still, Jefferson College, like many antebellum institutions, remained grounded in intense liberal arts instruction tailored to the wants and perceived needs of the region's aristocracy, as it was believed this type of learning provided academic rigor and deterred behavioral problems through strict scrutiny and pedagogical obeisance.[95]

Frenchmen, Americans, and the College Presidency

As the college progressed, the curricular structure was not the only aspect to change; the administration—namely the presidency—transformed as well. The first president of Jefferson College, nominated to the position at age forty by the institution's governing board, was Bernard Granet. He was highly educated and experienced in teaching boys and young men, had written several texts on education, and was, perhaps most importantly, French. Granet's wife was also engaged in education as the founder of a boarding school for young women on Royal Street in New Orleans.[96] Granet was well liked by the St. James Parish community, but regardless of his qualifications and pedigree, his tenure was short. In 1834, less than one year into his presidency, Granet resigned to help his headmistress wife expand her French Quarter school to include boys, albeit segregated from girls by means of a separate wing.[97]

After Granet resigned, the board selected a well-respected, although somewhat reluctant, French immigrant to head the Creole college. Claudius Crozet, former state engineer of Virginia, ex-army officer and Bonapartist, became Jefferson College's second president at the end of 1834.[98] As he was socially connected to Governor Roman, Attorney General Mazureau, and railroad president Dominigue Burthe, all supported Crozet and pressed for his nomination.[99] Prior to his arrival in the United States, Crozet had received a military education and served in Napoleon Bonaparte's army. With the return of Bourbon rule in France, Crozet sailed to America and served on the faculty at West Point in New York. His fame for participation in Napoleon's European campaigns made the French officer a notable figure at the academy.[100]

Born January 2, 1790, Crozet was the son of a wine merchant who lived and worked in the Rhône valley. He attended the Central Schools of France before gaining acceptance into the École Impériale Polytechnique (the same French institute Collège d'Orléans president Joseph Lakanal helped establish) in 1805. There, he studied artillery and engineering. In 1807, he applied to and enrolled in the Imperial Artillery School at Metz. After two years of advanced instruction, Crozet was commissioned as a second lieutenant in the first battalion of bridge builders for the Imperial Artillery Corps. On June 9, 1809, he entered the Grande Armée and was assigned to Napoleon's headquarters just in time to take part in the Battle of Wagram. Crozet spent the next three years in Holland constructing bridges for the French emperor. He was later promoted to Captain and assigned a regiment destined for Napoleon's ill-fated Russian invasion. But before Crozet could reach the emperor, he was attacked by members of the Russian military and held captive for two years. With the fall of Napoleon in 1814 and subsequent exile to the Mediterranean island of Elba, the Count of Provence became Louis XVIII, King of France and Navarre. Crozet was released from his Russian imprisonment and transferred to Paris for bridgework. However, when Napoleon escaped Elba in 1815 and began the "Hundred Days" campaign,

Louis XVIII fled to the Netherlands and Crozet was reassigned yet again. He and a group of French officers were reattached to Napoleon's armies and placed in charge of a garrison delivering gunpowder to Waterloo just before Bonaparte was defeated and overthrown. After the emperor's abdication and second banishment (this time to the island of St. Hélène, where he remained until his death in 1821), Louis XVIII was restored and Crozet requested leave from active duty. Dissatisfied with tumultuous revolutionary politics, he sailed to the United States.[101]

The ex-officer turned West Point instructor eventually tired of teaching and left New York to become Virginia's state engineer in 1823. He remained in this post for eight years before moving his family to the Deep South to become Louisiana's state engineer. While in Louisiana, Crozet designed an efficient method to clear streams and rivers, thus enhancing transportation. He also briefly ventured into railroad production and oversaw the laying of track not far from Jefferson College, but he became frustrated when the state legislature denied additional development plans.[102] Governor Roman, fearing that Crozet would abandon Louisiana, encouraged the disillusioned Frenchman to accept the Jefferson College presidency. Roman admired the Napoleonic soldier and learned much about the Frenchman, as the two had become neighbors and family. After relocating his own family to St. James Parish, Crozet permitted his niece, Anna, to marry the governor's son, Felix. Hence, the appointment of the state engineer to the college presidency was, at least to Roman, a clear choice.[103]

While serving as president, Crozet enforced military order similar to the regulatory structure Napoleon imposed on French lycées. Local planters lauded Crozet's stern administration because many considered their sons to be spoiled and in need of correction. To quote Simone de la Souchére Deléry, apropos the mollycoddled attitude of Louisiana's antebellum Creole youth: "How could it be otherwise with the Negroes' servility, always eager to please the 'little master,' and the Creole mothers' adulation?"[104] All in all, members of the ancienne population approved of Crozet in light of his former Bonapartist status. Unlike the conspiratorial regicide and former Collège d'Orléans president Joseph Lakanel, the new leader of Jefferson College was disinclined to trumpet his successes, past or present. Crozet's silence stemmed from a reluctance to discuss his service in the French armies. Evidence has been uncovered which indicates that following his military resignation, Crozet's opinion of Napoleon was severely dampened. When queried about the French emperor, he referred to Napoleon as a traitor to France and to mankind.[105]

To the dismay of Creole planters, not to mention that of Roman, Crozet's stint at Jefferson College proved short and wearisome. The French administrator often argued with his subordinates, particularly Hugues Jules de La Vergne, the college's treasurer, who was stalwart in his financial duties but curt and unyielding with the president. Dealing with tedious tasks and unable to engage in engineering projects, Crozet, once again, found education dull. Moreover, he had become disheartened with Jefferson College's isolated locale and often took long walks along the Mississippi River to break the monotony of rural life.[106] Tired of the Creole college, Crozet claimed ill health and removed himself from the presidency in 1836.[107] Leaving Jefferson College behind, he worked for a brief stint as a New Orleans drain surveyor until 1837, when he was renamed Virginia's principal engineer.[108] Despite his personal frustrations, the country life may have served Crozet better than urban New Orleans, as political turmoil erupted and rendered the city into three tenuous factions: one governed by French Creoles (the French Quarter), another lead by Anglo-Americans (the American District, otherwise known as Faubourg St. Marie), and a third, Faubourg Marigny, managed principally by immigrants. This municipal debacle spilled into the sugar parishes and ultimately tipped Louisiana's political scales in favor of the Anglo-American population.

Before Crozet left Louisiana for Virginia, feuding between French Creole, Anglo-American, and immigrant populations caused New Orleans officials to legally separate the city into three politically distinct districts. At the onset of the separation, French immigrants cast their votes in favor of Creole politicians. As a result, French Creoles, through numerical superiority, maintained authority and vexed Anglo-Americans. City funds were allocated to the maintenance of Vieux Carré streets, while roads in the American sector were neglected. American maintenance projects were so often overruled that the Anglo-American class demanded the separation of their residential and commercial district from the rest of New Orleans. The result was the sanctioned divide of New Orleans based on class distinction.[109] While New Orleans officials argued over governance, Louisiana's agriculture-based economic system teetered on the brink of collapse.

Crop surplus in the mid-1830s caused the price of US agriculture to plummet, resulting in the financial panic of 1837. Southern states in particular sustained damage from the economic crash. Individual fortunes were lost or severely injured, while tax-supported revenue spiraled out of control. Consequently, the federal government permitted states to cover "floating debts" with monetary endowments. In Louisiana, $355,000 was appropriated to these ends; however, owing to the state's legislative priorities, as well as its overextended educational system, the majority of federal monies were given to the College of Louisiana, Covington Female Academy, Franklin College, and Jefferson College. The remainder was applied to

Louisiana's sinking revenue. By July of 1839, the federal grant had been depleted.[110] At the onset of the collapse, Jefferson College was stocked with students representing both New Orleans as well as the preponderance of south Louisiana's sugar parishes. However, as sugar stock plummeted, enrollment dropped as well.[111]

While plantations grew in size and scope prior to the 1837 panic, so, too, did the need for credit to cover operation costs. New Orleans's banks extended enormous credit lines, and in turn planters willingly mortgaged their estates to expand their private empires. As the agricultural market swelled, Louisiana banks sold stocks to European investors and issued copious bank notes. When Europe withdrew from the American market due to speculative lending practices and decreased crop values, agriculture stock crashed, resulting in the largest US financial recession to date. Banks were unable to reimburse notes and land prices plunged. All the while, credit debt remained high.[112] The financial collapse caused several plantation bankruptcies, and in response to the economic downturn 166 plantations converted their main crop to cotton; however, cotton sales also dropped, which further diminished exports.[113]

In the midst of economic crisis, Thomas R. Ingalls was nominated to replace the ailing Crozet as Jefferson College's president. Ingalls, a New York native, had previously served as a faculty member at the rival Anglo-American College of Louisiana and had taught chemistry, pharmacy, and obstetrics at the Medical College of Louisiana in New Orleans. Though an American, Creoles considered Ingalls to be a capable educator who had the potential to keep Jefferson College afloat during the financial panic. As predicted, Ingalls experienced a successful presidential tenure.[114] According to Lusher, "the choice of the professor [Ingalls] placed at the head of the institution, gives us every reason to hope that it [Jefferson College] will prove worthy of the protection extended to it by the state and of the liberty of the citizens who have founded it."[115] Knowing that Ingalls had been a student of Crozet at West Point prior to his graduation in 1822 may have eased Creole apprehensions.[116] Whether or not the ancienne population approved of Ingalls's ethnic background, the college needed a leader who could steer the institution through the troubling fiscal calamity. Ingalls proved capable.

The American president began his administration by recruiting additional students and new instructors. Among those hired to teach at the college was Dominico Canova, the renowned lithographer, painter, and muralist. The artist, an Italian immigrant and son of a Supreme Court judge in Milan, was born in 1800. Canova immigrated to New York in 1825, and, in 1837, took up residence at Jefferson College to teach art and drawing lessons. In 1839, he opened a studio in New Orleans and was commissioned, among various projects, to paint ceiling frescos at St. Alphonsus Church in New Orleans's Irish Chanel. He also painted murals at plantations including St. Frusquin (renamed San Francisco Plantation in 1878 by then owner Achille D. Bougère) and Habitation Haydel (presently known as Whitney Plantation).[117] Stocked with a bevy of capable students and talented instructors, both planters and legislators credited the thirty-nine-year-old Ingalls for the college's improved condition: "Thanks to his [Ingalls's] cares its [Jefferson College] future prosperity was secured."[118] The legislature was confident that "Louisiana now had a college placed on a very respectable footing, and which led far to obtain a supremacy over all the universities of the Union."[119] By 1840, 238 students were enrolled with an increased tuition of $173.50 per academic term. In addition, students who took elective courses that required special services and equipment could expect up to $165 in fees. Surprisingly, these costs were far higher than those of other Louisiana institutions, yet enrollment continued to climb. Notwithstanding the economic collapse, fortune, as well as credit, returned. As before, high costs meant little to French Creoles.[120]

Along with tuition alterations, there came a slight transformation in religious ideals at the secular institution. Revised college statutes indicate that professors were required to not only teach their classes and maintain order, but were to "avail themselves of all suitable occasions to inform the students of their duties to God, their families, and their country."[121] Religion, it seems, was on the rise at Jefferson College. To enhance metaphysical education, Father Boué of St. Michael's Church was allowed to provide theological instruction to students in 1839.[122] In addition to this curricular change, the student body began to diversify as well. Starting in the 1840s, sugar parishes absorbed so many new German and Irish immigrants that the French-speaking residents constituted only thirty percent of the local population. As a result, French was no longer the dominant language.[123] Now college halls, formerly filled with French voices, were host to foreign tongues and accents. As enrollments continued to climb, the campus was in need of physical expansion and plans were drafted to erect an additional wing that would accommodate 200 pupils at an estimated cost of $40,000.[124]

With the completion of the new wing, enrollment rose to 264.[125] However, the toil of safeguarding Jefferson College through a period of national financial hardship weakened Ingalls. In March of 1841, Ingalls, alleging ill health as did his predecessor Crozet, announced his intention to resign the presidency at the end of the year. His replacement arrived in early 1842, another American by the name of Alexander Hill Everett.[126] Preceding his appointment at Jefferson College, Everett worked as the US minister to

Spain, and after 1829 he resided in Havana, Cuba, as a State Department delegate.[127] Despite his Anglo-American heritage, Everett won the admiration of sugar parish Creoles by displaying his literary talents and inclination to support liberal arts over all other curricula.[128] Born on March 19, 1790, near Boston, Massachusetts, Everett had studied at Harvard College and was briefly a member of the Whig Party. He served as the editor of the *North American Review* from 1830 until 1835 and was well acquainted with Henry Wadsworth Longfellow, the poet author of *Evangeline: A Tale of Acadie*. Roman and the college's board of directors found Everett's experiential and educational competency promising. Accordingly, they voted to hire the accomplished American and overlook his lack of French blood.[129]

Everett's modesty earned him additional Creole confidence. In a presidential speech, he confessed that he had no formal training in "the business of education." Instead of an educator or college president, he referred to himself as a "public servant" whose pursuit of literature had been a leisurely activity. Everett had traveled throughout Europe to study the expansion and progression of eastern civilization.[130] His European travels, literary background, and compliments to the Creole population were enough to secure him the affection of St. James Parish inhabitants. In a commencement speech, Everett flattered the crowd, avowing, "The gentlemen of Louisiana can never be otherwise than considerate and kind to a stranger, whom they have called from a distance, and from other occupations to devote the remainder of his life to their service."[131] The new president made a point to describe the fruitful fruition of transportation on the Mississippi River, noting that "the celebrated stream which flows by our side, at a few hundred yards distance, affords a more striking demonstration of the recent triumphs of physical science than could be given by volumes of description."[132] Everett depicted the evolution of river transportation from rafts to sailing vessels and finally to steam ships, but sensitive to Louisiana slavery ethos, he neglected to mention his abolitionist leanings.[133]

In his complimentary oration, Everett declared Jefferson College to be the finest institution of higher education in Louisiana. He also detailed his plans to increase the Creole college's academic standing. According to the new president, the campus had "a remarkably healthy and pleasant position on the banks of the Mississippi—fortunate in having secured the services of a corps of instructors, selected from among the best qualified Professors of the Old and New World."[134] Everett placed particular emphasis on state financial support, but also highlighted the need for adaptive curricula: "Without the security and protection afforded to the individual, by a good government, there can be no effectual attempt at general education. On the other hand, the best political institutions, if established by accident, in regions not prepared for their reception, by the previous diffusion of knowledge, decay and dwindle into nothing; or, perhaps, instead of working out their natural results, become themselves the fatal source of national distress and ruin."[135] The American administrator had hopes of developing a legal division at the college: "The establishment of a Law School, as early as may be practicable, is contemplated in the statutes, and will, doubtless, engage the attention of the Directors. . . . As soon as there shall be at the College a sufficient number of students, who have finished their courses of study, whether Classical or Special, a Law School shall be established, in which courses of lectures shall be given in public, criminal, and civil law."[136]

Further, Everett spoke at length about the need for courses in ethics and metaphysics. He described Christianity as being just as important as science in formal higher education: "It is our office, Gentlemen of the Faculty, who are entrusted, under the direction of the Board, with the immediate management of this institution, to facilitate, as far as we are able, the access to these treasures, to the youth who are committed to our charge."[137] Finally, the president informed students that they had a great opportunity before them. Students, Everett explicated, should not look upon their classes and studies as "irksome tasks." Quite the opposite, their time at Jefferson College, he exclaimed, could be the happiest of their lives.[138] To enhance physical activity and nationalism, Everett organized a cadet corps of forty students in accordance with his philosophy that "in a republic, where the defense of the country in times of emergency becomes the responsibility of private citizens, everyone is obliged to be prepared to answer the call of duty and patriotism." With the help of the college's governing board, Everett planned to slowly increase the cadet battalion. Apparently, in addition to religious instruction and the formation of a legal program, "the directors seemed to be thinking of turning their Creole college into a quasi-military institute."[139] Jefferson College had indeed experienced a period of administrational success and plentiful attendance, but ruin and dismay lay on the horizon.

Fire, Rebuilding, and the End of the First Regime

As Creole sugar planters continued to recover from the economic crash of 1837, they marveled at the growth of their prized educational institution. By 1842, under the administration of the meritorious president Everett, the college experienced its highest enrollments. However, on Sunday morning, March 6, 1842, Jefferson College, which stood as a testament to French Creole wealth, power, and prestige, was destroyed by fire.[140] Writing under the pseudonym René de Sennegy,[141] the Marist priest Aldric Lettin de la

Peychardière described the conflagration's speed as well as its influence on the community:

> Ten years, Jefferson shone in the heaven of the humanities and of the sciences; for ten years, it lived brilliantly, carefree and skeptical. Most certainly, yes, in those days of literary prosperity, the faculty was a precious jewel for Louisiana; the scholarly brood were not the delight of either the Church of St. Michael, nor of its pastors, nor even of the neighbors and the passersby. But, one day God grew tired and, while the Holy Office called the faithful to Church, while the students gave themselves up to laughter and games, on Sunday, March 6, 1842, at about 10 o'clock in the morning, the cry: Fire! Fire! resounded all at once in the vast establishment. Great emotion prevailed among the students and the teachers. Each one started to run. Vain efforts! The fire started by perhaps imprudent, perhaps vindictive hand, embraced the superb Athenaeum, and only a few hours sufficed the pitiless element to achieve its work. That night, immense ruins attested to the fact that the devastation was as rapid as it was complete.[142]

De Sennegy speculated that the fire was caused either by carelessness or malicious intent. Whether deliberate or not, New Orleans newspapers, including *The Courier*, reported that the fire broke out in the college's chemistry lab.[143] Three days after the destruction, members of the governing board held a meeting at the St. Louis Exchange Hotel in New Orleans to discuss petitioning the Louisiana legislature for rebuilding support. Reacting to the need for philanthropic donations from any supporters regardless of ethnicity, solicitation speeches were given in both French *and* English.[144] Fearing permanent closure as a result of students forgoing the smoldering remains for functioning academies, Everett ordered the faculty to continue instruction in makeshift lodgings near the ruined structure. The president's concerns were well founded, as other colleges advertised their ability to enroll Jefferson College's displaced students. A notice in the *Times Picayune* informed readers, "In consequence of the unfortunate accident which has happened to the College of Jefferson, the Board of Trustees of the College of Baton Rouge believe it to be their duty to inform parents and guardians that the latter is now in full operation, can receive one hundred and thirty pupils, and that nothing shall be neglected to attain the object of a liberal education."[145] Before long, the Creole college's governing board found a temporary site in New Orleans on the corner of Orleans and Dauphine Streets to hold classes.[146]

Making use of remnants from the first structure, the main college building, rebuilt on a grandiose scale in roughly two years, no longer sported fluted columns but large Doric columns and a central portico. This new structure, though shorter in length than the first, towered over the campus landscape with detailed Greek-revival features.[147] The college's new architectural style reflected an emerging trend in regional plantation structural design. With the influx of Anglo-Americans, sugar plantation architecture evolved beyond the subtly intricate Creole style plantation house. Late antebellum plantation homes were thick-columned mansions enveloped in an array of imported flora.[148] As Anglo-American influence permeated Louisiana plantation culture, Creole planters adapted their plantation homes to match or, in some cases, to surpass their American rivals. French Creoles combined opulent Greek revival architectural features with the layout of a traditional Creole plantation house.[149] Of those extant plantation homes between Baton Rouge and New Orleans, the majority were constructed in the 1840s and 1850s, a period when plantation architecture emulated American extravagance. Plantation homes of the mid-nineteenth century were fronted or surrounded by columns and wide galleries and contained a single hallway flanked by rooms, a design that was both imposing and functional, as it allowed a constant draft to waft through the house. Ventilation was essential due to south Louisiana's subtropical temperatures.[150] Similarly, Jefferson College was reconstructed with a prominent second-floor gallery and a series of large windows that could be opened to improve airflow.

As with the first structure, no architectural records related to the new edifice have been uncovered. Nonetheless, architectural historian Samuel Wilson Jr. drew a correlation between the rebuilt college and buildings designed by James Gallier Sr.[151] Born in Ravensdale, Ireland, Gallier spent his childhood studying at the School of Fine Arts in Dublin. In 1820, he continued his education in Dunbalk and gained practical architectural experience while employed with regional construction firms. Seeking better working conditions, Gallier sailed to New York in 1832, and in 1834 the young Irishman journeyed to New Orleans.[152] Wilson's assertion that Gallier designed the rebuilt Jefferson College is unlikely. According to Gallier's autobiography, he spent the first half of 1842 sailing back and forth between Louisiana and Cuba, working on plans to construct a hotel similar to the New Orleans St. Charles Hotel. Plans for the Cuban hotel ultimately fell through, but Gallier himself explained that due to his travels, he had little time for other projects.[153] Although he may not have contributed to the reconstruction of Jefferson College, Gallier is remembered as the architect responsible for designing the iconic New Orleans city hall fronting Lafayette Square on St. Charles Avenue. The government building was later renamed

Gallier Hall in memory of the famous architect who died at sea in 1866 when his ship, the *Evening Star*, capsized off the coast of Cape Hatteras.[154]

It is possible that the rebuilt college's architect was James Harrison Dakin. Owing to a series of uncovered cash advance receipts, it has been established that Dakin designed and constructed several plantation homes along the gold coast. For example, the Bringier family of St. James Parish commissioned Dakin to build their domiciles at both Ashland (also known as Belle Helene) and Bocage plantations.[155] Not only were the Bringiers well connected with neighboring Creole planter families, they were also associated with Jefferson College, as their son Amedee had studied there.[156] Dakin, born 1806 to James Dakin Sr. and Lucy Harrison, launched a productive career as a partner in the New York architectural firm of Town, Davis, and Dakin. In 1833, he designed the Rockaway Marine Pavilion in Long Island before moving to New Orleans in 1835 to work alongside brother Charles Bingley Dakin. Built overlooking the Atlantic Ocean, the pavilion hotel spanned 230 feet, stood two stories tall, and possessed a colonnaded porch and portico. Overall, the design looks strikingly similar to the Greek Parthenon.[157] Architectural plans for the pavilion included a massive two-tiered cupola, which was discarded in the final draft. The completed Marine Pavilion resembles rebuilt Jefferson College, so it is possible that Dakin applied the pavilion's features to the Creole institution's reconstruction designs. The New York native went on to design prominent Louisiana structures such as St. Patrick's Church in New Orleans and the Old State Capital Building in Baton Rouge. Dakin passed away on May 13, 1852, while living in the capital city. His body was reportedly returned to New Orleans and interred in the Girod Street Cemetery, but when the cemetery was deconsecrated and deconstructed in the 1950s, records were scattered and the architect's remains were lost.[158]

If the first college edifice embodied Jeffersonian ideals, the second reflected the sumptuousness of plantation affluence. Simone de la Souchére Deléry described the new Jefferson College structure as a "long building with white columns, less a school in appearance than a rich planter's mansion."[159] The grandeur of the new construction was palpable; so, too, was the incurred debt.[160] Even though bank reliability and sugar finances had stabilized by 1845, south Louisiana planters, in 1846, experienced yet another financial blow as the federal government drastically reduced sugar tariff protections. Later readjusted to better profit planters, sugar tariffs were, in short, the dominant form of protection that reduced inordinate competition between Louisiana and Caribbean plantations. By forcing the US to rely predominantly on internal sugar production, Louisiana's economic viability was maintained. By decreasing the tariff, however, sugar planters were left financially vulnerable and understandably leery of precarious investments, which included financing an indebted institution of higher education.[161] Still, many had contributed to the Jefferson College rebuilding effort, but their monetary gifts were not enough, and credit was used to finish construction. Given the financial irregularity of the late 1840s, French Creoles, both in New Orleans and in the sugar parishes, came to grips with the need for American monetary support and began to discuss the reunification of New Orleans's three governmental districts. By 1852, Louisiana officials brokered a reunification deal. The Anglo-American faction agreed, but would only comply if the largely American suburb, then known as Lafayette, was included in the merger. With the suburb's annexation, Anglo-Americans possessed a total urban population that surmounted the combined Creole and French immigrant populace, therefore offsetting ancienne population political authority. As Christina Vella describes it, "Reunification was the beginning of the end of Creole domination."[162] With the college fire and New Orleans's political comportment leaning toward the Americans, Creole families returned to the traditional practice of sending their sons abroad to finalize their education. One such Jefferson College student was Gabriel Aime, son of wealthy sugar planter Valcour Aime. Gabriel had attended the Creole college before and after the fire. Not long after reenrolling, Valcour decided it was time to send his son to Europe for the "grand tour."[163]

By the time reconstruction efforts were complete, attendance at Jefferson College had diminished to seventy-eight students.[164] When classes started in the fall of 1844, student numbers had further dropped to fifty-two boys and young men, ages ten to nineteen.[165] To make matters worse, the college's president Alexander Everett resigned in March, owing to his want to pursue a diplomatic career in China. Hastily, the governing board selected a president from the faculty, F. Chevet. Chevet had dutifully served as the institution's instructor of drawing and mathematics, had published the results of chemical experimentations in the *American Journal of Science and Arts* as well as the *Annals of Electricity, Magnetism, and Chemistry*, and had compiled budgetary reports detailing the college's financial resources to the Louisiana legislature. Though familiar with the college's internal workings, he had little experience as a leader.[166] Burdened by considerable debt, a shrinking student population, and the loss of the celebrated Everett, Jefferson College continued to struggle.[167] In an attempt to provide financial support to the ailing institution, the Louisiana legislature gifted $10,000 in 1845 under the condition that the governing board relax entrance requirements, increase the number of indigent students educated free of charge, and dispense with the requirement that pupils supply silver goblets and utensils.[168] The board reluctantly agreed, but the legislature's financial assistance was a one-time contribution. State legislators had grown tired of supporting a feeble set of colleges and sought to create a new seminary of learning in Pineville, Louisiana, that would outdo all previous higher education endeavors. While Jefferson College grappled with its financial situation, the American College of Louisiana was closed in 1845. The East Feliciana Parish institution had experienced years of dismal enrollment. Students in central Louisiana were scarce and quality teachers were hard to find. With the decision to close the floundering American college, all students were dismissed, as were the faculty and staff.[169]

Confusing the College of Louisiana's history is the account of its purchase by Centenary College of Mississippi, then located in Brandon Springs, Mississippi. In 1839, Centenary College was chartered under the direction of a Methodist-Episcopal minister, Reverend Benjamin M. Drake. The Mississippi Conference of the Methodist-Episcopal Church not only oversaw Protestant churches and missionary efforts in both the titular state as well as the state of Louisiana, it was also the governing authority behind Centenary College. In 1841, Centenary College opened, but student numbers were insufficient to adequately fund the institution. As a result, Methodist administrators sought a new location that would better serve the Mississippi Conference's educational plans. Learning of the defunct College of Louisiana campus, the Methodist-Episcopal Church purchased the property for $10,000, moved Centenary College to the empty facilities in 1845, and renamed it Louisiana Centenary College. In 1846, the institution's name was changed to Centenary College of Louisiana.[170] Lee Morgan explains this title modification and its import in his book *Centenary College of Louisiana, 1825–2000: The Biography of an American Academy*: "The trustees voted to change the name of the institution from 'Louisiana Centenary College'—it had been known by this name for a year—to 'Centenary College of Louisiana.' They also resolved quickly to declare the graduates of both Centenary College in Mississippi and the College of Louisiana in Jackson to be alumni of 'Centenary College of Louisiana' . . . an action which showed clearly that they considered the two colleges to have 'merged.' From that time on, the date 1825 has been cited as the year of the founding of the institution known today as Centenary College of Louisiana."[171] Centenary College's governing board believed that by allowing former College of Louisiana graduates to claim the Methodist institution as their alma mater, the institution would accrue alumni dollars and claim a founding date of 1825, despite the fact that the College of Louisiana had formally closed and returned its charter to the state legislature in 1845. As time passed, the renamed Centenary College of Louisiana suffered much like the Anglo-American College of Louisiana. To offset low enrollments, women were admitted in 1900; however, the institution continued to decline. In 1905, the board of directors acquired land in Shreveport, Louisiana, and moved the campus there, where it continues to exist as a Methodist liberal arts college. In 1979, the original College of Louisiana campus in Jackson was added to the National Register of Historic Places, but not before several buildings had been demolished. The Louisiana Office of State Parks currently manages the site, which includes a single dormitory wing, a professor's cottage, and a Civil War–era cemetery.[172]

Though Creoles had momentary cause to celebrate the College of Louisiana's demise, Jefferson College soon suffered a terminal blow. In addition to the closed College of Louisiana and the rebuilt, yet financially weakened, Creole college, the recently established Franklin College in Opelousas, Louisiana, proved to be a similar burden to the state legislature, as it failed to mature beyond that of a preparatory school. Again, professors and gold coast residents tried to save Jefferson College. President Chevet, along with St. James Parish representative Robert Nicholas, beseeched the legislature to endow Jefferson College with monies set aside to construct the Louisiana State Seminary of Learning (later renamed the Louisiana State Seminary of Learning and Military Academy, present-day Louisiana State University). Similar petitions from both Franklin College and Spring Creek Academy in Rapides Parish were put forth but were denied. Weary of supplying funds to a sickly set of colleges, the Louisiana legislature terminated all support in 1845.[173] With this cessation, Jefferson College was doomed to close. As Lusher lamented, "It is obvious that the 'Retrenching Act' had sounded the death knell of the state institutions known as colleges."[174] Indeed, as funds were redirected to create the Pineville seminary, chances for long-term college survival in antebellum south Louisiana diminished.

In 1846, the board of directors, hoping to keep Jefferson College afloat, sought a Catholic religious order to save the institution. The Society of Jesus was contacted in hopes that Jesuit priests would purchase the campus and assume control. The order's fathers and brothers were indeed interested as their own college, St. Charles College, was located in the inaccessible and poorly populated community of Grand Coteau. The St. James Parish academy offered the Jesuits a tempting solution due to its geographic location and river access.[175] In 1847, Father Jean Baptiste Leon Maisounabe, stationed near the Creole college at St. Michael's Church, was able to survey the campus and advise his superiors regarding a potential purchase.[176] Born May 10, 1805, in Bayonne, France, Maisounabe had been appointed to oversee Jesuit missionary and educational activities in the Deep South. Writing to his provincial in Lyon, the French Jesuit explained, with some amount of trepidation, that "Jefferson [College], which we're considering buying; there's great pressure on me to do this; I dare not make a decision because of lack of manpower to maintain this establishment." Maisounabe had, in fact, previously requested additional priests and brothers from northern Jesuit provinces to enhance St. Charles College but had been denied.[177]

After in-depth consideration, Maisounabe wrote again requesting definite answers to specific questions. Regarding Jefferson College, he asked, "1. Should we buy the place? 2. if we do, should we hold on to Gr. Cot. [St. Charles College, Grand Coteau]? 3. What resources will we have to pay for it?" Acting without direct orders, Maisounabe "decided in favor of buying," and, in September, laid out his plans to purchase: "I visited Jefferson College [and estimate the cost] at 80,000 [Spanish] piastres. I think my estimate is higher than it would actually cost. After pondering over all that I've explained, I made my election according to St. Ignatius' Rules and decided on buying the place. I asked the Bishop [Antoine Blanc] to handle the transaction personally or through some agent of his own choosing; I set 25,000 piastres as the maximum that could be offered, which is to include the library, the physics and chemistry laboratories, and two hundred iron beds which are in the building. To pay off this debt I'm banking chiefly on the college's success; and I have no doubts about its success if we get some good teachers, especially Englishmen."[178] Maisounabe's hopes were high, but his ambitions exceeded the Society of Jesus's ability to staff Jefferson College. The Jesuits were concerned about the future prospects of St. Charles College and Spring Hill College in Mobile, Alabama, as both were experiencing financial difficulties. In addition to the Jefferson College prospect, there was the possibility of opening a college in Baton Rouge (the College of Sts. Peter and Paul). Together the new Baton Rouge college and the St. James Parish institution would, in Maisounabe's opinion, adequately replace the remote St. Charles College. As well, the priest dearly wanted a church, rectory, and college in New Orleans (the Church and College of the Immaculate Conception). In response to Maisounabe, the French provincial denied his request and opted for the Baton Rouge and New Orleans projects.[179]

On September 29, 1847, Maisounabe justified the order's decision not to buy Jefferson College: "Shortage of money is forcing me to call a halt for the time being to both our N. Orl. [New Orleans] and J. [Jefferson College] projects. . . . I'm writing the Bishop asking him to hold up the entire proceedings with regard to Jefferson; prudence obliges me to do this because of the uncertainty of our establishment at N. Orl. [New Orleans]."[180] Although the Jesuits decided against the St. James Parish facility, they never forgot it. In October, Maisounabe wrote,

> After all the thought I've put into it, I don't intend to forget all about Jeff. Coll. now. In fact, the way I feel, my conscience wouldn't be at rest if I didn't do whatever I could to bring about the success of our twofold project (Jeff. And N. Orl.) or at least prepare the way for its success. But before getting myself involved too deep in such serious business, I'd like to have a clearer picture of the whole financial situation and what I can expect in the way of subjects. Perhaps I'll have time to receive instruction from His Paternity before Jeff. Coll. is sold. Every day I beg God in my prayers for this grace—that He may bring about the failure or success of our plans depending on which is more in accord with the greater glory of God.[181]

Because the Jesuits did not purchase the college, it was closed in January of 1848, auctioned to the Citizens Bank of Louisiana for $23,000, and placed under the protection of caretakers.[182] Sugar planters looked on as their college, along with its buildings, lands, and all instructional material, became property of the New Orleans–based bank.[183]

Despite Jefferson College's closure, a small testament to its antebellum existence was printed that visually depicts the institution within Louisiana's agricultural expanse. In 1848, John La Tourrette published *La Tourrette's Reference Map of the State of Louisiana: From the Original Surveys of the United States, which Show the Townships, Sections, or Mile Squares, Spanish Grants, Settlement Rights & C., Also the Plantations with the Owners Names Engraved Thereon.* In an attempt to accurately portray the size and scope of Louisiana's cotton and sugar industry, La Tourrette rendered Jefferson College between neighboring sugar plantations directly across the Mississippi River from the Roman family estate. As La Tourette submitted his map for printing mere months before the college closed, it was included.[184] This map has survived, albeit in limited quantities. One copy has been framed and hangs in the first-floor hallway of Houmas House Plantation in Burnside, Louisiana, fourteen miles from the former college site.

From 1848 until 1853, Jefferson College slumbered, sans students and instructors, while local foliage grew wild throughout the grounds; still, one Creole kept watch over the campus, searching for an opportunity to restore it. A. B. Roman, after his second term as governor, was employed as an agent for and oversaw the 1848 transfer of Jefferson College to the Citizen's Bank of Louisiana.[185] In time, Roman fostered the institution's reawakening under a different title and French immigrant administration. Even though the Creole college's first regime ended in failure, its antebellum existence illustrates ancienne population desires to overshadow Anglo-Americans while preserving social class hierarchies in the face of ethnic rivalries. After Jefferson College's revival as Louisiana College, sugar parish Creoles experienced a first-hand example of the thorny relationship between themselves and the française étrangers.

3

The Forgotten Louis Dufau

If you wish to have permanent colleges in your own native state, you must depend on yourselves, and not at all upon any one man.

—Louis Dufau

In the wake of Jefferson College's closure and the 1837 financial crash, Louisiana economics improved. By the 1850s, French Creole planter affluence was at its height.[1] However, the Creole college stood empty. As ancienne population wealth recovered, talk of republican ideals and secessionist thought were ablaze in the sugar parishes. All the while, French immigrants continued to arrive in New Orleans. Many of these immigrants had relocated to Louisiana on account of the 1848 French Revolution. As members of the française étrangers moved out of New Orleans and settled along the Mississippi River, Creole planters received news worthy of note. In the fall of 1853, St. James Parish residents learned that a charter had been issued for a college to be located in the city of Convent.[2] But where was this new institution to be housed? Before long, the question was answered. A French immigrant by the name of Louis Dufau had purchased the Jefferson College campus with plans to reopen it but under a different name. A proclamation, published in bold typeface by the migrant himself, informed locals of the forthcoming institutional resurrection, title change, and administrational shift: "LOUISIANA COLLEGE, Now Situated In The Parish Of St. James, In The Extensive And Commodious Buildings, Formerly Known As The Jefferson College. PRESIDENT: LOUIS DUFAU."[3]

On September 5, 1853, the notary public of Orleans Parish approved Dufau's purchase of the gold coast campus from the Citizens Bank by means of a $20,000 line of credit. The acquisition included all books, furnishings, and laboratory equipment.[4] Like many French immigrants, Dufau, born

1816, lived in New Orleans, an exile from his homeland. Before he arrived in the United States, he had received a master of arts degree from the University of France, specializing in French history, languages (including French, Spanish, and Latin), and literature. Dufau taught at the Lycée Louis-le-Grand, managed a preparatory lycée in Paris, and had been a member of the French Society for the Abolition of Slavery.[5] He was not only a political activist and educator; he was an author. In 1840 he published a French grammar text, *Manuel des Verbes Français, Réguliers et Irréguliers* (Complete List of French Verbs, Regular and Irregular).[6] One year later he produced a French geography book, *Atlas Historique et Géographique de la France Dupuis les Temps les Plus Reculés Jusqu'a non Jours* (Historical and Geographical Atlas of France from the Oldest Time until Today). In 1842, he coauthored a governmental history of France, *Précis Historique des Régences en France, Précédé d'une Introduction et Suivi des Pièces Justificatives et Documents Historiques et Législatifs* (Historical Handbook of the Regencies in France, Preceded by an Introduction and Followed by Documentary Evidence and Historical and Legal Documents), with a fellow Parisian scholar, F. Solar. After these first three books were printed, Dufau, in 1843, penned an argumentative volume: *L'Université et Ses Adversaries, Quelques Observations Sur le Libelle Intitulé "le Monopole Universitaire, Destructeur des Lois et de la Religion"* (The University and Its Adversaries, Some Comments on the Libel Entitled "The Monopoly of the University, Wrecker of Laws and of Religion"). This particular book detailed, in Dufau's estimation, the many enemies of education. Chief among these was organized religion. Though Dufau was Catholic, he disliked hierocracy and claimed it hindered formal instruction by condemning ideas that might question the Church's authority. While in Louisiana, he singularly evoked enough hostility between himself and local clergy that historian Caryn Cossé Bell labeled him "a bitter opponent of the Catholic Church."[7] Dufau's opposition to religious hierarchy might have been the result of a life lived in revolutionary France.

While in Bordeaux, Dufau met and courted Cecile Street. Eleven years his junior, she was the daughter of Timothy Benedict Street of South Carolina and Angelique Leroy from Saint Domingue. After a short courtship, Dufau and Cecile married on October 10, 1843, and had three children: Jules (b. 1844), Camille (b. 1848), and Marie (b. 1851).[8] Perhaps believing his vocational background might better serve a city that—in appearances, at least—welcomed and employed French immigrants, Dufau moved his family to New Orleans in 1848. Upon their arrival in Louisiana, Dufau, his wife, and their children took up residence in Freetown, a small community on the west bank of the Mississippi River in Jefferson Parish.[9] New Orleans, though new to Dufau, was familiar to Cecile, as members of the

Street family owned property and resided in the city.[10] In order to financially provide for his relocated family, Dufau tutored, served as the editor of a political journal titled *Moniteur du Sud* (Southern Monitor), and gave public lectures on the history and language of France under the auspices of the University of Louisiana.[11] Newspaper advertisements informed the public that lecture admittance cost "five Cents per sitting, or by subscription for One Dollar per month" and took place in either the reading room of the Cabinet de Lecture on Exchange Alley, between Conti and St. Louis Streets, or at New Orleans's First Municipal Council building. Solicitations also announced that "Louis Dufau will give lessons in French Literature and History, either in private families or academies of young ladies and gentlemen." In 1849, Dufau worked in the French department of Orleans High School. Former Jefferson College English professor Erastus Everett governed the school located on Esplanade Avenue. Everett, a native of Princeton, Massachusetts, and an 1836 Dartmouth College graduate, left New Orleans in 1855 to teach in Brooklyn, New York. In 1875, he became chair of Greek and Latin at Rutgers University and remained so until 1879.[12]

Dufau showed promise as an educator, but he also revealed signs of an easily agitated temper. On May 1, 1849, the *Daily Picayune* reported that Dufau and the First Municipal Council doorkeeper quarreled over a public lecture. The doorkeeper, Mr. Duncan, was away from his post tending to family members afflicted with cholera and was therefore unable to unlock the council building for patrons to attend one of Dufau's lectures. Infuriated with Duncan, Dufau sent a letter of complaint to the municipal council claiming that the doorkeeper had deliberately denied him and his attendees entry. The contents of the letter were more than enough to upset the doorkeeper. As journalists at the *Daily Picayune* recounted, "Mr. Duncan characterized the communication . . . as impudent, and moved to resend the resolution previously passed, granting the use of the Council room for the lectures."[13] Dufau was now short one lecture hall, which must have put a damper on his pocketbook since he was using the venue to sell copies of his 1840 book *Manuel des Verbes Français, Réguliers et Irréguliers*.[14] Republished in 1851 by Jacques-Felix Leliévre in New Orleans, the reprint included an introduction that pandered to French Creoles and Anglo-Americans alike. Studying the French language, as was elucidated in the book's new prologue, was "encouraged by all of the people who understand the usefulness and the beauty of a language that the American B[enjamin] Franklin considered rightly as the proper addition to a solid and distinguished education. Such works cannot but fully succeed in Louisiana where French is, and will remain for still a long time the native language of the majority."[15]

Apparently, Dufau's introduction sat well with Leliévre, as he agreed to disseminate the book. Indeed, Leliévre had established a reputation for himself by reprinting local authors' texts in French alone, which he believed would sustain Louisiana's linguistic heritage.[16] Echoing Leliévre, Dufau concluded that he had created "a useful work, not only for the Americans who want to learn French, but also for those who, while they speak it, may desire to know it well."[17] While Dufau's books did much to increase his standing as an educator, he was not without editorial criticism. His articles in the *Moniteur du Sud*, as well as his public lectures, were denounced as halfhearted, ignoble means to increase his social position. Charles Testut, a New Orleans literary critic, railed against Dufau's flippant praise of socialism and lack of imagination. This critic, in his 1850 book, *Portraits Littéraires: Nouvelle Orléans* (Literary Portraits: New Orleans), began with a simple yet powerful statement: "There is nothing remarkable about M. Dufau's articles." This disparagement was the commencement of a long, heated review. In Testut's words,

> His writings are most often tepid which undermines the conscientious work of the critic. M. Dufau doesn't have what a newspaper of today demands: he doesn't have the liveliness, the easiness, the strength, nor the eyesight required by such type of literature. As for the content, it is useless to speak about it.
>
> M. Dufau tried also to be a columnist. He is there so heavy, so boring that he would do better, we believe, to renounce this useless activity: science does not replace imagination.
>
> The only merit of the work of M. Dufau consists in a relatively good choice of articles, in a laudable quality of language, and most of all in the mission that the *Moniteur du Sud* has chosen: to defend the homeland in a foreign country.
>
> We have said above: "As for the content it is useless to speak about it." Here is why:
>
> When he came here, M. Dufau, in too much of a hurry to become known, invited people to readings about history. It was a method like any other. But, at that time—still close to us—M. Dufau attacked, harassed, fought with all his strength what he now defends and preaches as much as he can. Last year, socialism was for him an object of ridicule, of bad jokes, a fantastic utopia. Today, it is the lighthouse toward the future, it is the Holy Ark which will save the people from the shipwreck.
>
> Thus, the excessive desire to become well known sent first M. Dufau in the path of the Right. Then, suddenly, one day, the need for a

> little more publicity, and the ambition to show off took him over, and sent him in the path of the Left. *Sic itur ad . . . nihilum!* [thus one goes to . . . nothingness]
>
> If M. Dufau had remembered his La Fontaine, particularly the fable "The Hare and the Tortoise," he would not have taken the wrong path, he would have "hurried slowly." If he would have checked his talents, he would not have created a newspaper, and he would have been well recognized by the party that fits him.
>
> We heard—and we believe without hesitation—that M. Dufau is a good teacher. In fact going from the known to the unknown, it appears to us that he has the serious qualities demanded by the profession of educator.
>
> So, as M. Dufau knows very well, generally, it is not a good idea to have more than one iron in the fire. He will recover easily from journalism, an occupation that does not fit him, while enjoying the successes he will certainly obtain, with time, in the honorable and difficult profession of educator for which he appears gifted.[18]

The message was clear: give up journalism and stick to teaching. Whether a result of Testut's denigration or not, Dufau set his political writings aside, for a time, to publish academic catalogues and newspaper advertisements for his new French Quarter college.[19]

Notwithstanding the public criticism, on October 1, 1851, Dufau opened a one-building college located at 313 Dauphine Street.[20] He titled his New Orleans academy Louisiana College but managed it much like his former lycée in France.[21] Several French immigrants, new to the United States, relied on educational experiences gleaned from Europe for employment. Chiefly, they became accountants, bankers, clerks, college instructors, doctors, lawyers, merchants, and private tutors. Few (roughly six percent) became planters.[22] Similarly, Dufau relied on his teaching and publication experience to direct his efforts in the Crescent City. Sympathizing with his fellow française étrangers, he hired other refugees, including Jean-Sylavain Gentil, to teach at his French Quarter college. An ardent republican and journalist, Gentil had been arrested as a result of Louis-Napoleon Bonaparte's (Napoleon III) censorship of the press. Now, in New Orleans, he taught courses in history and romance languages.[23] Touting a bevy of European educators, Dufau's college, founded as a private institution devoid of state support, was well followed by the city's populace. Newspapers, including the *Daily Delta* and the *Daily Picayune*, printed favorable reviews of Louisiana College. In fact, New Orleans locals were more than satisfied to send their sons to Dufau's institution.[24]

A College of His Own

During the months of February and March, a column in the 1852 *Louisiana Courier* provided details about Louisiana College, as well as Dufau's teaching skill acquired in Paris and Bordeaux. The article implies that Dufau originally hesitated to establish a college in New Orleans because competition might arise between such an institution and the city's many recently opened public schools. Five years prior, on May 3, 1847, the Louisiana legislature passed a "free school act" that fostered the creation of tax-supported primary schools.[25] Dufau may have doubted his ability to attract students away from these no-cost institutions given that his academy, inclusive of preparatory curricula, required tuition dollars in order to persist. *The Daily Picayune* reported, however, that prominent New Orleans's families who remained unlisted urged Dufau to open a college for the instruction of the city's youth. Resolute to please these "prominent families," Dufau pushed forward, converting his stance from hesitant to zealous. The French immigrant announced that his French Quarter institution would rival those of Europe and the American North and become a "Grand College." English, "being the national language," was used in all classes, and curricula fully adhered to a liberal arts model. Dufau, now a college president in his own right, taught French, Greek, and Latin, while Orton A. Dempster, formerly of the University of Edinburg, taught English. Louis Elkins served as the original mathematics instructor (later replaced by a Mr. Torrey), a Mr. Saint Marc of the University of France taught physics, and William Pitt Riddell, a graduate of both Yale College and the University of Louisiana, taught chemistry. A man named Castellon governed the Spanish department, while F. Emion taught primary French. Finally, a Mr. Gaunt presided over penmanship. The college also offered lessons on the history of France, the Middle Ages, and the United States. As well, students received a healthy dose of bookkeeping, geography, and political economy. Fees for boarding students were set at twenty-five dollars per month, while half-boarders paid fifteen. Not only were boarding students expected to supply a bed frame, mattress, and washstand, they were required to provide a mosquito bar and netting, a necessity for those who wished to diminish the possibility of contracting yellow fever.[26]

Despite his reputed negative perception of organized religion, Dufau advertised that he was "convinced that all training, in order to be solid and good, ought to rest in a basis at the same time moral and religious." Yet no classes, other than a general philosophy course, were geared toward the metaphysical.[27] Perhaps his reference to religion and morality was meant to appease those parents who preferred to enroll their sons in colleges and schools undergirded by faith. While no evidence has been uncovered to confirm or deny the aforementioned, Dufau's stance on religion wavered. This vacillation may have resulted from the death of his fifteen-month-old daughter, Marie, in 1852.[28] Whether her demise influenced Dufau's spiritual considerations remains uncertain, but his loosened stance on religion and religious instruction is noted in several Louisiana College advertisements and academic catalogues.

In addition to administrative and instructional duties at Louisiana College, Dufau also taught English and French at Miss S. S. Hull's Young Ladies' Institution at 159 Carondelet Street.[29] Conceivably, the extra income may have helped support Dufau's family and maintain his French Quarter college. Even so, these advertised educational activities did much to enhance his public image as a prominent teacher. In fact, Dufau, like his French immigrant peers, became rather well known in New Orleans. French immigrants, mostly living as middle-class citizens, considered themselves equal to Creoles. Members of the ancienne population may have collaborated with said persons over politics and education, but they rarely considered them to be aristocratic equals. Still, these française étrangers, being expatriates, brought ideals of nationalism across the Atlantic and aligned themselves with Creoles through a mutual understanding of the region's French cultural synthesis. Moreover, French immigrants understood ancienne population desires for a highly stylized hierarchy and helped reinforce the tiered social system where well-to-do Creoles maintained upper-class citizenry.[30]

While Dufau's Louisiana College advanced, newspapers alleged that the building "was situated in a quiet, airy and healthy portion of the city."[31] These claims were poorly founded as, in 1853, the deadliest recorded bout of yellow fever struck New Orleans, devastating student enrollments.[32] From the onset of hot, Southern, summer days until the first frost in late October or early November, yellow fever, or "yellow jack," as it was commonly referred to, plagued New Orleans. Transmitted by the mosquito *Aedes aegypti*, victims of the disease suffered high fevers, bodily weakness and pain, constipation, chills, internal and external bleeding, and damage to the liver, kidneys, and spleen, which rendered the flesh ashen yellow.[33] Yellow fever so often occurred in Southern port cities that by the 1840s, physicians referred to the American South as the unhealthiest region of the United States, with New Orleans sitting atop a list of sickly cities.[34] Benjamin H. Trask estimates that from 1849 until 1858, "twenty to twenty-five thousand people died in New Orleans" from yellow fever alone. Between the 1853 months of May and November, no fewer than 9,000 people died from the dreaded disease.[35]

After the 1853 epidemic, Dufau made the expensive decision to relocate his college to a healthier locale. He learned that yellow fever remained a rare occurrence in rural St. James Parish, presumably due to the dispersed population.[36] Aware of the parish's relative health, as well as the presence of an empty college campus, Dufau saw the gold coast as an encouraging possibility.[37] Upon contacting the notary publics for both Orleans and St. James Parishes to enquire about former Jefferson College, Dufau was greeted by a longtime supporter of Creole education, André Bienvenu Roman. Subsequent to his second term as governor, Roman was appointed notary public and recorder of deeds and mortgages.[38] He also represented the Citizen's Bank in its dealings with large real-estate ventures at home and abroad.[39] Retiring from the governor's office in 1843, "the two Houses of the Legislature voted [Roman] thanks for the manner in which he had discharged his duties, and the citizens of New Orleans entertained him at a public dinner."[40] As the notary public for St. James Parish, Roman approved the sale of Jefferson College to Dufau and agreed to promote Louisiana College.[41]

Despite a growing unease surrounding the française étrangers, especially since so many had been elected to high-ranking offices, Creoles understood the need to count all French-speaking men as political brethren. Roman himself regarded immigrants fleeing the 1848 revolution with apprehension, as he assumed that several were involved with the Parisian insurgence.[42] Still, one of Roman's closest educational counterparts had been Claudius Crozet, former Napoleonic military officer and president of Jefferson College. Roman, relinquishing his trepidation, saw Dufau's proposition as an opportunity to revive the Creole college. With contracts and a $20,000 credit line approved, Dufau acquired the campus and presented himself before the Louisiana legislature on April 28, 1853, to secure a charter for his soon-to-be relocated institution. The legislature consented, and with Governor Paul Octave Hébert's signature, the charter was made official.[43] Concerning the removal of Louisiana College to St. James Parish, John R. Ficklen explains that "in October, 1853, one of the New Orleans journals announced that the College of Louisiana, which had been established on Dauphine Street at great expense by M. Louis Dufau, had been removed to St. James Parish, and had taken possession of the buildings of the ex-Jefferson College."[44] Roman, in support of Louisiana College, persuaded a few Creole sugar planters to purchase shares and serve on the college's governing board. These planters included, but were not limited to, James D. Denegre, Alexis Ferry, Septime Fortier, and Valcourt LaBarre.[45] Although a French immigrant presided over the college, Creole planters in effect controlled the institution.

St. James Parish residents, external to the new governing board, were also interested in the revival of their plantation college. Five years silent, the campus was reanimated to the great expectations of Louisiana's rural and urban populations. The *Daily Picayune* reported,

> Jefferson College had like Rome its decline and fall, . . . [but] at length the old life is to return to the place; the busy studies and joyous sports of boyhood are soon to have dominion where so long a desolate silence has reigned, and a new generation of the youth of Louisiana will occupy the place where so many hundreds of their predecessors now young men engaged in the struggles and cares of life, passed their college days. We speak of this at length because so many of our readers and so many of the young men of the State will take a peculiar interest in the subject. Well then, Jefferson College is to resume its old glories when it was the favorite of State Governors and Secretaries, and bankers and planters, and mighty men of that ilk. The officers and the students of Louisiana College . . . are to be transferred to the Jefferson buildings and the classes will open.[46]

Alcée Fortier, historian and Creole descendent, portrayed Louisiana College's opening with equal gusto: "It was the old Jefferson College resurrected."[47] Vernon Elmer Kappel contended that Fortier was confused, thought Louisiana College was actually Jefferson College, and never acknowledged Louisiana College as an institution separate from the former. To quote Kappel, "Should not the learned historian of Louisiana, Alcée Fortier, know this [the relocation of Dufau's Louisiana College to St. James Parish—not the reopening of Jefferson College itself] to be fact?"[48] Regardless of the marked difference between the new Louisiana College administration and that of former Jefferson College, one could argue that Fortier did indeed know the college had changed hands but was heralding a return to the familiar, an aristocratic planter social class reunited with its own educational institution. This may have been what the Creole historian meant in the preceding quote.

After relocating, Dufau advertised Louisiana College's healthy geographic location stating that St. James Parish "was remarkably fortunate during the late [yellow fever] epidemic."[49] Pleased with his purchase, the college president enacted a new curricular structure to attract greater student numbers. This alternative itinerary was notably different from Dufau's French Quarter college: "The Course in education in the establishment [Louisiana College] is both Superior and Elementary. There is, consequently, a primary course and a collegiate course."[50] In fact, Dufau's St. James Parish institution had three academic departments: collegiate, practical, and preparatory (the collegiate and practical were grouped together in the "superior" division).[51] Curricula, although divided into thirds, were meant to last eight years with students progressing through the preparatory department before entering one of the institution's higher branches. Although more advanced, the collegiate and practical courses bound students to an overwhelming series of liberal arts courses pertaining to grammar, history, languages, literature, and, in the final year of instruction, sciences including botany, chemistry, civil engineering, geology, and physics.[52]

Regarding language, Dufau himself described English as his second tongue, but even though he preferred to speak and write in French he chose to do so in English, as it was the national patois. As such, it was the language of instruction at Dufau's St. James Parish college, just as it had been in New Orleans.[53] The decision to move away from the bilingual pedagogical method employed at the former Jefferson College may not have bothered French Creoles too much, as toward the end of the antebellum era they had begun to tolerate English as the preferred language for business and education.[54] Linguistic predilections aside, Dufau intended students to start young, progress through the preparatory department, matriculate through the collegiate or practical course of instruction, and extend their stay beyond the bachelor of arts degree if they wished to acquire a master's degree.[55] Accordingly, the St. James Parish college was filled with boys, not young men. Such enrollment practices were more akin to the preparatory lycée model Dufau was familiar with, not the "college" Creoles knew and wanted. Just as disconcerting, the so-called practical branch was anything but practical, as it was a duplicate of the collegiate course of study. According to a Louisiana College student ledger from 1853, linguistics (English, French, Greek, Latin, and Spanish) monopolized all levels of academic study.[56] Not only were classes overtly focused on linguistics, daunting oral exams followed. On the last Saturday of each month, faculty assessed students in front of onlookers drawn from the general public.[57]

Dufau failed to recognize that long-lasting and single-minded linguistics education was not what Louisianans desired. He was, however, well aware of the need to attract and retain students, as he often published advertisements concerning his regular presence in New Orleans and willingness to accompany students to the relocated academy. "Families desirous of sending their sons to this institution [Louisiana College] are respectfully informed that the President of the College will be aboard the steamer *Gipsey* leaving New Orleans, on Sunday, the 6th . . . at 9 o'clock A.M. for the purpose of taking charge of the students who may be confided to his care."[58] In addition to securing students, Dufau, in order to enhance retention, enforced residential rules. While located in New Orleans, he had admitted boarders, half-boarders, and nonboarders alike. Once relocated to Convent, Dufau claimed that he was "convinced that it [was] impossible to obtain in an establishment made up of outside students and boarders, the same result and success of that of an educational establishment where only boarders are admitted."[59] Whether or not students were pleased with the boarding requirement, all were required to live on campus and adhere to a disciplinary structure just as fierce as the regulatory mandates of old Jefferson College.[60]

The rules and regulations of Louisiana College also dictated that students participate in military training to enhance obedience. Those who did as they were told received rewards; those who defied instruction accumulated demerits.[61] The severity of Dufau's sugar parish college outstripped the relative benignity of his French Quarter institution. In addition to aligning Louisiana College with the harsh directives of former Jefferson College, Dufau maintained that students provide silver goblets and utensils.[62] He not only catered to local wealth and want for stern instruction, he appeared to accept the regional practice of chattel servitude. Regardless of his prior membership in the French Society for the Abolition of Slavery, Dufau allowed slaves to mend and wash students' clothes. If local families sent slaves to attend to their sons' needs, the president acquiesced. However, Dufau forbade slaves from entering the college buildings "without special permission."[63] Though the French immigrant kept quiet about his former membership in the abolitionist society and adopted regulatory conventions resonant of the Creole college's earlier administration, he encountered a series of debilitating circumstances that made his stay in the gold coast difficult.

Trouble in the Sugar Parish

Severe regulations did little to endorse Dufau's reputation as a positive pedagogical figure, and his mandate that students reside on campus hindered enrollments. Prior to the yellow fever epidemic of 1853, the French immigrant's New Orleans college enrolled a maximum of 274 students, with 227 attending the preparatory department.[64] Conversely, his St. James Parish institution enrolled 185 students, with 162 attending the preparatory department.[65] With disproportionate enrollments and severe debt, Dufau was in a financial quagmire. To address the lack of funds, tuition at relocated Louisiana College was set at $240 for the 1854–1855 academic year.[66] With the addition of obligatory fees for laboratory and instructional materials, the cost of attendance exceeded $300.[67]

To generate extra capital, Dufau converted the president's house into a hotel for guests, visiting professors, and parents wishing to call upon their children. The price for one room, lights, and fuel for the fireplace was twenty dollars a month or a dollar and fifty cents per day. Guests, "none but the quiet and well bred," were charged fifty cents for breakfast, seventy-five cents for dinner, and twenty-five cents for supper.[68] Dufau even tried to save money by employing Louisiana College graduates as lower paid instructors.[69] Along with poor college finances, Dufau lacked private donations. Indeed, the French immigrant found few financial supporters to champion his educational cause and help repay the college's debt: "We, the undersigned, wishing to assist the said Louis Dufau in establishing a valuable and permanent institution of learning in Louisiana, do hereby agree . . . to pay the several sums annexed to our respective names, one half on the first day of May, 1855, the other half on the first day of May, 1856, to the Citizens' Banks of Louisiana in New Orleans."[70] Those who signed the agreement were Alex De Clouet, James D. Denegre, Frs. Dugue Jr., Alexis Ferry, Adrien Frere, Septime Fortier, Governor Hébert, Valcourt Labarre, A. B. Roman, former Louisiana Supreme Court Associate Justice Edouard Simon, and the college's physician, Dr. Damaré. Of these eleven, six were the original shareholders and governing board members listed on the college's 1853 charter. Dufau also informed the public that persons wishing to become shareholders could do so by contacting him or any member of the board.[71]

Apart from Dufau's solicitations, however, no evidence has been uncovered to suggest that anyone, apart from those listed in the academic catalogue, bought shares in Louisiana College. Likewise, there are no documents to verify that the original shareholders followed through on their promises to assist in paying the $20,000 debt. If poor financial support was not daunting enough, Dufau also received staunch critique regarding curricular decisions. Historically, Dufau was perceived as irreligious or, at the very least, antitheocratic, but just as he did in New Orleans, Dufau countered his dogmatically secular stance and insisted that his sugar parish institution was based on moral and religious foundations.[72] As the 1854–1855 Catalogue of Officers and Students states, "Religion in the College is placed on the same foundation that the constitution of the United States places it. Each student, therefore, is left free in exercising or enjoying his own religion, according to the dictates of his conscience. Yet, all students from Catholic families are requested to attend church on the Sabbath, and all students from Protestant families are requested to attend Biblical Exercises in the College on the Sabbath. These two classes embrace all students of the College—unless the parents or guardians shall notify the President, that they wish their sons or wards, to be regarded as exceptions."[73] The college president even permitted a priest to prepare Catholic students for First Holy Communion and notified parents that the religious beliefs of all enrolled were respected.[74] Including firm rules and stern military training, religion may have served as an additional means to enforce positive student behavior. Yet the allowance for personal spiritual practices did little to hamper condemnations hurled against Louisiana College for its predominant lack of religious instruction. As in Dufau's New Orleans institution, theology courses were excluded in his St. James Parish college.[75]

In addition to pedagogical criticisms and pitiable funding, Dufau's personal troubles were myriad. On November 10, 1855, the morning edition of the *Daily Picayune* reported, "We regret to learn, through the officer of the steamer *Capital*, the death of Mr. Louis Dufau, President of Louisiana College."[76] Being made aware of his premature obituary, Dufau contacted and corrected the tabloid. The false report was retracted in the newspaper's evening edition and attributed to a miscommunication that confused Dufau with Mr. Bac, a Louisiana College instructor who had recently passed away due to an unidentified illness. The *Daily Picayune*'s editors, keen to correct their mistake, apologized for any alarm or damage the inaccurate statement may have caused parents, students, or the college itself. The retraction went so far as to explain that, although Bac died as a result of fever, "there was no other case of sickness at the College, nor in the immediate neighborhood, and that everything was getting on well."[77] Still, the false notice was an ill omen of forthcoming tribulations.

As well as correcting false reports of his demise and contending with a barrage of condemnations concerning Louisiana College's secular nature, Dufau was also engaged in disputes with local clergy. After Dufau opened his St. James Parish college, Father Auguste Simone Paris of St. Michael's Church "made Dufau see the possibility and even the necessity of saying Mass at the college, at least on Sunday."[78] Per his conversations with Paris, Dufau allowed Catholic students to attend services at St. Michael's but only once a week. Dufau was satisfied with the arrangement until Father Claude Anthony Tholomier replaced Paris as the church's pastor. Unlike Paris, Dufau did not care for Tholomier. Clearly, the feeling was mutual, as the priest and college president began a series of bitter disagreements that almost erupted in a legal suit. Tholomier informed Dufau that he owed 250 piasters for the reservation of church pews for college students. Dufau grumbled about the exorbitant price but complied and paid the required sum. In response, Tholomier demanded more money, roughly 700 piasters. Dufau countered that his previous agreements with Father Paris never included discussions of such high pew prices.[79]

Dufau complained that his students were forced to sit with slaves after they were denied pews designated for white parishioners and "that it was only at Paris' invitation that they had occupied the little choir chapels and that this was only until someone came to say Mass at the college." In short, he refused to pay the additional amount. In a letter to Archbishop Antoine Blanc in New Orleans (on July 19, 1850, the port city had been raised from a diocese to an archdiocese by Pope Pius IX, and, with its elevation, Blanc was promoted from bishop to archbishop),[80] Dufau exclaimed that, in no short terms, "Tholomier wants a scandal, and money."[81] Though failing to mention his oppositional purview of Church hierarchy, Dufau explained to the archbishop that as an educational administrator responsible for multiple Catholic students, he had a duty to provide for their religious edification, albeit external to the college campus. Any disgrace resulting from a legal battle between Louisiana College and the Catholic Church might, Dufau explained, prevent students from attending Mass. In closing, Dufau queried Blanc, "Would it be right to deprive the faithful of divine services?"[82] The archbishop, disturbed by the squabble, encouraged both parties to settle the matter amicably. Dufau, opting to avoid the courts, suggested arbitration with the college's physician, Dr. Damaré, representing both himself and the priest of St. Michael's Church.[83] To aid his defense, Dufau recorded events as he remembered them and sent a copy to Tholomier:

> With a view of reconciliation and to avoid the scandal of a lawsuit, Dufau had asked Mr. de Laneuville, then Mr. de Poorter and Antoine Webre to see Tholomier. The latter had been asked to propose an arbitration. Dufau would name two arbiters; Tholomier would name two. . . . Instead of replying categorically to these gentlemen Tholomier accused Dufau of breach of faith, etc. Tholomier has proved once more that he has neither logic nor reason. Dufau cannot let Tholomier's assertions pass in silence. 1. It is completely inaccurate that Tholomier sent several persons to settle the affair amiably. First he sent an unknown person to present his bill. Then Dufau received a visit from Dr. Damaré who, after discussing matters of mutual concern, brought up Tholomier's name. Dufau immediately stopped him, telling him that he wished to keep Damaré's friendship. 2. Tholomier spoke of the Mr. Bac affair and said that if the deceased had only Dufau as security, he would have refused him a miserable burial of 15 piastres. Dufau hopes he will not die while Tholomier is pastor of St. Michael. Tholomier seems to forget that he has already received 350 piastres from Dufau which he gave freely without any contract between them. Does Tholomier think his services at St. James [Parish] are worth 700 piastres? 3. Tholomier mentioned that Dufau refused to withdraw a promise of 100 piastres which the late Auguste Theriot had placed in Dufau's hands for the establishment of a college at St. James. It is the first time that Dufau learned that Tholom-

> ier had pretentions of being the owner of this gift. Dufau acknowledges that he owes 100 piastres to Theriot's estate but he does not see how Tholomier can collect this sum. Even if the deceased had made this gift to be applied to some pious work Tholomier would have to explain it before the courts and show that he had not extorted it from a dying man. 4. Tholomier has left the idea that he acted only on orders and advice of his superior. Dufau cannot believe that the Bishop [Blanc] would have advised Tholomier to have recourse to the courts instead of arbitration. Dufau is going to write the Bishop about this. Tholomier accuses Dufau of being a man of bad faith, etc; Dufau accuses Tholomier only of caring too much for money.[84]

After reading Dufau's testimony, Tholomier transcribed his interactions with the college president and shared Dufau's account with Damaré. Having reviewed the transcript, Damaré was sorely displeased. In his estimation, Dufau was unjustly vilifying the parish priest. The physician vocalized his discontent by calling Dufau a "wretch!"[85] In response, Tholomier wrote the following and sent a copy to Archbishop Blanc but did not provide Dufau with a facsimile:

> It is not fitting for Tholomier to dictate the reply to the letter Mr. Louis Dufau wrote to Blanc about him. Tholomier had read it before Blanc through one of his parishioners. Tholomier told the truth every time he talked to Blanc with the intention of informing him of the state of things between Dufau and himself. Dufau assumed too much merit in stating that he was on such good terms with Father Auguste Simon Paris. On Tholomier's first visit to Dufau, Dufau spoke disparagingly of Paris and Paris told Tholomier that he had never made the promises which Dufau claimed but had always said Dufau was to make arrangements with the pastor who would come. However when Tholomier arrived at St. Michael, Dufau asked him nothing about what he owed for the seats occupied by his students. So Tholomier had to lay claim to it in a letter dated in August and written at the advice of Father Vital Gille, S. J. To this letter Dufau replied with insulting remarks face to face. Neither Blanc nor the Jesuits were spared. Several weeks later Dr. Damaré was sent to apologize and make any arrangements Tholomier would like and assuring Tholomier in a letter that he [Dufau] would accept them. Tholomier's stupidity is that he did not keep that letter. Tholomier had said in his letter in August that Dufau was to pay 250 piastres each year for his students, and that wishing to remain free he did not wish to commit himself to go such and such a day or any number of times to the College but

> would do everything his position called for and that, gratis. In spite of all these agreements, Dufau wrote to Blanc that without any special contract he had sent 150 piastres to the pastor of St. Michael not as a debt but as a gratuity. The truth is that Tholomier had been obliged to ask Dufau for it at least six weeks after their agreement. As for the other 200 piastres which Dufau says he gave as a gratuity, it was his students who gave it in gratitude for preparing them for First Communion. Dufau owes Tholomier 350 piastres.[86]

Further fueling animosity between Dufau and the St. Michael's priest, Tholomier published his account of Dufau's treachery in local newspapers. Upon discovering Tholomier's declaration, Dufau requested proof of the priest's allegations. Tholomier, however, could not produce sufficient evidence, and the matter was dropped. Louisiana College's president emerged triumphant but begrudged. Dufau then severed ties with St. Michael's Church and stated that he "wish[ed] very much to have nothing more to do with Tholomier."[87]

Aside from clerical criticisms, Dufau struggled to increase enrollments. Not only was Louisiana College lacking in wealthy shareholders, the institution's collegiate and practical branches were poorly stocked. Since the close of Jefferson College, Creole planter families had reaccustomed themselves to sending their elder sons abroad for higher education.[88] Despite the propinquity of Louisiana College, planters cared little for the institution in general. Dufau's attempts to create a college that matched the academic rigor of northern colleges and universities had failed.[89] What enrollment existed quickly declined due, in large part, to sugar planters' indifference to provide for struggling colleges and schools. Annoyed with paying mill and poll taxes to support public institutions created by the 1847 free school act, plantation-owning Louisianans pressed the state legislature to decrease funding for instruction that, in planters' opinions, only served the regional poor. Wealthy Creoles had no desire to send their sons to "pauper institutions" and continued to rely on private tutors to furnish their sons' rudimentary education. As rich planters held power in the Louisiana legislature, political concern revolved around appeasing the upper class while the lower classes struggled to attain legitimate forms of state subsidized education. Given sugar planters' desire to cut property taxes, especially as it concerned their sizeable plantations, capital reserved for parish schools was radically reduced in the Louisiana Constitution of 1852.[90] Money from local backers had also dried up as St. James Parish Creoles, no longer interested in assisting the tempestuous Dufau, turned their backs on the ill-fated Louisiana College, despite its status as a private academy.

In an attempt to increase student numbers, Dufau paid for several newspaper advertisements. In July of 1856, the *Daily Picayune* printed the following: "Louisiana College, located in the buildings formerly occupied by Jefferson College, Parish of St. James. The Louisiana College is now the property of a company of stockholders, among whom we observe Louis Dufau, A. B. Roman, James D. Denegre, L. Labarre, Fr. Dugue Jr., Dr. Damaré, Septime Fortier, Alexis Ferry, Alex Declouet, Edouard Simon. It has been charted by the state of Louisiana, and received University privileges in March last. It is administered by a Board of Trustees, composed of Mesers. Louis Dufau, President ex-officio; James D. Denegre, L. Labarre, Alex Declouet, Fr. Dogue Jr., Alexis Ferry, Septime Fortier."[91] Dufau may have hoped that showcasing the names of supportive Louisianans would influence attendance. To further increase his contact with Louisiana educators and school leaders, Dufau acquired a position on the State Teachers Association's publication subcommittee. This group was charged to "report upon the expediency of establishing a monthly state journal of education . . . devoted to the educational interests" of Louisiana.[92]

As a member of the publication subcommittee, Dufau made the acquaintance of Joseph S. Copes, a respected physician from Delaware. Copes moved to New Orleans in 1849 and was well connected in Louisiana society. He practiced medicine at Charity Hospital in New Orleans, was involved with the Academy of Sciences and the New Orleans Medical Association, was a member of the University of Louisiana's administration, and served as Superintendent of the Second District of Public Schools in New Orleans.[93] Before moving to Louisiana, Copes lived in Mississippi and was a founding member of the Mississippi State Agricultural Society. While there he helped establish Sharon Female College and sat on the board of directors for Oakland College (site of present-day Alcorn State University). Dufau, seeing an opportunity to establish a beneficial relationship with this medical doctor and superintendent, asked Louisiana College's one-time chemistry instructor, William Riddell, to invite Copes to attend the August 1856 commencement ceremony. Riddell had taught for Dufau from 1851 until the spring of 1852 in New Orleans. The chemist informed Copes that three Louisiana College students would most likely receive the bachelor of arts degree and that two would probably receive the master of arts degree.[94] Dufau may have asked Riddell to write Copes, as the former Louisiana College instructor was now employed alongside his older brother, the eminent New Orleans physician John Leonard Riddell, at the University of Louisiana. A few days after Riddell sent the aforesaid letter, Dufau, on parchment emblazoned with the rural college's image (the phrase "Louisiana College" adorned the main portico), wrote to Copes to request his attendance.[95] Whether or not Copes attended the sugar parish commencement is not known. Nevertheless, attempts to increase enrollment at Louisiana College and garner social, as well as financial support, proved unsuccessful. In retaliation, Dufau crafted a livid commencement speech targeted at the local populace.

Grappling to repay his $20,000 debt, an agitated Dufau lectured to the commencement crowd and, irritated with the lack of support, pronounced that Louisianans sent "more than a thousand" of their sons away to the North or to Europe instead of patronizing regional education—namely Louisiana College. "I may be mistaken, gentlemen," he scolded, "yet I am under the impression that you lack public spirit in regard to the education of your children! . . . The best proof of it is, that you send your children to be educated among them [Northerners and Europeans], while I have never heard, that they send in return their children to be educated among you!"[96] Dufau continued his condemnations: "How many persons in Louisiana, intelligent and respectable people indeed! Who say, whenever they see the beginning of an institution of learning, 'very well, let them succeed . . . but in the mean time, let us send our youth to Kentucky, to Virginia, to Maryland, or to New England! . . . Do you call this consistency? Do you call this patriotism?'"[97] Preaching to the crowd, Dufau emphasized that he had done all he could as a private citizen to maintain a college devoted to the public.

Now, Dufau asserted, it was up to the public to help. "To found a college on a permanent basis, the community ought to be interested in its success," he declared, "and the President ought to be rather the agent of the community than the proprietor of the College. . . . Thus, people of Louisiana, if you wish to have permanent colleges in your own native state, you must depend on yourselves, and not at all upon any one man."[98] In publishing his 1856 commencement speech, Dufau verbally and symbolically linked Louisiana College to the state. He even placed the Louisiana symbol of a pelican nursing her chicks on the address's cover and claimed that Louisianans historically and readily relied upon French immigrants to educate their sons. Dufau pronounced that as his college was named after the state and led by a Frenchman, it deserved community support. In conclusion, the college president bemoaned, "I do not know if the fatal destiny of old Jefferson College is reserved to Louisiana College. I hope not. All that I can say and promise for my part is to stand firm and unmoved to the last moment."[99]

No longer confident that public support alone could prevent Louisiana College's demise, Dufau contacted a former affiliate, the University of Louisiana, and brokered a deal that rendered his college a feeder institute subservient to the New Orleans university. The agreement between Louisiana College and the University of Louisiana appeared in the September 21, 1856, issue of the *Daily Picayune*:

> The University of Louisiana has connected with the Louisiana College, which established by act of the Legislature, has been for two years under the charge of Mr. Louis Dufau, long and well known here [New Orleans] as a teacher. The college has been in operation for two years in the buildings of the old Jefferson College, about sixty miles up the coast, and where so many of the young men of our State received their education under the able administration of President Ingals, Everette, . . . and others. Why it fell through we can not say; but pity it is, it did. Its auspices and prospects were of the fairest, and we know that hundreds of those who have many pleasant remembrances of the fine old buildings, large play grounds, and other appurtenances of this extensive establishment, to say nothing of its many able professors, some of whom are still with us, will unite in the hope that the "Old Jefferson," that once promised to be the leading college of the State, may yet, one day, be restored to its pristine renown, vigor and usefulness.[100]

In spite of Dufau's ranting and articulation agreement, Louisiana College closed in the fall of 1856.[101] Records indicate that after Dufau secured the original credit line of $20,000 (payable in four yearly installments of $5,000), he sold his French Quarter college property. Monies from this sale may have been used to cover relocation expenses and to help pay the first loan bill. Dufau, by means of capital generated from student tuition, fees, and other miscellanea, was able to pay the 1855 bill; however, when he could no longer reimburse the debt on Louisiana College, the campus was returned to the Citizens Bank.[102] With no viable employment alternatives in St. James Parish, Dufau returned to New Orleans. Upon his return, however, a series of personal conflicts distanced him from the field of education. Not only had Dufau failed to maintain a college of his own, after returning to the Crescent City he alienated himself from Louisiana leaders responsible for the progression of regional education. As a result, his and Louisiana College's record in the pedagogical history of Louisiana was diminished.

Return to New Orleans, Return to Provocation

Realizing that Louisiana College would close, Dufau made arrangements with the University of Louisiana's governing board to assist in establishing a college department and an affiliated preparatory school in New Orleans and relocate the majority of his younger students to said department.[103] Devoid of his St. James Parish and original French Quarter college properties, an alliance with the University of Louisiana was Dufau's last means of retaining a leadership role in Louisiana's educational scene. Recorded in the University of Louisiana's board of directors' minutes, Dufau "propos[ed] to discontinue the Louisiana College situated in the Parish of St. James, and bring with [him] as many of the students as [he] can to enter the university."[104] By the end of the mid-1850s, the university possessed an academic "college" department comprising both upper-level courses and a preparatory grammar school in addition to a well-staffed law school and medical program (in 1847 the Medical College of Louisiana merged with the University of Louisiana to create said program).[105] In 1850, Claudious Wistar Sears, an instructor in the preparatory division, was promoted to the rank of mathematics and natural philosophy professor in the college division.[106] This West Point–educated instructor was not only elevated in rank; he was given the college department's deanship and charged to recruit students. In 1856, Sears presented a plan devised by Dufau and former Louisiana College instructor Riddell to enroll approximately one hundred and forty students—the majority being transfers from Dufau's sugar parish college.[107]

The college department (housed in the university's west wing) operated financially independent from the University of Louisiana, but was beholden to the institution's governing board.[108] In 1856, an academic catalogue was published that described the department and its leadership. Sears, the catalogue explained, oversaw the college department as a whole while Dufau governed the grammar school.[109] Though apprehensive about operating both preparatory and college-level courses alongside law and medical programs, the board acquiesced and admitted, "The fact . . . that these gentlemen are willing to undertake the [college department] enterprise at once . . . relieves the difficulties in the way of making the literary department of the university a private concern to a great deal."[110]

On December 23, 1856, Dufau was installed as principal of the grammar school and professor of moral sciences.[111] Along with Dufau's appointment, the catalogue reported that the preparatory division was seeking "competent professors."[112] Without delay, Dufau hired several former Louisiana College instructors to teach in the grammar school.[113] With the college department established, boys from nine to fourteen years of age well versed in reading and writing were allowed to enroll. The preparatory division boasted a rigorous set of language and philosophy classes and educated students to either enter Sears's college courses or an "active business life."[114] Once again in New Orleans and holding a position of some authority, Dufau attempted to reenter the city's social circles. On March 5, 1857, Dufau invited Copes to attend the department's annual examinations.[115] As before, no documents have been uncovered to indicate whether or not Copes was present at the event. Copes's attendance aside, Dufau, now situated at a respected university, made efforts to forge fraternal relations with fellow faculty and was permitted to join the newly formed University of Louisiana Mu chapter of Phi Kappa Sigma—a fraternity initially consisting of both students and faculty founded at the University of Pennsylvania in 1850.[116] But where success followed Dufau, so, too, did disagreement, disappointment, and failure.

Not long after his appointment to the University of Louisiana faculty, Dufau began to argue with and complain about his colleagues. In a letter to the dean of the university's faculty, Dufau quibbled about foul language used by professors and how it debased the institution. He chastised his peers for not adhering to "proper" meeting etiquette when discussing university business and lambasted Sears for acting as if he, not Dufau, was principal of the grammar school. In a final outburst, Dufau reminded the university's faculty and governing board that "at least two thirds of the students of the collegiate department were formerly students of Louisiana College, and the future prospects of the collegiate department, in our opinion, depend upon the actions of the late president of Louisiana College."[117] In addition to criticizing his peers and superiors, the French immigrant met with discord over his dual appointment as principal and professor. In a letter to the governing board, Dufau haughtily explained that the university's preamble permitted no other university official, save the president, deans, and department principals, to serve as both administrator and professor. He cited policies that reinforced his right to collect money from students, enforce disciplinary measures, and administer the grammar school as he saw fit.[118] Dufau was fast becoming a nuisance to the university's upper administration.

In addition to administrative arguments, Dufau fought with his former Louisiana College colleague Riddell, who now served as both treasurer and professor of chemistry and natural sciences in the University of Louisiana's college department.[119] Although Dufau and Riddell agreed it was best to transfer Louisiana College students to the new department, Dufau believed that Riddell tarried in his duties and failed to garner the board's approval to hire instructors formerly employed at the St. James Parish college. To compensate for Riddell's shortcomings, Dufau approached the board, received the appropriate approval, and sought out his Louisiana College peers to teach in the grammar school. Weary of dealing with perceived incompetence, Dufau no longer wished to work with Riddell. The grammar school principal complained that Riddell made managerial decisions without his consent, spent department money sans permission, and caused an air of confusion and dissent among the faculty regarding Dufau's leadership. He also claimed that Riddell neglected his classes and hampered the departments' overall success. Riddell disagreed, issued a public letter describing his negative opinion of Dufau, and attempted to declare a legal injunction against the grammar school principal in the fourth district court.[120]

Although the injunction was not granted, the grievance between Dufau and Riddell was so troublesome that a university policy had to be created to deal with issues arising between the two: "All disagreements between Louis Dufau and W. P. Riddell (if any shall arise) shall be immediately . . . referred to the Board of Administration, where decisions shall be final and binding."[121] Nevertheless, Dufau believed his former colleague had grievously injured the college department. Endeavoring to mentally banish Riddell while aggrandizing his own attempts to create a lasting educational institution, Dufau avowed, "I have considered the . . . [Riddell] matter as of a secondary importance, having always and constantly in view to build up a permanent basis an institution of learning in Louisiana, this being now the only reason for which I have been induced to quit the parish of St. James, and to place myself and the college which I have been connected under the patronage and government of the . . . University."[122] Dufau may have meant well, but due to difficulties experienced at the University of Louisiana the college department had, by the end of the 1850s, failed to attract sufficient student numbers.[123] By then, the university's board had tired of Dufau and sought ways to rid themselves of the French immigrant.

On February 15, 1860, the governing board rescinded Dufau's building lease. Compounding existing issues, the board learned that Dufau, after consulting with Sears in an attempt to increase funds, had sublet a portion of the college department wing to local social clubs and vendors (an ice cream merchant, in particular) without their approval. Ultimately, the board decided to close the department and associated grammar school. Along with their decision to terminate the college department, the governing board made a point to mention their belief that Sears was a competent and attentive teacher who richly deserved the reputation of his practice despite his recent misstep. Likewise, Riddell was not repudiated for his quarrelsome relationship with Dufau. The chemistry professor remained at the university until he accepted the post of assistant state geologist in Texas. Dufau, however, was not mentioned nor was he retained as a university professor.[124] In all, the college department and grammar school lasted three fitful years.[125]

Free of the trappings imposed by educational administration, though not of his own volition, Dufau returned to public life and became an active member in a New Orleans Scottish Rite Masonic lodge. Louisiana masons had long feuded over which organizational rite should govern regional lodges. These arguments were intensified by ethnic rivalries between south

Louisiana's predominant social classes: Creoles supported the French Rite, Anglo-Americans upheld the York, and French immigrants fostered the Scottish.[126] As immigrant freemasons, members of the française étrangers rallied around republican beliefs resonant of both revolutionary France and the Masonic triumvirate: "Liberty, Fraternity, Equality."[127] Responding to this combatant rift, the Grand Lodge of Ancient Freemasons of South Carolina attempted to assert control over the Southern jurisdiction of lodges.[128] As regional power struggles gave way to Anglo-American dominance, the York Rite emerged as the predominant configuration for Louisiana's Masonic lodges.[129]

In the late 1850s, James Foulhouze, a high-ranking New Orleans mason, founded a separate division of Scottish Rite freemasons titled *Supreme Council of 33ds for the Sovereign and Independent State of Louisiana*. Foulhouze's plan to unite Louisiana masons under a single rite fell flat because numerous Louisiana lodges external to New Orleans refused to pay homage to the new Supreme Council. As a result, New Orleans lodges stood alone with Foulhouze at the helm. Dufau, ever desirous to increase his social standing, volunteered to serve as the newly formed council's grand secretary and published a revised set of principles and rituals based on the Scottish Rite.[130] Dufau's aid in separating Scottish Rite lodges from the Southern jurisdiction aligned him with several prominent New Orleanians; however, the secession of Masonic lodges from the larger fraternal order provoked hostility and heaped infamy on the ex-educator.

By the latter half of the 1850s, New Orleans Scottish Rite masons were banned from lodges around the world and labeled "irregular masons."[131] Such was the case for Louis J. B. Brunett, a sea captain initiated into New Orleans's Louisiana Masonic Osiris Lodge, Number 8. While traveling, Brunett was denied admission to lodges in Gibraltar, London, Marseilles, New York, and Paris. In New York, he was informed that his Masonic "diploma was like a counterfeit banknote, and that the signatures of Messrs. J. Foulhouze and Louis Dufau made the parchment worthless."[132] In 1859, the New Orleans's St. Andrew Masonic Lodge, Number 5, petitioned the Grand Lodge of the State of Louisiana for reinstatement into the national order. For this restoration to occur, New Orleans masons needed a scapegoat. None would do so well as the Scottish Rite grand secretary, Dufau. The New Orleans lodge accused the French immigrant of purposely misinterpreting English orders issued by the Anglo-American-controlled Louisiana State Masonic Lodge. Dufau, the lodge claimed, not only tampered with information concerning dues and property ownership, but he also intentionally altered directives related to the installment of the York Rite over the Scottish. The New Orleans lodge declared that he did both willfully to separate the port city from the greater fraternal order and thereby enhance his personal prominence. Lodge members begged for readmittance, claiming that Dufau, Foulhouze, and other influential masons led them astray. "The facts speak for themselves," St. Andrew's lodge petitioners wrote; "you are well aware of the share Messrs. Lisbony and Dufau have taken, since three years, in the unfair war wag[ed] against your Worshipful Body. Recent events have proven to you that they were over-scrupulous about the means by which they could attain their object, in sowing anarchy and discord in an Institution whose bases are Peace, Harmony, and Charity!"[133] Following the former college president's fall from Masonic leadership, Dufau's second daughter, Camille, was diagnosed with scarlet fever and died at the age of eleven. Unseated and burdened by the loss of a second child, Dufau's freemason superiors offered him some consolation and allowed him to remain a member of the Polar Star Lodge, Number 1.[134]

Not permitting the loss of several educational positions, Masonic degradation, and the death of two children to dampen his pursuit of social acclaim, Dufau found another cause to champion. As the Civil War dawned, Dufau joined the Union Association of New Orleans. As he had with the Masons, the French immigrant became the organization's secretary. The purpose of the Union Association was to uphold the federal constitution and promote the reconciliation of Louisiana to the United States. As an article in the *New York Times* explained, "The terms of membership of this Association [were]: First, citizenship of the United States; second, citizenship of the state of Louisiana; third, to declare the intention to become a citizen of the United States; fourth, the taking of the oath of allegiance to the United States, and signing this Constitution." The Union Association sponsored daily raisings of the United States' flag, Union balls, and public speeches from antisecessionists. One such assemblage in New Orleans attracted a staggering 15,000 attendees.[135]

Despite the Union Association's growing predominance in south Louisiana, Dufau's role as secretary was brief. Less than one year after joining the coalition, Dufau died suddenly and without the level of renown he so desperately desired. On July 12, 1862, the *Daily Delta* reported, "Louis Dufau, the Secretary of the Union Association of this city, departed this life yesterday at about 4 o'clock p.m. at his residence in Freetown. The Union Association, and all his friends, are invited to attend his funeral, which will take place from Foyer Maconnique Lodge, on St. Louis Street, between Exchange Alley and Royal Street, at 4 ½ o'clock This Afternoon."[136] Dufau's brothers-in-law, Gustave Leroy and Octave Morel, presided over the funeral.[137] Although the *Daily Delta* listed the date of Dufau's death as July 12, the French immigrant actually passed away on July 11.[138] New Orleans archdiocesan cemetery records document that Dufau, age forty-five, died as a result of an "irritation of the brain." Given the onset of an exceptionally high fever, head and body aches, vomiting, and expedient fatality associated with nineteenth-century diagnoses of brain irritations, Dufau may have died of complications stemming from meningitis.[139] An indicator of the illness's fatal speed lies in the fact that the Union Association had no time or—given the need to perform Dufau's funeral—simply neglected to remove the secretary's name from the association's July 13 newspaper advertisement regarding a public reading of the Declaration of Independence.[140]

After her husband's death, Cecile Dufau relocated to Houston, Texas, but maintained ownership of her and her husband's Jefferson Parish home. In 1893, Cecile tried to sell her Freetown property to a Mr. Edward Jones for the sum of $1,150. However, in 1895, public notary John Baptiste Lyman recorded Jones's inability to make payments and Cecile's agreement to release him from all contractual obligations.[141] Documents describing Cecile's life are scarce. It is believed that she never remarried and remained in Houston until her death on May 1, 1902, at the age of seventy-eight. That year, Cecile was interred in the Harris County Glenwood Cemetery, 350 miles from her husband's remains.[142]

Cecile's life, wed to the fractious Dufau, is poorly recorded. What she contended with as the wife of a man who courted hardship can only be speculated. Still, records indicate that Cecile was listed along with Dufau on property deeds associated with his French Quarter college. She also agreed with her husband's decision to purchase the former Jefferson College, and her consent is recorded in the 1853 act of sale signed by Dufau.[143] Possibly, Cecile's support indicates a need for her husband to succeed and generate capital to support their Louisiana-based family. On the other hand, the junior Dufau, Jules, experienced the failure of his father's educational ventures firsthand. Not only was he present when Louisiana College closed, he was one of the many students transferred to the University of Louisiana's college department and was in attendance when it was terminated. It seems that where his father taught, Jules attended. After Dufau's death, Jules enrolled at Straight University on Esplanade Avenue and was later employed as a business clerk before moving with his mother to Texas. In 1908, Jules passed away. He is buried near Cecile in Glenwood Cemetery.[144]

Although Louis Dufau and his Louisiana College existed, very little has been written about either. What has been recorded is often incorrect. Kappel lists several nineteenth- and early twentieth-century historians who omitted Dufau from Louisiana's larger narrative of educational history. These historians either judged the French immigrant and his ephemeral college to be insignificant or confused them with the antebellum Anglo-American

College of Louisiana.[145] Such historians, many of whom are descendants of the same Creoles who abandoned Dufau while in St. James Parish, include Henry C. Castellanos, Edwin Whitfield Fay, Alcée Fortier, Charles Gayarré, John Smith Kendall, Henry Rightor, Charles Testut, and Edward Larocque Tinker. Even the famed Catholic historian Roger Baudier excluded Dufau and credited the Creole planter Valcour Aime for founding Louisiana College at the Jefferson College site.[146] Although the aforesaid historians erred, Louisiana College did exist. In 1858, two years after Dufau's egress from St. James Parish, the well-known *Norman's Chart of the Lower Mississippi River* was published. Printed on this map, apart from the copious diagrams of plantation estates, the word "college" is listed next to a small black rectangle, representative of the empty Louisiana College buildings. Unlike *La Tourrette's Reference Map of the State of Louisiana*, a full title, in this case "Louisiana College," is not given. Even though Dufau's college had closed, the map's designers, Marie Adrien Persac and Benjamin Moore Norman, felt it necessary to indicate the campus's location not far from St. Michael's Church.[147]

The Creole college remained closed until Louisiana sugar planters chose to revitalize the only educational institution dedicated to their heirs. However, to return the campus to its former glory would take considerable financial resources. Such wealth existed and was owned by Valcour Aime, a planter who had become king among sugar barons. Contained within the college's Reconstruction–era academic catalogues is a short institutional history. Again, Dufau remains absent. In his stead, Aime and A. B. Roman are inaccurately credited for opening Louisiana College. The catalogue indicates that the institution closed as a result of financial difficulties and that Aime purchased the campus for $20,000, constructed a Gothic Chapel proximate to the main building, and dedicated it to his children, Félicie and Gabriel.[148] This account is partially correct. Certainly, Aime played an important role in the college's post-Dufau revival, but not while the French immigrant was present in St. James Parish. Dufau's administration, similar to the foreign French governance of the Collège d'Orléans, reflects a conspicuous degree of tumult between the ancienne population and Louisiana's française étrangers. Though Creoles used the latter for instructional purposes, they denied them social class equality. If anything, the legacy of Louis Dufau is one of an ambitious French immigrant who embodied societal and educational ideologies resonant of revolutionary France. By means of college administration and instruction, Dufau attempted but failed to launch himself into Louisiana's high society—an aspiration that, for many, was wholly impossible.

4

Return of the Sugar Barons

Valcour Aime, you are the Louis XIV of Louisiana.

—Unknown

While visiting Valcour Aime's plantation, a French nobleman (according to Michael Jean Fortier) effused, "Valcour Aime, you are the Louis XIV of Louisiana. The beauties of your plantation remind me of the Chateau de Versailles."[1] As one of the wealthiest French Creole sugar barons in south Louisiana prior to and immediately after the Civil War, Aime was renowned for his plantation success and is credited as having been one of the first planters to refine granulated sugar. In fact, a sample of his refined sugar was awarded a grand prize at the 1853 New York industrial exposition. Aime's refining innovations substantially contributed to the industrialization of the gold coast, and as a direct result, he amassed a fortune.[2] In addition to his financial success, Aime served as a charter member for the original Jefferson College, and three years after Louis Dufau left St. James Parish, he became a seminal figure among sugar planters responsible for the college's revival. Aime is not only remembered for his contributions to the sugar industry but also for influencing the Creole college's religious foundations. The life and exploits of Aime have been recorded, and to some extent embellished, thus exemplifying this particular sugar baron as a pertinent example of French Creole society and culture.

Born Francois Gabriel Aime in 1797, he was the youngest son of St. Charles Parish denizens Francois Aime II and Marie Félicité Fortier. When he was an infant, a nurse referred to him as "Valcour," and the sobriquet was retained.[3] After Francois's death in 1799, Valcour and his older brother,

Michel, were sent to live in New Orleans with their grandfather, Michel Fortier II, owing to their mother's lengthy European sojourns. (Marie had traveled abroad to consult physicians about her poor eyesight.) When Marie died in 1806, Michel Fortier became the legal guardian of Valcour and his brother, and had both educated by private tutors.[4] As teenagers, the brothers served under Captain René Trudeau, Troop of Horses, St. Charles Parish, in the War of 1812, from 1814 until 1815. Taking his leave of the military, Valcour Aime, now twenty-one, married Josephine Roman in 1819.[5] Josephine's brother was none other than André Bienvenu Roman, the future two-time governor of Louisiana. The newly married couple resided at the Roman plantation in St. James Parish with Josephine's widowed mother, Marie Louis Patin. All the while, Aime managed the plantation for his mother-in-law until her death on October 14, 1830. After Marie Louis's death at the family's New Orleans townhouse on Dumaine Street (known today as Madame John's Legacy), Aime offered to purchase the sugar parish estate from the eldest Roman heir, Jacques Telesphore Roman, in 1836. Telesphore sold the plantation and bought land adjoining his former property where he built Bon Séjour plantation.[6] Aime profited from his expenditure and constructed a grand home to replace the former Roman edifice. The new house with its decorative gardens became regionally known as "Le Petite Versailles."

Aime and his wife were fruitful in their marriage. Between 1819 and 1830, the Creole planter fathered five children: Edwige (b. 1819), Josephine (b. 1820), Félicité Emma (b. 1823), Félicie (b. 1825), and Gabriel (b. 1826). After Edwige's birth, Aime sold his inherited St. Charles Parish property to his brother Michel in 1820 in order to expand his sugar empire.[7] By 1829, he had implemented steam engines and vacuum-panning equipment to increase the productivity of his mill, dubbed the "St. James Sugar Refinery."[8] Aime's refining ventures were successful, and he grew in wealth. In turn, other planters, including A. B. Roman, learned from Aime and profited from his knowledge. As wealth increased between the Aime and Roman families, so, too, did marriage opportunities. Creoles considered interfamily marriage a form of heritage preservation. In this tradition, the Aime daughters, Edwige, Félicité, and Josephine, married Fortier and Roman cousins, produced more than a few heirs, and closely bound the families.[9] Accordingly, the Aime and Roman families sold or exchanged goods and properties with each other.[10]

During the 1830s and 1840s, Aime expanded his plantation home, adding two-story wings, Doric columns, and marble floors.[11] In 1842, he began planting an immense English garden, which later included an artificial lake, river, and water feature.[12] The lake contained an island surmounted by a brick fort with a functioning cannon. In reference to Napoleon Bonaparte's second exile, the island was casually referred to as St. Hélène.[13] To embellish his garden, exotic plants were imported, bridges were constructed to span the miniature river, and a Chinese pagoda was built atop a flower-covered hill with a brick-lined grotto underneath.[14] Encircled with columns, Aime's mansion consisted of sixteen rooms, including a banquette hall, bedrooms, private parlors, a children's dining room, a library, a large interior marble staircase, two exterior wooden staircases, and a sizable courtyard.[15] Such extravagant living conditions were, according to Roulhac B. Toledano, the result of riches accumulated by successful sugar harvests, granulation, and sales. "Sugar," Toledano explains, "had made this way of life possible, and Aime spared neither personal effort nor money to increase the quality of his cane and the quality and amount of refined sugar."[16]

To improve production methods, Aime sailed to Cuba in 1845 and inspected the country's sugar industry. He also sent assistants to England to study alternative manufacturing techniques. He even instructed his son, Gabriel, while touring Europe to acquire syrup samples and investigate refineries where sugar was produced from beets. Aime's research won him notoriety and increased his assets. By 1852, the Aime plantation and refinery consisted of 9,500 acres and 215 slaves, an estimated value of $700,000.[17] Bound to slave labor like his planter peers, Aime was firmly entrenched in a plantation hierarchy that prescribed the use of white overseers to manage daily affairs. Overseers were imperative to sugarcane farming, reaping, boiling, and granulating. Slaves alone, per Aime's and his peers' purview, could not be trusted with such hefty responsibilities.[18]

Although Aime was a product of the slave-owning sugar elite, those who visited his bountiful estate and English gardens, sometimes traveling on the Creole planter's personal riverboat dubbed the *Gabriel Aime* in honor of his son, held him in high regard.[19] One such admirer was Eliza Ripley. Born in Lexington, Kentucky, on February 1, 1832, Ripley moved to New Orleans with her family at the age of three. As an adult, she wrote a series of short stories that recounted her south Louisiana upbringing. These narratives were first printed in the *Times Democrat* and later reproduced in her 1912 book, *Social Life in New Orleans: Being Recollections of My Childhood.*[20] In addition to producing narratives of New Orleans's weddings, dances, and daily life, Ripley recounted a visit to Aime's St. James Parish plantation. She described the grand Aime house and its rooms filled with magnificent furnishings, as well as the sprawling gardens. Arriving by steamboat, Ripley was escorted across the levee and into the embrace of Aime's daughter Félicie. Together, the two surveyed the entirety of Le Petite Versailles.

> Félicie and I, with a whole escort of followers, explored the spacious grounds, considered the finest in Louisiana. There was a miniature river, meandering in and out and around the beautifully kept parterres, the tiny banks of which were an unbroken mass of blooming violets . . . [the] tiny stream . . . was spanned at intervals by bridges of various designs, some rustic, some stone, but all furnished with parapets. . . . There were summer houses draped with strange, foreign-looking vines; a pagoda on a mound, the entrance of which reached by a flight of steps. It was an octagonal building with stained-glass windows, and it struck my inexperienced eye as a very wonderful and surprising bit of architecture. Further on was—a mountain! covered from base to top with beds of blooming violets. A narrow, winding path led to the summit, from which a comprehensive view was obtained of the extensive grounds, bounded by a series of conservatories. It was enchanting. There I saw for the first time the magnolia frascati, at that date a real rarity. . . . Next morning the *Belle Creole* [riverboat] was due, and our visit to fairyland was drawing to a close. . . . We gingerly walked the plank, in single file. The boat backed up to give her leeway, and once more for a moment we were in full view of the house. Two figures fluttered handkerchiefs from the balcony, Mme. Valcour and Félicie waving a last adieu—alas! . . . at the rear end of the boat I had one more last glimpse of Valcour Aime's plantation.[21]

Aggrandizing Aime and his plantation, several, albeit questionable, tales surround the ostentatious life of this particular sugar baron. According to legend, the Creole planter once won a $10,000 wager by producing an entire meal, including coffee and cigars, using only the "fruits" of his plantation. Another rumor states that Aime played host to a member of French nobility. Some versions of the story identify the nobleman as Louis Philippe I, but this is unlikely given that Phillipe was already France's monarch at this point in history and was contending with his country's unstable economy. Other renderings name the guest as "a French nobleman" without exacting nomenclature. In accordance with this tale, Aime served the nobleman a sumptuous dinner on gold place settings. Marc M. Matrana narrates two renditions of this story: "[Aime] latter threw these [place settings] into the Mississippi River, as no one was worthy to eat after the royal guest. Another version of the story states that Aime threw the solid gold place settings into the river during the Civil War, rather than have Union soldiers steal them. Still, if any of these stories are totally historically accurate," mused Matrana, "they illustrate Aime's vivacious, often flamboyant personality. They also exemplify his position as a planter-celebrity, a status earned during his era but that even today exists in popular regional lore."[22] If anything, these stories embody Aime as a quasi-mythical, larger-than-life figure in the minds of modern Louisianans.

While Aime dedicated immense attention and energy to his gardens, home, and plantation, his family remained paramount, and he showered them with gifts befitting their rank. To his daughter Edwige, wife of Florent Louis Fortier, Aime acquired and gave Richbend Plantation. Aime also purchased neighboring St. Joseph Plantation and gifted it to his daughter Josephine and her husband, Alexis Ferry, as a wedding present. Similarly, Aime financed the construction of Félicité Plantation for his third daughter, Félicité Emma. She and her husband, Septime Fortier (Florent's brother), bore fourteen children while living at their estate. These three daughters, all of whom married cousins, lived a short distance from Le Petit Versailles. Aime's fourth daughter, Félicie, married another cousin, Alfred Roman, son of A. B. Roman. She and her husband had three children and resided at the Cabanocey Plantation until they moved to Paris in the 1850s.[23] The Aime family doted on Gabriel in particular. As Aime's namesake and heir, he took special care to provide Gabriel with a life of privilege, education, and travel.

Aime wanted his son well educated but also wished to support his brother-in-law, plantation-peer, and state governor. Well aware of Roman's involvement with and devotion to nearby Jefferson College, he enrolled Gabriel at the Creole institution. With the college's destruction by fire in 1842 and subsequent reconstruction, Gabriel reenrolled, but he was soon sent abroad for further education via life experiences, entertainment, and the "grand tour" of Europe. Gabriel's travel diary recounts a series of adventures any wealthy antebellum planter's son might have experienced. In Europe, Gabriel spent most of his time in Paris but also traveled to Antwerp, Arles, Bologna, Bordeaux, Brussels, Florence, Genoa, Grugi, London (where he reviewed a filthy sugar refinery per his father's instruction), Marseilles, Naples, Pisa, Rome, Toulouse, and Tours. While touring the continent, he studied architecture, art, and culture without the accompaniment of a chaperon or tutor. Before he could venture beyond France, however, he was delayed due to a case of gonorrhea. It was not unusual for young men to engage in illicit sexual activities during the nineteenth century. Parents may not have approved, but in general they did not admonish their unmarried sons for expending "sexual energies" with prostitutes.[24] Under doctors' orders, Gabriel remained in bed and submitted himself to treatment. In his diary, he recorded physicians' visits and the remedies they prescribed. These medicines consisted of nausea-inducing herbal capsules and urethral injections of lead acetate diluted in water.[25] Though deemed poisonous in the twentieth century, lead acetate was a common tonic for a host of illnesses in the nineteenth century.[26]

Believing himself cured, Gabriel continued his European escapades. He attended theatrical performances, visited the Coliseum and Vatican in Rome, explored the ruins of Pompeii, climbed Mount Vesuvius, and traveled to see the Leaning Tower of Pisa. Stopping in Marseilles on his twenty-first birthday, Gabriel glimpsed the prison fortress Château d'If made legendary in Alexandre Dumas's novel *The Count of Monte Cristo*. Throughout his journeys, he kept in touch with his father, occasionally sending the senior Aime syrup samples from European sugar refineries.[27] Gabriel also sent gifts to his other family members, including a golden, chiming clock. As can be expected, his family (particularly his sisters) spoke incessantly about Gabriel's adventures to friends and relatives.[28] Clearly, the heir to the Aime fortune was deeply loved. After his grand tour, Gabriel traveled to Europe twice more, once in 1850 and again in 1853. In 1854, at the peak of Aime's wealth and contentment, Gabriel was on his way home to St. James Parish. However, in order to return to his family, he had to journey through the Caribbean, Cuba, South Alabama, and New Orleans, all of which were experiencing the dreaded yellow fever.[29]

Death, Religion, and War

Even though Aime's prosperity allowed him to construct an opulent home and gardens, they provided little comfort when, on September 18, 1854, Gabriel passed away. Docking along the Gulf Coast during the annual epidemic, Gabriel traveled to Bladon Springs, Alabama. This small rural town was known for its hot springs and drew people to what were considered the "curative powers" induced by mineral baths. Bladon Springs' residents had constructed a therapeutic spa and hotel to increase revenue by way of tourism in 1838.[30] LeBreton explains that Gabriel "had been back from Europe for weeks, but stopped on his way to Bladon Springs to rest from the trip and avoid if possible the heat to which he had grown unaccustomed since his prolonged stays abroad."[31] Louisiana history enthusiast Glenn Falgoust hypothesizes that Gabriel, after arriving in the fever-infested coastal South, was bitten by a mosquito and became infected. Realizing he was sick, the young Creole visited Alabama's hot springs in hopes of being cured. This optimism, Falgoust contends, accounts for Gabriel's prolonged stay in Bladon Springs in spite of its 250-mile distance from Le Petite Versailles.[32]

When he ended his spa treatment, Gabriel traveled to New Orleans, remained one night in the fever-ridden city, and journeyed to the Aime plantation the next morning. Once he reached Le Petite Versailles, Gabriel indicated that he felt ill, but instead of allowing his family to provide care, he retired for the evening desirous of a full night's sleep. The next day, Gabriel's fever was worse, and in two days' time he was found dead in his room.[33] His obituary, which highlighted the demeanor and refinement associated with those who possessed the ancienne population pedigree, was printed in the *Daily Picayune*:

> DEATH.—We regret to learn the death, after a short and sudden illness, of Mr. Gabriel Aime, in the parish of St. James, at the residence of his father, the well known, intelligent and universally esteemed planter, Mr. Valcour Aime. We knew this young man well, and sympathize deeply with his afflicted family and wide circle of friends. He was about thirty years of age, unmarried, and united by the ties of blood, like most of our Creoles, to many of our old and most respectable families. His amiable disposition, generous heart, and high sense of honor, endeared him to all who knew him, and many indeed and deep will be the sighs for his untimely fate. It is but a few weeks since we saw him tall, well developed, manly formed swelling with youthful health and vigor, and his easy polished, courteous manner lending additional grade to his pleasant words and lively, smiling countenance. If any one bid fair to live a long and honored life, surely it was he. Inscrutable are the ways of Providence! Before its decrees we must bow the head in silent submission, for manhood in all its pride and strength is but as the grass before the mower's stroke.[34]

Understandably, Gabriel's demise left an ineradicable mark on the Aime family—his father in particular. In his personal diary, Aime writes, "Let him who wishes continue. My time is finished."[35] His plantation journal echoes this sentiment. The September 18 entry interrupts the daily cane growth, sugar production account, and weather report with "V. Aime has retired from active life." His son-in-law Florent Fortier continued the journal and managed the St. James Sugar Refinery.[36] Two years after Gabriel's death, his mother Josephine passed away, and in 1859, while residing in Paris, his sister Félicie took ill and died.[37] Bereft of his gregarious attitude, Aime sought religious reclusion that consisted of intense Catholic devotion, fasting, and material denial. When not at prayer in his garden grotto, the once vivacious planter could be found at the nearby St. James Catholic Church or visiting the large brick and stone Aime family tomb in the adjacent cemetery.[38] Alone at his plantation home, Aime pored over his son's travel diaries, reading them again and again. In the margins the melancholic sugar baron scrawled, "A tear for you all" and "May you, my son, who art in heaven with your kind mother ask God to forgive me so that I may join you."[39]

Before and after his son's death, the master of Le Petite Versailles had maintained strong familial and business ties with his four sons-in-law.[40] As well, Aime kept in close contact with his neighbor, A. B. Roman.[41] Roman, in addition to his positions as the state's former governor and notary public for St. James Parish, served as Aime's personal lawyer. Knowing that Aime had no male heir to whom to bequeath his estate, Roman encouraged him to buy the defunct Louisiana College and reopen it under its genesis name, Jefferson College.[42] Roman may have persuaded Aime by reminding the French Creole that he was one of the original 1831 charter members and therefore had some responsibility for the region's educational needs.[43] Either way, Aime bought the college at auction for $20,000 in 1859. After purchasing the campus, Aime ordered the construction of a Gothic chapel near the institution's main building. The chapel, built between 1859 and 1861, was dedicated to his deceased children, Félicie and Gabriel, and consecrated to the Sacred Heart of Jesus. For some time, the chapel contained a reproduction of Bartolomé Esteban Murrillo's painting *Madonna and Child*.[44] James Gallier Jr., son of New Orleans's Gallier Hall architect, is purported to have designed the chapel. A design sketch, credited to Gallier Jr., has been preserved and is stored in the Southeastern Architectural Archive at Tulane University. The drawing resembles the Gothic chapel but is not an exact match. Therefore, it cannot be said with certainty that Gallier Jr. was the building's architect.[45]

Although he owned the college, Aime had no intention of leading it. Rather, he ceded the property to his four sons-in-law, Alexis Ferry, Florent Fortier, Septime Fortier, and Alfred Roman (Alfred had returned to Louisiana after Félicie's death).[46] Aime requested that his sons-in-law create a corporation under the name Jefferson College and sell shares at a cost of $100 until $10,000 was attained. He believed that it would cost at least that much to properly endow the institution. Dividing the managerial control of his personal shares equally (although not gifting them) among his sons-in-law, Aime empowered them to serve as the college's collective governing authority.[47] Even though records do not indicate whether the $10,000 was acquired in full, Aime's sons-in-law enticed local sugar planters to purchase stock with the potential of sitting on the college's newly reconstituted governing board. With Aime's appointment of Charles Oscar Dogué as Jefferson College's president on August 10, 1859, the Creole college reopened and enrolled fifty students. Dogué, a Louisiana journalist, lawyer, and poet who had studied at the Collège St. Louis in Paris, enforced liberal arts instruction, stern discipline, and a tuition price of $300.[48]

Aime made a point to indicate that he would not profit from Jefferson College. Any income resulting from his shares was relegated to Dogué in order to improve the institution.[49] As a result of Aime's purchase and delegation of power, control of the college was effectively restored to French Creole planters. As a letter from Aime to the Jefferson College governing board elaborated, "The directors will have the full right to adopt the rules and the discipline which will be observed at the college, and to establish the whole educational system, to name the president, and following the advice of the president, to name definitely the teachers and the Catholic chaplain who will become part of the institution."[50] The directors were also responsible for the election of new board members. By the time a new college charter was issued in March of 1861, the administrative board included a bevy of names: B. Avegno, E. Babin, Oliver Blineau, S. W. Bringier, Seraphin Brow, Antoine Carriere, T. Cottman, Robert W. Davis, Albert Duffell, Victor Forstall, Fergus Gardere, T. R. Gaudet, J. L. Generes, J. W. Godberry, Joseph LeBourgeois, W. H. Letchford, N. Melancon, Jules Mossy, John Oliver, A. B. Roman, C. Roselius, C. M. Sheppard, Ben Toledano, B. S. Webre, and the college's president, Dogué.[51]

Directed by Aime's sons-in-law, shareholders could cast one vote per share. However, no shareholder could cast more than twenty votes regardless of the number of shares owned.[52] Despite the stipulated $100 fee per share, a stock receipt, signed by Dogué and issued on October 31, 1859, points out that shares were sold for as little as $50.[53] Although the governing board wavered in collecting funds, they were effectively empowered to manage all aspects of the college. Aime, however, made one crucial change to the institution, one that would deny Jefferson College its secular heritage: the wealthy recluse dictated that the governing Christian denomination at Jefferson College would be Catholicism. South Louisiana's clergy, including Archbishop Antoine Blanc, supported the institution's religious conversion. The archbishop praised Aime's decision, and, in an attempt to support the worldly Creole college turned Catholic academy, the prelate suggested potential instructors.[54]

After Jefferson College reopened, reports circulated describing the campus's healthy and welcoming location. St. James Parish residents and visitors alike commented on the reinvigorated institution and its beautiful chapel surrounded by thriving plantations and hearty sugarcane fields. One traveler, passing through the sugar parishes, reported, "Two miles below the Convent [of the Sacred Heart] is Jefferson College. . . . The buildings are roomy, substantial, and in thorough repair. . . . There are about 50 students attending. . . . There are six sugar refineries [in the parish], most of them conducted on an extensive scale: those of Valcour Aime, E. J. Forstall, L. Roman, B. Lapice & Bro, Mrs. Winchester and E. C. E. Mire."[55] While just begun, the revived college was, for the time being, relatively well supported; nevertheless, planter capital would not sustain the college for long because in 1861, the American Civil War commenced. Sugar barons, their families, and heirs were embroiled in a series of battles that forced Louisiana planters, once more, to abandon Jefferson College.

Before 1860, planters had been reticent to entertain secessionist thought. Sugar parish inhabitants understood that the largest market for their cash crop rested with tariff-bound consumers residing in the upper Mississippi Valley and Atlantic seaboard. Disunion, to sugar planters, meant the loss of tariff protection and the frightening possibility of unchecked competition with Caribbean sugar plantations.[56] But after Abraham Lincoln's election as president on November 6, planters changed their antisecessionist tone.[57] As historian Ted Tunnel explains, "Lincoln's victory forced sugar planters to weigh the benefits of remaining in the Union, against, as they perceived it, the possible or even probable loss of their slaves at some future date."[58] To be sure, Lincoln believed that slavery was immoral and should be eliminated. Although branded an abolitionist by critics, Lincoln never identified as a member of the movement, nor did he assume the label.[59]

While Louisiana legislators, led by Governor Thomas Overton Moore, rushed to secession, some upper-class Creoles, including A. B. Roman, disagreed. In 1861, Roman voted against Louisiana's departure from the United States. Thirteen Louisiana parishes also voted to remain in the Union. Of those thirteen, denizens in the gold coast parishes of Ascension, Assumption, St. John the Baptist, and St. James voiced their disinclination to secede.[60] An ordinance of secession, however, was adopted on January 26, 1861, and Louisiana became the sixth state to leave the Union.[61] While the decision was not unanimous (113 votes in favor, 17 against), Creoles envisioned secession as an opportunity to reinvigorate French custom and the potential removal of the Anglo-American irritant. To enhance this supposition, Creoles, during the Civil War, published copious French journals to strengthen Louisiana's historical *lingua franca*.[62] Roman, now serving as one of three commissioners chosen to confer with the federal government and secure the confederacy's peaceful separation from the United States, traveled to confer with Lincoln. The US president, however, refused to discuss the matter.[63]

Turning from their antisecessionist vote, St. James Parish Creoles took up arms and organized military companies. Aime himself donated $500 to each parish military company and organized Company A of Louisiana's 30th Infantry Battalion, aptly named "Valcour Aime's Guards."[64] "The French and Creole elements," as John D. Winters explains, "set the pace in volunteering for [Confederate] service. Other foreign groups followed, but with far less noise and ostentation. Irish and Germans outnumbered all others, but lacking funds, they attracted less attention to their service. However unheralded, the lowly Irishman and German contributed far more to the cause of the South than most of the richly attired French and Spanish troops who made a fine display on the parade ground but often absented themselves under fire."[65] As war waged on, the number of enlisted Creoles increased, but when New Orleans fell to Union forces in April of 1862, the militant spirit of Confederate-minded sugar barons fell with it. Planters could do little as federal soldiers looted plantations, and gunboats bombarded refineries and mansions alike.[66] The Mississippi River, as Charles P. Roland tells, had been "friend and benefactor in time of peace [but] had betrayed them in time of war and had borne the invader into their midst."[67]

When word of invasion reached the sugar parishes, Anglo-Americans fled with their slaves and as much material property as they could carry to Texas, Mexico, or further into South America. Conversely, the majority of French Creoles remained, preferring to face the oncoming onslaught than flee. Those planters who stayed experienced less property damage than those who abandoned their plantations. With no one at home to address or receive Union troops, soldiers ravaged deserted plantation houses.[68] Like his ancienne population peers, Aime, owner of Jefferson College and impresario of Le Petite Versailles, remained at his plantation as disaster approached. While enduring the onslaught of gunboats and federal troops, the Aime family experienced yet another bout of calamity. After the fall of New Orleans, Aime's first daughter, Edwige, died. Defeated and depressed, Aime stood on the river levee watching cannon balls damage neighboring plantations while his barns and stables were ransacked.[69]

With the onset and protraction of the Civil War, many colleges and universities in the American South suffered as students and professors enlisted in the Confederate army. In September of 1861, Governor Moore decreed that all males between the ages of eighteen and forty-five enlist in the militia. As a result, most of Louisiana's institutions of higher education were disabled. Likewise, due to a lack of students and the arrival of Union troops, Jefferson College closed for a third time in 1862.[70] Immediately after the bombardment of Donaldsonville, the main college hall was converted into a barracks for northern soldiers who burned, stole, or severely damaged furniture, library books, and scientific equipment.[71] As LeBreton details, "The [college] buildings suffered cruelly at the destructive hands of the soldiers. Horses and mules fed where aristocratic young Creoles had sat at academic desks."[72] With the passing of the Emancipation Proclamation, planters slipped into an emotional miasma. According to Lillian Bourgeois, "The millions of dollars invested in slaves were a total and immediate loss. Worse yet, there were no hands to till the soil to recover the losses. Fortunes were swept away, plantations changed hands, and weeds grew where sugar once thrived."[73] Religion offered some consolation, but depression worsened as formerly affluent members of the ancienne population surveyed their ruined estates.[74]

With New Orleans subject to Union control, General Benjamin Franklin Butler was placed in command of the city. Adhering to Lincoln's orders, Butler sought to increase public morale, though he was not completely successful. He arranged food supply distributions, street cleanings, and promoted public displays sponsored by the Union Association, of which former Louisiana College president Louis Dufau was secretary. However, Butler was short-tempered and had little patience for confederate sympathizers. Despite the general's shortcomings, Lincoln believed that the reinvigoration of New Orleans, with its exports of cotton and sugar, would hasten Reconstruction.[75] Thus, Butler required sugar planters' cooperation if he was to restore order in war-torn Louisiana. To appease planters and restart a heavily damaged agricultural market, Butler tried to reconcile former slaves and slaveholders in a new labor scheme.[76] Before Butler fully realized his plans, however, General Nathaniel Prentice Banks replaced him as chief administrator for south Louisiana. After Banks's arrival, planters approached the newly installed general and asked for help controlling freedmen who were, in their opinion, running roughshod over the sugar parishes. Though Lincoln had precluded the parishes of Ascension, Assumption, Jefferson, Lafourche, Orleans, Plaquemines, St. Bernard, St. Charles, St. James, St. John, St. Martin, St. Mary, and Terrebonne from the Emancipation Proclamation due to Union occupation, slaves fled plantations when news of the executive order reached the gold coast. To address the issue, Banks, in 1863, imposed a wage system where plantation owners paid ex-slaves a pittance to perform labor.[77] Building on Butler's labor scheme, Banks instituted a contractual wage system that required freedmen to return to their former masters' plantations and help restore sugar production.

Under contract, ex-slaves received food, basic medical care, and shelter—little more than was afforded to them prior to emancipation. In addition, they were paid an average salary of three dollars per month or five percent of all proceeds rendered from annual crop yields (i.e., sharecropping). Relegated back to the fields, freedmen were prohibited from leaving plantations without permission.[78] To increase labor efforts, Banks issued General Order Number 23 in 1864, which forced freedmen to seek out contracted employment. Those who did not engage in contractual work were made to toil on abandoned plantations. To Banks, labor was a public duty whereas vagrancy was a punishable crime.[79] As antebellum forced labor gave way to Reconstruction sharecropping, planters manipulated an already corrupt employment system to create a new form of indentured servitude.[80] Planters approved of Banks's labor system but were displeased when he announced plans to develop educational institutions for freedmen.[81] In 1864, Lincoln instructed Banks to summon a constitutional convention to discuss, among other things, tax-supported instruction for both black and white children. This new constitution, while prohibiting the expenditure of public dollars on religious colleges and parochial academies, promoted the establishment of freedmen's schools.[82] By the end of 1864, there were nine such schools along the banks of the Mississippi River with an average attendance of 2,400.[83]

Although Banks was responsible for creating a string of African American schools, his efforts to increase sugar production faltered. Between 1861 and 1865, the number of sugar plantations decreased from 1,291 to 200 with a gross profit reduction from $194 million in 1861 to approximately $25 million at the end of the Civil War.[84] These numbers did little to improve Creole spirits, as "former slaveholders," according to William Kauffman Scarborough, "emerged from the war and Reconstruction an embittered class."[85] Though some planters recovered their prewar wealth, others were financially ruined, and virtually all were enraged over the rights bestowed on the black population.[86] After St. James Parish troops returned to the gold coast, a play was staged at Félicité Plantation with Aime and his remaining family members in attendance. One actor, mourning the demise of the Old South, stated: "It is necessary despite the devil for the world to go on; youth must be amused but it would be wise to do without luxury—to do everything in a simple manner—that is the only way it is possible to continue."[87] The actor's comments did little to amuse, however, as all were aware that their way of life dependent on slave labor had come to an end.[88] As Southern Reconstruction advanced, antebellum social class hierarchies unraveled. French Creole planters would, for decades to come, look back on their prewar lifestyles and idealize "the past as a 'grand and lordly life.'"[89]

Accommodating a new social order was indeed difficult. Many failed to adapt and lost their fortunes to bad investments or plantation neglect.[90] Sharecropping allowed the Louisiana agrarian system to persist, and plantation owners remained largely in control; however, French Creoles' firm hold on the sugar industry had been shaken. By 1869, a host of Louisiana plantations had been purchased by individuals foreign to the ancienne population. Planters, such as the Irish-born John Burnside, bought lands formerly owned by deceased or bankrupt Creoles.[91] As sharecropping became standard practice, sugar planters saw a possible return to glory. Others, however, viewed sharecropping with suspicion, believing that freedmen would not serve as quality workers.[92] Although sharecropping became commonplace, emancipation had irreparably altered the sugar industry, as well as regional social class hierarchies. The enactment of the Thirteenth Amendment did more than abolish slavery; it provoked a wave of nationalization that encouraged French Creoles to demarcate themselves as citizens of the nation rather than Louisiana alone. Creoles, now seeking racial rather than ethnic fortitude, aligned themselves with Anglo-Americans to bolster and unify the "white race" against large populations of liberated African Americans.[93]

During and after Reconstruction, "Louisiana's sugar-planter elite held firmly to a social, racial, and economic agenda where black labor stability lay at the axis of their business success and class position."[94] French Creole planters, in ever increasing numbers, moved out of the sugar parishes to metropolitan centers. Those who remained managed what was left of their plantations, all the while retaining haughty airs that separated Creoles from Acadians. Toward the end of Reconstruction, these particular strata persisted. Creoles, with their large homes stocked with finery, were the antithesis of Acadian families who lived in small, sparsely furnished, yet brightly painted houses. Both groups, however, were active in community gatherings, festivals, and music. Large families and close-knit neighbor networks characterized Acadians, while peer associations and smaller family units were associated with French Creoles. As the Creole class diminished in the latter half of the nineteenth century, the Acadian population grew.[95] In time, Acadian laborers replaced the preponderance of Louisiana sharecroppers. The postwar economy failed to decrease costly land prices, and, as a result, poor Acadians living in the sugar parishes could not afford to purchase significant quantities of land. Thus, Acadian-owned plantation capital remained difficult to acquire.[96] As Reconstruction came to a close in 1877, efforts to improve African American rights halted. Democrat redemption replaced Reconstruction victories including the ratification of the Fifteenth Amendment, which gave African Americans the right to vote. In this new era, Southern politicians rallied against racial equality. In 1883 the Supreme Court nullified the Civil Rights Act of 1875, and white Southerners, French Creole or otherwise, reentered a world where racial supremacy was again the accepted norm.[97]

As class and racial strife festered, south Louisiana's educational structure, despite attempts to modernize, lagged behind northern schools, colleges, and universities. Post-bellum teachers in Southern states were accused of remaining true to the secessionist cause. Instructors who taught Confederate history in lieu of US history and trained their students to taunt federal officers with jeers of "Yankee scum" were labeled anti-Union and replaced with pro-Union teachers. Nationalistic instructors adhered to loyalty requirements and quelled rebellious behaviors by forcing pupils to participate in patriotic activities such as singing the national anthem.[98] "Old teachers of questionable loyalty were fired, and new Union supporters were hired. Pro-Southern schoolbooks were purged, and northern schoolbooks replaced them."[99] In addition, the educational structure in New Orleans was remodeled to resemble the Boston system with a centralized governing bureau and city-wide superintendent.[100] These administrational revisions influenced instructional language, too. With the collapse of the Confederacy and the dawn of Reconstruction politics, English wholly replaced French as the language *de jure* in Louisiana's schools, colleges, and universities.[101]

J. Carlyle Sitterson highlights that sugar planters, even after the Civil War, were loath to discontinue their children's education regardless of the legal "schoolhouse" vernacular. The lean years of Reconstruction encouraged planters to send their male children to institutions such as Johns Hopkins University in Maryland or the University of Virginia, while their daughters were enrolled in women's academies located in Baltimore or New Orleans.[102] These same planters, many of whom were Jefferson College shareholders, were unable to take up the task of reopening the Creole institution due to their weakened financial condition.[103] In reality, the cost of tuition elsewhere was less than the expense required to reopen the sugar parish college. Aime emerged as one of the few French Creoles to escape destitution. As such, his relatives, particularly the Fortier and Roman families, avoided impoverishment. Nonetheless, sugar planters across south Louisiana were obliged to restructure their livelihoods in accordance with a drastically changed cultural and economic climate reliant on sharecropping and Acadian labor.[104]

Segregationist Plans and Catholic Priests

With the advent of emancipation, a newly freed black population surrounded plantation owners. Both French Creoles and Anglo-Americans, although coping with this recently liberated, uneducated populace, took comfort in the fact that although they were free, ex-slaves possessed little power to enact any lasting social or political change.[105] Still, freedmen proved fiercely motivated to increase their intellectual competency and that of their offspring. Acting on behalf of the South's former slaves, in 1865 the federal government created the United States Bureau of Refugees, Freedmen, and Abandoned Lands, otherwise known as the Freedmen's Bureau, to fund, govern, and regulate black schools. Arriving in Louisiana, bureau members discovered that private schools serving African Americans had opened under the supervision of both black and white teachers, many of whom were local Louisianans, while others had traveled from the North to support the education of ex-slaves.[106] During Reconstruction, the Freedmen's Bureau opened 126 institutions for black students throughout south Louisiana.[107]

Most freedmen's schools were established in the metropolis of New Orleans. Nevertheless, similar educational institutions existed in the sugar parishes. Taking note of rural black schools, planters did little to support them, rebuked education taxes set by General Banks, balked at northern carpetbagger aid, and focused their reconstruction efforts internally. As Howard Ashley White explains, "A people who had seen their social and economic system overthrown were more interested in repairing their fortunes than in promoting education among their former slaves."[108] Planters were wary of freedmen's schools and loathed the possibility of selling or leasing confiscated and abandoned plantation properties to support black education. The Confiscation Acts of 1861 and 1862 permitted the seizure and sale of properties (slave, land, or otherwise) that belonged to disloyal persons in federally occupied territories to better the Union.[109] Toward the end of the Civil War, Union generals Absalom Baird and Edward Richard Sprigg Canby, both involved in the administration of educational institutions governed by the Freedmen's Bureau, proposed to use confiscated properties, or funds generated from their sale, to perpetuate freedmen's schooling.[110] By the start of 1865, the federal government had seized roughly 96,000 acres of land in Louisiana alone—the majority of which was located in the gold coast.[111]

Adding insult to injury, rumors circulated in St. James Parish that Union officials schemed to convert the empty Jefferson College facilities into a freedmen's academy.[112] The notion was more than sugar planters could handle. René de Sennegy described the situation in his 1877 text *Une Paroisse Louisianaise: St. Michel du Comté d'Acadie* (A Parish in Louisiana: St. Michel in the County of Acadia): "Ah! Vengeance has sometimes rare refinements of cruelties. Jefferson College was [a] Creole flower, raised for the sons of family; it was the free university par excellence; the man of Washington smiled at that thought, and he cried with mockery: 'Let the slave take the place of the master; let us transform the white dwelling into an institution of the emancipated.' That odious idea had been scarcely sensed when the creole blood trembled with indignation; it was the bitter dregs of the chalice. 'No, it will not be thus; we shall save from this last infamy the roof which shelters our childhood, the dwelling built in our later years, the Athenaeum which should give to our children science, letters and the arts, must it be for that to despoil us of our last might.'"[113] General Banks was certainly aware of the college's existence, as it was one of the only educational institutions labeled on a regional map he himself commissioned. The map, *New Orleans to Vicksburg: Prepared by Order of Maj. Gen. N. P. Banks*, provides rich information related to battlefields, bayous, lakes, forts, river mile points, roads and railroads, sawmills, towns, timber acreage, and so on.[114] Jefferson College was a point of reference included on this and other Civil War maps due to its central location between Baton Rouge and New Orleans, as well as its proximity to the river and service as a military barracks.[115] French Creoles, concerned that Banks or the Freedmen's Bureau would confiscate and racially repurpose their long-time college, acted to intercede; however, in order to succeed, the Creole institution had to be legally removed from state and federal oversight.

Despite their "reversal of fortune," sugar planters had no intention of allowing former slaves a college among ruined plantation homes and burnt cane fields. To avoid the transformation of Jefferson College into a school for freedman, Aime "bought out the shares of the other stockholders."[116] After acquiring his peers' shares, Aime discussed the situation with A. B. Roman and decided to offer the college to the Archbishop of New Orleans, Jean-Marie Odin. If Odin were to accept, the college would then become sectarian, and, according to Louisiana's revised 1864 constitution, government control would dissolve.[117] Archbishop Odin, however, refused to accept the college. Instead, he instructed the planters to "go to the vicarship of St. Michael [the church in St. James Parish]; see Father [Henri] Bellanger, agree with this religious; you have there at your doors all that you could desire. The Marist Fathers have proven themselves in France as educators. Their College of St. Chamond, of the Seyne and of Riom[,] are held in great esteem."[118] Like the Sisters of the Sacred Heart and the Society of Jesus, the Society of Mary had gained notoriety in the sugar parishes for their missionary endeavors.

The Marist order had arrived in Louisiana at the onset of the Civil War and assumed control of St. Michael's Church at Odin's request.[119] First conceived in 1816 by Father Jean-Claude Colin and approved by Pope Gregory XVI in 1836, members of the Society of Mary were no strangers to war and revolution. A young Colin, born in 1790 and ordained a priest in 1816, had experienced the intense political and cultural turmoil of revolutionary France in which Catholic clergy were severely limited in their abilities to proselytize and say Mass. French churches and monasteries were stripped of adornment, which was sold to the public. As well, Catholic priests were forced to swear allegiance to the French government and demonstrate their loyalty to Paris over the pope in Rome. Those who refused to pledge fealty faced imprisonment, deportation, or death.[120] Religious societies, even before the revolutions of 1789 and 1799, were suspicious of France's secular government. In 1762, France expelled the Society of Jesus over concerns that the religious order had become too politically powerful. In 1773, Pope Clement XIV assented to European monarchist demands and suppressed the Jesuit order worldwide.[121]

One hypothesis holds that the Society of Jesus's removal served as a catalyst for the Society of Mary's establishment. As Marist historian J. Coste explains, Colin, subsequent to his ordination, listened as clergyman Jean-Claude Courveille recounted a spiritual revelation in which Mary, the mother of Christ, conveyed her want for a new religious order that would fill the void left by the Jesuits.[122] With the formation of the Society of Mary, Colin wrote its guiding constitution, which according to Coste "was basically that of the Society of Jesus."[123] Like the Jesuits, the Marists found themselves engaged in teaching and educational administration not long after the order's informal creation. In 1829, Colin was appointed superior of the College of Belley. It was there that the founding idea behind the Society of Mary was formed, thus linking the new religious order with education from its inception.[124] Busy with writing and college supervision, Colin sent priests to Oceania islands for missionary work. He also accepted six schools in France for the Marists to oversee. In the words of Donal Kerr, the time for education in France and abroad "seemed ripe." François Guizot, France's Minister for Public Instruction, informed a more religiously lenient government led by Louis Philippe I that education should include moral and religious curricula. An agreement was reached and "Guizot's Law" was approved on June 28, 1833, thus reinstating religious instruction at Parisian colleges and schools. As the Marists traveled to the United States, they encountered various Jesuit fathers and brothers since Pope Pius VII had restored the Society of Jesus in 1814.[125] Working alongside Benedictines, Dominicans, Jesuits, Sisters of the Sacred Heart, Ursuline Nuns, and other Catholic orders, the Society of Mary took hold in south Louisiana.

In 1864, Aime, Roman, and Jefferson College's board of directors adhered to Archbishop Odin's advice and invited Marist Father Henri Bellanger to meet and discuss the Creole college issue at the Roman family plantation. Bellanger, born August 23, 1823, in Angers, France, had professed his vows of ordination at the age of twenty-three. He was sent to Louisiana in 1863, and, after his arrival, was placed in charge of St. Michael's Church.[126] Once assembled, the group devised a contract to transfer Jefferson College's administration in its current war-torn state to the Society of Mary for a period of six years. Father Bellanger, however, did not approve of the contract's temporal constraints and declined the offer. Bellanger believed the physical circumstances were too daunting to overcome in such a short period of time.[127] Extensive war damage left the college bereft of essential materials and walls were marred with graffiti. In a 1906 address to Jefferson College's students and alumni, John L. Peytavin described the scene: "Like the dissolute Romans of old, they [Union soldiers] used the steel bodkin to engrave their ribald verses and curses on the . . . walls of the college buildings."[128] Disappointed with the meeting's outcome, Roman ended the discussion. Four months later, he engaged in negotiations with Father Abbe Marion of St. James Catholic Church on the west bank of the Mississippi River. Marion, a diocesan priest, accepted Roman's contractual terms and informed Odin that he would undertake the college's reopening. The archbishop forbade Marion to take on the task, as the archdiocese could not afford the necessary repairs.[129]

In a conversation that included Odin, Marion, and governing board member Florent Fortier, the archbishop stated, "No, dear sir, . . . No; I know too well what such charges cost the diocesan administration; never will one of my priests take a like burden while I live. But, I repeat to you, you have the Marist Fathers, speak to them; that is what you need. Father Bellanger is at this moment in the city [New Orleans], do you want me to summon him to the Archbishop's palace, and you will hear in my presence?"[130] Odin understood the financial strain a damaged college would place on his Episcopal See's financial and human resources. On the other hand, the Society of Mary, an economically independent order, might be able to tackle the costly venture. Though Fortier agreed to meet Bellanger once more, the priest could not be found. As such, Marion and Fortier left New Orleans on a steamship bound for St. James Parish.[131]

By chance, Bellanger happened to be aboard the same vessel as the aforementioned two. Having located the Marist, Marion and Fortier revisited the subject of transferring Jefferson College to the Society of Mary. Not long after their ship arrived in Convent, Fortier and Bellanger met again at St. Michael's Church to review a revised act of transfer.[132] De Sennegy recounts the fortuitous encounter thus:

> By a stroke of Providence, the shareholder and the Marist Father [Bellanger] left New Orleans, carried by the same steamer. They met as simple acquaintances, they exchanged reciprocal politenesses, then the conversation veered imperceptibly toward the pending question [regarding Jefferson College] and the discussion was carried on with such loyal spirit on both sides, that a few days after the meeting on the Mississippi, a beautiful morning of May, Mr. Florent Fortier presented himself at the vicarage of St. Michael['s Church], an act of transfer in hand. Placing the document before Father Bellanger: "Read and sign,"

> he said to him, "and right away, the affair has dragged too much." The religious then obtained a moment of reflection. Twenty-four hours were allowed him; he profited by them to go down to New Orleans, to visit His Grace . . . Odin, and without doubt also a notary. The act was found good in its form and tenor, and on May 6, 1864, Father Bellanger accepted it in placing his signature on the contract.[133]

Aime, pleased with the agreement, donated his college shares, valued at 6,433 piasters, to Bellanger.[134] Overseeing the exchange, the four Jefferson College Corporation members, Alexis Ferry, Florent Fortier, Septime Fortier, and Alfred Roman, facilitated the transfer of shares to the Society of Mary by means of a notary act on condition that the priests consult and collaborate with the governing board and all external shareholders, maintain the campus's physical location, and improve the college to become a preeminent institution of higher education in Louisiana.[135] As soon as the Marist fathers and brothers acquired the college, federal troops departed. Lillian C. Bourgeois explains that this military egress was the result of friendly relationships between the United States and French governments. Given that the French Marists held no link to the United States or the Confederacy, Union troops were ordered to vacate the campus and relocate to northern St. James Parish.[136] No longer host to military personnel, the institution became known as St. Mary's Jefferson College, the first Marist college in the United States.[137]

As dictated in the official transfer agreement, "the Marists [would] engage themselves on the pain of losing their rights . . . to establish as soon as possible and to maintain always a collegiate institution of the first class for the classical and religious education of the young men of Louisiana."[138] Thanks to the ceded shares, the Marists emerged as the predominant institutional shareholders.[139] External shareholders were displeased that the Marists had majority control but brought no formal complaint against the Society of Mary, so long as the Creole governing board held sway over the institution's curricular administration, controlled faculty pay, and could veto any unsavory actions taken by the Marists.[140] To keep shareholders informed of the financial stability of St. Mary's Jefferson College, the religious order was required to disseminate annual budgetary and organizational reports.[141] Regardless of these administrative limitations, the Creole college now existed, in effect, as a Catholic college, free from state or federal supervision. Since the Louisiana Constitution prohibited the expenditure of public funds to support sectarian education, sugar planters had succeeded in their endeavor to remove Jefferson College from the clutches of the troublesome Confiscation Acts and Freedmen's Bureau.[142]

Notwithstanding the historic exclusion of African Americans from formal religious education, Catholic leaders, including Father Bellanger, found ways to provide catechetical and intellectual instruction to St. James Parish's enslaved inhabitants prior to emancipation. All the same, longstanding legal regulations regarding the religious edification of slaves hindered these efforts. The *Code Noir*, first published in 1685 and later reprinted in 1724, ordered Louisiana officials to baptize slaves in accordance with Catholic doctrine. Plantation owners were instructed not to task enslaved people on Sundays and religious feast days. As Claude F. Jacobs and Andrew F. Kaslow clarify, "Since baptism did not call for emancipation, slave owners did not object to the rite."[143] With the publication of *Appeal to the Colored Citizens of the World* by David Walker in 1829, Louisiana politicians crafted new legislation that outlawed the education of slaves in general. Walker's manuscript called for slaves to rise up and fight for liberation. Fearing a possible rebellion, Southerners kept slaves illiterate rather than have them learn to read and become empowered by antislavery texts like Walker's.[144] Nevertheless, Catholic leaders began providing basic theological instruction to the black population of St. James Parish in 1845.[145] Frank M. Uter adds, "During the day they [Sisters of the Sacred Heart] would visit the cane fields and pass out holy cards to African-American field workers and encourage them to stay faithful to their Catholic faith or seek [religious] instruction."[146]

These religious educational efforts proved poorly organized and yielded little results. Nevertheless, since slaves were exposed to plantation owners' spiritual orientates, and as the bulk of south Louisiana's slaves were reported to be Catholic, they attended Mass on a regular basis. Mass, however, was divided along the color line as slaves were forced to sit or stand in the back of the nave while white patrons sat in pews proximal to the alter—an unmistakable differentiation of racial strata in a religious setting.[147] Further emphasizing the import of this situation, Randall M. Miller explains that "sharing a common Catholicism . . . between master and slave . . . did not forge a sense of common social purpose."[148] As if the segregation of Mass was not enough, some white churchgoers forcibly prevented slaves from approaching the altar for communion, regardless of the priest's intervention to impart the sacrament.[149] For much of Louisiana's clergy, slaves were, after all, Catholics deserving of the Holy Eucharist.

Conversely, some clergymen were all too accustomed to the South's racially bifurcated milieu. These priests remained indifferent to, refused, or in some cases rebuked religious instruction to slaves.[150] To make matters worse, Catholic orders managing colleges and schools in post–Civil War Louisiana worried that enrolling freedmen would cause white families to withdraw their children and thereby financially damage religious institutions.[151] At the close of the 1866 Second Plenary Council of Baltimore, Archbishop Odin requested Reverend Mother Anna Shannon, superior of the Convent of the Sacred Heart and St. Michael's Academy in St. James Parish, to open a school for the children of former slaves.[152] Receiving Mother General St. Madeline Sophie Barat's blessing, the Sisters of the Sacred Heart established St. Joseph's School for freedmen in 1867.[153] By the end of the nineteenth century, the Sisters of the Blessed Sacrament, led by Mother Katherine Drexel, enlarged St. Joseph's School to accommodate increased enrollments.[154] During this period of black parochial school development, New Orleans's public schools experienced state-imposed integration, a phenomenon that disappeared prior to the end of the nineteenth century.[155] Despite such progress, St. Mary's Jefferson College retained its racially segregated stance.

Regardless of the college's religious transformation and subsequent disconnection from government influence, sugar parish shareholders remained fervent in their want to control St. Mary's Jefferson College, in spite of the Marist presence. In short, the Society of Mary remained, for good or ill, subservient to the governing board per the original transfer agreement despite the number of shares owned by the order. Indeed, "embroiled or meticulous [lay] share-holders could create difficulties; in a word, the Society [of Mary], although mistress of the greater number of [share] actions, did not feel at home." Then again, de Sennegy prophesized, "A religious order, faithful to its high mission, is always sure of the future; it possesses life, it can wait."[156] Tolerating officious shareholders, the Marists improved the campus and watched for an opportunity to take full control of the Creole college.[157]

Marist Control, Creole Decline

With the transfer complete, St. Mary's Jefferson College opened on July 12, 1864, with twelve students in attendance: A. Bourgeois, I. Crane, J. Crane, J. Damaré, E. Duffel, J. Gauthreaux, A. Jourdan, N. Landry, J. Malarcher, P. Martin, A. Robert, and V. Seré.[158] To aid the Marists in reopening the college, local planters, though financially taxed by war and Reconstruction, pooled funds and donated $7,000 to the institution's governing board. Even Mother Shannon provided monies to support the Marists.[159] As a new Catholic institution, the college needed tuition dollars to support all operations. Tuition during the antebellum Jefferson College regime was costly. So, too, was tuition throughout the Louis Dufau and Valcour Aime periods. Though the Civil War had depleted south Louisiana fortunes and left an indelible need for affordable local education, the Marists maintained tuition at $150 per term ($300 per year) with a $10 entrance fee.[160] Understanding the lack of capital fortitude behind Confederate dollars, the Society of Mary shrewdly insisted on US currency as the only acceptable method of payment. Bent on crafting a premier educational institution per the governing board's insistence, the Marists needed substantial capital if they were to succeed. As a result, tuition remained high throughout Reconstruction. Nevertheless, the Marists were, at least partially, considerate of Louisiana's decreased financial standing and reduced some affiliated costs. Instead of requiring students to supply costly silver utensils and goblets like the original Jefferson College and Louisiana College administrations had done, the priests permitted their wards to purchase inexpensive electroplated items.[161]

Costs aside, de Sennegy, a Marist priest and fierce critic of Dufau's secular Louisiana College, must have beamed to witness, in his words, "the Marists purif[ying] with holy water the Athenaeum."[162] Illustrating the transformation of Jefferson College from an institution guided by the hand of man to one led by Christ, the celebrant de Sennegy wrote: "The crucifix appeared hanging over the chair of the professor, speaking to all of work and sacrifice; the sweet image of the Mary Immaculate showed itself in the exercise and schoolrooms, cheering the eyes. The university school became a residence of peace, of intellectual labor and of prayer. No more somber bars no more forced incarceration, a park, solid masses of greenery, fields planted with young, productive trees. Behind a Christian institution morose foreheads and philosophic promenades of a sage of 12 to 15 springs! No, for these young spirits games are needed, laughter, the fanfares and the salutary distractions which rest the intelligence, capture the heart for an instance and fortify the body."[163] Unhindered by minuscule attendance, the Marists dutifully collected tuition dollars and advertised admission ages ranging from eight to fifteen years. They also mentioned the proximity of their college to the Convent of the Sacred Heart, the use of lands adjoining the campus to cultivate vegetables, and the addition of new oak allées directly in front of the main building. These allées extended beyond the original antebellum oak allée and terminated before the banks of the Mississippi River.[164]

With the Creole college's reopening, sons of local war-torn sugar planters, who had previously enrolled in northern or European colleges out of luxury, now attended St. Mary's Jefferson College out of affordability. Sugar planters welcomed annual tuition bills in lieu of the overwhelmingly expensive grand European tours and sought ways to exponentially reduce attendance prices when possible. For example, after gifting his Jefferson College shares to the Marists, Aime placed a condition on the new administration: that his male relations would be given an education free of charge.[165] Subsequently, several of Aime's kindred received instruction "on the house."[166] Substantiated by Michael O'Brien, "Good bloodlines might forgive a small shortage of ready cash. . . . Having a famous member in the family helped; politicians were acceptable . . . but Founding Fathers were best. Merchants were fine, though planters were a touch better, while having both in the family was advisable, as insurance."[167] As late as 1897, the Fortier family reminded the priests of Aime's contributions to the college, as well as their promise to educate his relatives at no cost.[168]

Though Aime remained a recluse, he godfathered many parish children (usually boys) who were, in turn, granted a free education at St. Mary's Jefferson College. As LeBreton writes, "No doubt the well intentioned parents knew that the godfatherhood of Valcour Aime carried with it, not only the distinction of having him as a godfather, but . . . assured education at . . . Jefferson [College]. Whether he suspected it or not, Valcour Aime continued to serve in the capacity whenever he was invited to do so."[169] Aime not only provided for parish children, he supported the St. James Catholic Church, not far from Le Petite Versailles, by donating a pair of large silver candlesticks, a set of Stations of the Cross, and two painted depictions of the Holy Family.[170] The aging "Louis VIX of Louisiana" was indeed generous with his remaining wealth. Old, bearded, and clad in austere garments far beneath his social rank, Aime was conscious, if not outwardly reflective, of the declining Creole aristocracy.[171]

Signaling the ancienne population's demise, André Bienvenu Roman, former Louisiana governor and founder of Jefferson College, breathed his last on January 26, 1866. While walking in New Orleans's French Quarter, Aime's trusted brother-in-law and lawyer, without warning, fell to the ground and died.[172] The *Daily Picayune* described his life and loss as follows: "One of the worthiest and most honored sons of Louisiana went to his rest on Sunday last. Ex-Gov. Roman died suddenly, while passing from the house of his son-in-law, Mr. E. F. Lavillebeuvre, in Dumaine Street. He had been in his usual health, and was about to pay a visit. A few yards from his home he fell, and died instantly, evidently without pain or consciousness of the shock. The remains were taken from the home in Dumaine Street, on Monday afternoon to the steamer *Lafourche*, to be transferred to St. James Parish, and deposited in his family tomb."[173] Roman's complete obituary, albeit verbose in recounting his relations and gubernatorial administrations, does not list his connection to Jefferson College, his dealings with Dufau's Louisiana College, his promptings of Aime to reopen the institution under its original title, or his aid in transferring the college to the Society of Mary.[174]

Whether Roman's death came as a shock to Aime is unknown, as no records illuminate the sugar planter's thoughts on his brother-in-law's demise. Aime, however, was not long without Roman, nor did he witness the complete ruin of the Creole elite. Shortly after Christmas 1866, Aime contracted pneumonia and passed away on New Year's Day 1867.[175] The death of this legendary French Creole was publicized in several Louisiana newspapers. The *New Orleans Daily Crescent* somberly announced his passing:

> The death of Valcour Aime—The Painful announcements of the death of this old and excellent citizen reached us at a moment when the CRESCENT office and its attaché were in the midst of the annual New Year's festivals. Mr. Aime was so well known and so well esteemed throughout the state, that the news of his demise spread a gloom upon our festivities. Valcour Aime was one of the few survivors of that staunch old generation of creoles who to the last retained their traditions of their Gallie ancestry; chivalrous to the verge of romance, brave as the Paladins, generous to prodigality. We regret exceedingly to see the "*ancienne regime*" [*sic*] so rapidly passing away. Mr. Aime was born of one of the most ancient and aristocratic of the old French families that settled in the territory of Louisiana. His family connections are spread throughout the state, and number such names amongst them are Roman, Ganucheau, etc. He died at the age of three score and ten, upon the family estate, which he had done much to beauty, and surrounded and soothed by weeping relatives and friends. Generosity and hospitality were never more beautifully illustrated than by Mr. Aime amid the vicissitudes and calamities of the late war. His plantation in St. James was the synonym of bounty to hundreds and hundreds of fugitives and sufferers, unknown to him, perhaps, but who can never cease to cherish his memory, and honor his name and his family.[176]

After Aime died, members of his family sold portions of his grand estate to repay debts. On October 9, 1867, Aime's schooner, the *Gabriel Aime*, was auctioned in New Orleans.[177] Four years later, the *Daily Picayune* advertised the sale of Le Petite Versailles to John Burnside.[178]

Originally from Northern Ireland, Burnside traveled to Virginia as a teenager and worked as a grocer's clerk. He excelled in commerce, moved to New Orleans, and became a partner in a mercantile operation. Eventually Burnside bought the firm, renamed it "John Burnside and Company," and began purchasing sugar plantations. Among his many plantation acquisitions, the Irishman procured and took up residence at Houmas House in Ascension Parish. Burnside was just as successful in the sugar industry as he was in commercial business. In 1861 alone, his plantations produced 7,652 hogsheads of sugar and set a new production record. Unlike other south Louisiana planters, Burnside returned to financial prominence after the Civil War. As Scarborough explains, "Burnside gave neither financial or moral support to the Confederacy." Instead, the shrewd businessman conserved his resources and emerged as the region's leading sugar baron. Upon his death in 1881, Burnside left an estate worth $8 million.[179]

In life, Aime may have been given to the same disdain for Irish and German immigrants as his French Creole peers. If so, the very idea of an Irishman owning and operating Le Petite Versailles and the St. James Sugar Refinery might have repulsed him. Nevertheless, Aime's efforts to revive Jefferson College succeeded. Grateful for his donation, the Marists said Masses for the sugar baron before and after his death but did little to publicly acknowledge his contributions. Complaining of the Marists' "lack of tribute" to the deceased Creole, Aime's descendant, Alcée Fortier surmised in 1894, "It would be strange indeed if the Tulane University were ever to forget Paul Tulane. Why have the present owners of St. Mary's Jefferson College apparently forgotten Valcour Aime, the pious man, the friend of education and of the poor?"[180] (The University of Louisiana was renamed Tulane University of Louisiana in 1884 to honor the philanthropic donations of Paul Tulane, a retail and real estate mogul from Princeton, New Jersey.)[181] Responding to Fortier, the Marists printed descriptions of Aime's patronage in historical accounts of St. Mary's Jefferson College.[182] Louis Dufau, however, was left out of these narratives. In fact, when St. Mary's Jefferson College opened, the institution's stationary bore the same campus effigy employed on Louisiana College's stationary. Yet the "Louisiana College" name etched above the main portico had been scratched out, leaving only traces of its lettered existence. In time, the campus letterhead was substituted with an overlapping "M" and "A," a traditional symbol representing the Marist order. Later, these initials were replaced with the Marist shield above the Latin motto *Sub Mariae Nomine* (In Mary's Name). The shield and motto were surrounded by the phrase "Jefferson College, Convent, La."[183] As Catholicism and Marist control waxed, Creole authority further faded from the scene. Though Fortier had memorialized his ancestor, Aime's social class, much like Dufau's educational legacy, was not to last. According to Toledano, "Valcour Aime survived his wife and three of his children as well as the Civil War. Yet he did not survive to see his way of life, the whole social and economic structure under which he had thrived, crumble. . . . The subsequent decline of his family's fortune and his properties parallel the collapse of Louisiana's Golden Age."[184]

Sans Aime and Roman, St. Mary's Jefferson College persisted as its new

Catholic affiliation resonated with St. James Parish's changing ethnic population. Inflated numbers of German and Irish immigrants, Catholic born and bred, had taken up residence along the gold coast, as French Creoles, still reeling from war, fell back on their faith for spiritual stability. Accordingly, a religious college in an increasingly Catholic region was welcome. As such, enrollments increased and St. Mary's Jefferson College prospered through Reconstruction with a mixture of religious and liberal arts instruction.[185] By the end of the 1860s, enrollment had risen to 102 with the majority of students originating from south Louisiana (Algiers, Assumption Parish, Donaldsonville, Lafourche Parish, Marksville, New Orleans, Plaquemines Parish, Point Coupee, St. James Parish, St. John the Baptist Parish, Terrebonne Parish, Thibodeauxville, and Vermillionville). Other pupils came from Mobile, Alabama; Scottville, Texas; the Mississippi Gulf Coast; and as far away as France and Italy. Among the Louisiana student ranks were Gabriel Fortier, Joseph Roman, Onesime Roman, and Sidney Roman, all relatives of Valcour Aime and A. B. Roman.[186] As well, Marist efforts to enhance instructional materials were underway. Colonel Samuel H. Lock, professor of natural philosophy at the Louisiana State Seminary of Learning and Military Academy, visited the Marist college while conducting a geographical and topographical survey of the region. Locket commented on the institution's curricula and pedagogical items and described the campus as "well kept, abundantly supplied with physical and chemical apparatus, and with a very fair beginning of a geological and mineralogical cabinet."[187]

The Catholic administration of St. Mary's Jefferson College appeared relatively lax compared to the overtly firm disciplinary structure of antebellum Jefferson College and Louisiana College. According to one advertisement: "The system of government [at the college] is mild and parental, yet firm in enforcing the rules of established discipline."[188] Sporadic in former administrations, religion now took precedence. Students, regardless of faith or denomination, were expected to attend religious services; however, the Marists made a point to mention, "no influence is ever exercised over their minds in regard to their religious persuasion."[189] Notwithstanding the college's open yet insistent stance on religion, the Society of Mary ornamented the campus with Catholic art, statues, and symbols. The student body also observed religious feast days in honor of Catholic saints.[190]

The existence of a Catholic college amid Reconstruction-era sugar plantations no doubt provided a sense of increased community for French Creoles. If their sons were educated together, perhaps a vestige of the old aristocracy would be preserved. On the other hand, St. Mary's Jefferson College's new slogan might have offended what was left of ancienne population pride. Academic catalogues throughout the 1870s were inscribed with "Non Coronatur Nisi Legitime Ceravert."[191] This maxim, meaning "Not crowned, except he strives lawfully," may have reminded Creoles of their slow social decline from the lofty heights of their antebellum quasi-Parisian hierarchy. Still, the former elite took some pleasure in knowing that *French* Marists composed the college's faculty, served their sons dutifully, and had sent for additional priests and brothers from France to help with instruction.[192]

A newspaper advertisement dated August 12, 1865, sheds some light on the overall academic structure of the Marist institution: "Opening of Jefferson College. They [the Marists] offer elementary, commercial and high school. This includes some college work."[193] Certainly, students needed preparatory courses if they were to partake in college-level classes. This model emulated the traditional European six-year college structure, attracted a greater range of student numbers, and, presumably, fostered retention. It was assumed that if students began their preparatory work at the college, they would remain until graduation. But, unlike Dufau's administration, enrollment numbers were relatively well dispersed across upper- and lower-level academic departments.[194] In time, three branches of study existed at St. Mary's Jefferson College: classical, commercial, and preparatory (the former two composing the "college department").[195] The preparatory curriculum consisted of arithmetic, Christian doctrine, English, geography, history, penmanship, reading, and spelling. The college branch furthered preparatory curricula via enhanced liberal arts instruction, while the commercial branch was "designed to fit young men for mercantile pursuits."[196] To facilitate experiential learning in the commercial branch, a portion of the main college building was transformed into a working bank "fitted out with the usual bank furniture to give the commercial students practical experience of all the details of the business." The college bank was, in actuality, a physical duplicate of era banks complete with grated cashier windows and inlaid wood panels.[197]

As the Louisiana legislature had empowered the new Catholic institution to grant college degrees, high school diplomas, and literary honors, students were examined twice a year to test their academic progress.[198] Perusing the college's academic catalogues, readers were exposed to a litany of course topics as well as religious symbols, including the hand of God pouring oil into a lit lamp and a cross-intersected crown. Not only did religious images adorn catalogues; palmettos, common to Louisiana's wetlands, framed pictures of class and dorm rooms.[199] Catalogues, produced by the *Morning Star Print* (the Catholic press of New Orleans),[200] informed parents and pupils alike that St. Mary's Jefferson College "sees to it that they [the students] are taught how to fulfill their duties to God, their country, their fellowmen and themselves,"[201] and that the Marists adhered to a highly classical curriculum with a particular emphasis on science and commercial arts.[202] In addition to classical, commercial, and scientific courses, students were expected to engage in routine theological instruction.[203] Overall, the college maintained an educational paradigm heavily influenced by the liberal arts model of late nineteenth-century France, but promoted a clear intellectual divide between commercial and classical instruction.[204]

Progressing from the preparatory department, students who adhered to the classical branch were prescribed courses in anatomy, botany, chemistry, English, geography, geology, Greek, history, hygiene, Latin, mathematics, mental philosophy, mineralogy, physiology, penmanship, and religion.[205] Those in the commercial branch attended courses in bookkeeping, business arithmetic, commercial law, commercial forms, mercantile correspondence, penmanship, shorthand, and typewriting.[206] Beyond the aforementioned, students could partake in German and Spanish language courses, participate in the brass band, or receive private violin or piano lessons for an additional eight dollars.[207] Unlike students in the classical branch, young men enrolled in the commercial division were exempt from Greek and Latin.[208] Commercial students were meant to become members of a "business class" of corporate managers and accountants, unlike male members of the genteel sugar planter echelon.

Whether students followed the classical or commercial branch, they were eligible to receive the arts baccalaureate or bachelor of science degree, depending on curricular background. In order to meet the necessary qualifications for graduation, students had to complete "four [college level] subjects daily, five times a week, through the entire course of four years."[209] At the end of their final year, students submitted themselves to comprehensive oral and written examinations. Those who did not obtain a minimum score of 75 percent were denied graduation. The master of arts degree was granted to those already possessing an arts baccalaureate from St. Mary's Jefferson College, sans additional course requirements upon the submission of an "essay of merit" to the college president. If the president and members of the faculty agreed that the essay illustrated advanced learning, the master's degree was conferred.[210]

From the onset of the Marist administration, curricula rarely changed. Granted, a commercial branch was added, but at first it greatly resembled the classical branch with only slight variations as concerned scientific courses. In time, science and mercantile curricula were expanded, but the college never advanced to match the academic prowess of peer colleges and universities. A late nineteenth-century Louisiana biographical and historical series described the scholastic progress of St. Mary's Jefferson College as follows: "That fine [college] building is now occupied by an ecclesiastical order, whose efforts, doubtless praiseworthy, do not approach the standard of education [originally] projected by Governor Roman."[211] How Roman would have judged the Marist administration remains a matter of speculation, as he had died sixteen years prior.[212]

With the passing of Aime, Roman, and other formerly wealthy Creoles, the Society of Mary was left relatively alone to administer their Catholic college as they pleased. Indeed, the Marists had entrenched their institution in a now deeply religious portion of south Louisiana and marketed it accordingly. Newspaper advertisements announced, "This ancient establishment, agreeably situated at COLLEGE POINT, one of the most eligible and picturesque spots formed by the windings of the Mississippi, combines, with the best sanitary conditions, all the advantages, which can make it pleasant and attractive. Whatever the causes may be which contribute to the salubrity [*sic*] of the place, it is a well-known fact, that no malignant fever ever broke out within its walls, and the pupils have always been noticed for their healthy appearance."[213] The Marists, concerning the overall "health" of College Point, had spoken too soon as in 1873, three instructors—all priests—died of typhoid. Although no students succumbed to the disease, parents withdrew their sons, and attendance plummeted. As a result, both lay shareholders and the college's governing board reacted in the negative and prepared, once again, to close the institution.[214]

After the typhoid related deaths, Reverend Father Pére Favre, Superior General of the Society of Mary, assessed conditions at St. Mary's Jefferson College. Favre must have been astounded at what he found: sickness, death, low enrollments, and financial instability. As the Marists could do nothing to sway the governing board's decision to close, Favre deemed the college's state to be desperate. In order to save the ailing institution, the priests opined, they had to dispense with Creole control and hatched a plan to remove the college from planter authority.[215] Shareholders, after reviewing the institution's financial condition, resolved to declare bankruptcy. Such pronouncements were unsurprising, as the majority of Creole sugar planters, despite their Catholic heritage, rarely supported the Church financially.[216] According to Miller, "Southerners generally had no reputation as great [financial] givers, and Catholic southerners were no different."[217] Disregarding moribund support, the Society of Mary refused to abandon their educational work in St. James Parish.[218]

Weary of contending with Creole dominance, the Marists demanded $44,000 in deferred wage payments from the governing board. The board, due to a lack of funds resulting from the college's physical rehabilitation, proved unable to generate the fee, and the Society of Mary sued.[219] As de

Sennegy explains, "By a providential meeting . . . an assembly of shareholders was held, the burdens which were imposed upon them [the Marist Fathers] for twelve years were clearly exposed; these [Creole] gentlemen unanimously recognized their inability to provide for them [the Marists]."[220] As members of the board could not counter the legal suit, the parish sheriff seized the college and all associated property in February of 1867 and sold it at auction to the highest bidder. Before the college could close, the Society of Mary put forth a bid of $10,000 and won unopposed in March of that same year.[221] No longer obliged to observe the demands of regional shareholders, the Marists focused on their own initiatives.[222] After years of reorganization, they drafted a new college and religious organizational charter in 1870; however, the document remained unsigned for an additional five years because church and missionary duties delayed the order from incorporating itself as a legitimate, state-acknowledged instructional and spiritual corporation.[223]

In 1875, the Marists finally established their charter, which denoted the seat and domicile of the new corporation to be fixed to St. Mary's Jefferson College.[224] A portion of the document, "An Act to Incorporate the Society of the 'Fathers of the Society of Mary,'" reads as follows:

> We, the undersigned and successors [The Society of Mary], desiring to associate ourselves for a literary, scientific, religious and missionary purpose within the State of Louisiana, and to enjoy the powers and immunities of a corporation or body politic, so that such a purpose may be the more effectually accomplished; do hereby by virtue and in pursuance of section six hundred and seventy-seven of the Revised Statutes, adapted by the General Assembly of the State of Louisiana in the year one thousand eight hundred and seventy, create, organize and establish in the Parish of St. James, in the State of Louisiana, a corporation or body politic under the name and style of "The Fathers of the Society of Mary" for the purpose of diffusing knowledge, promoting learning and morality, instructing youths in the liberal arts and sciences and in missionary work in the Towns and Parishes of the State of Louisiana.[225]

Absent lay management, Henri Bellanger was elected both president of the college and of the Society of Mary's Louisiana corporation (Father Stephen Chaurain replaced Bellanger as college president in 1865 after he, Bellanger, stepped down to serve as pastor of Holy Name of Mary Church in Algiers. Leaving Louisiana for Lyon, France in 1878, the first Marist president of St. Mary's Jefferson College expired in 1879 and was buried in the Cemetery of Loyasse).[226] Taking stock of their winnings, the Marists estimated the worth of their newly established corporation, including all college property, buildings, and land, at $180,000. In order to keep the new corporation removed from planter control, the charter concluded with a telling statement: "In case of his [Father Bellanger's] death, resignation, removal or otherwise, the Board shall by and with the consent of the superiors of the order of the Marist Fathers appoint his successor or successors, to be chosen among the members of this corporation."[227] In short, the acting presidential authority thereafter was to be a member of the Society of Mary and no one else.

The once secular Jefferson College, now a religious institution free of Creole governance, spurred regional Catholic zeal. De Sennegy, chronicler of religious and social events in the Louisiana gold coast, wrote in response, "Come then, O you who are hungry and thirsty for truth and science, come under the peristyle of the Athenian temple, listen to the voice of the Catholic and Marist priest, announcing that the true God has overthrown the gods, arresting for an instant your intelligence on the ancient and hazy legends of the Greek and pagan world, uncovering for you their pure lines, beautiful and noble, and conducting you to evangelical poems in which the Christian appears entire with his tendernesses, his joys and his sorrows; come to the forests, under the great oaks, to listen to the voice of the infinite, of nature and of God."[228] Before Jefferson College's transfer to the Society of Mary, south Louisiana sugar barons had done much to reopen and remove the institution from state and federal control. As a result, Creoles avoided the racial conversion of their plantation college. Passing through Reconstruction, the Society of Mary overcame planter rule and emerged as the seminal authority of a college repurposed to serve God through education. In the years to come, the religious order, dedicated to the Blessed Mother of Christ, had to contend with decreased rural populations, increased demands for scientific and commercial curricula, the Americanization of southern colleges and universities, institutional competition with secular private and public higher education, and the devastating Mississippi River flood of 1927.

Figure 1: "Jefferson College Aerial," c. 1870. General Photograph Collection, Marist Houses and Communities, Jefferson College, Cabinet 189, Drawer 1. Image courtesy of the Archives of the Society of Mary, Atlanta, GA. Rebuilt main college building (center), matching porters' lodges (front of main building), Alumni Hall (left of and behind main building), Thouvenin Hall (right of and behind main building), campus chapel (right of Thouvenin Hall), gymnasium (right of chapel), lay-faculty residence (right of gymnasium), president's house/college hotel (bottom, left).

Figure 2: F. Girsch, "Governor A. B. Roman Portrait," c. 1850s. Louisiana Image Collection, Box 49: Portraits, Folder 38, Louisiana Research Collection, Tulane University, New Orleans.

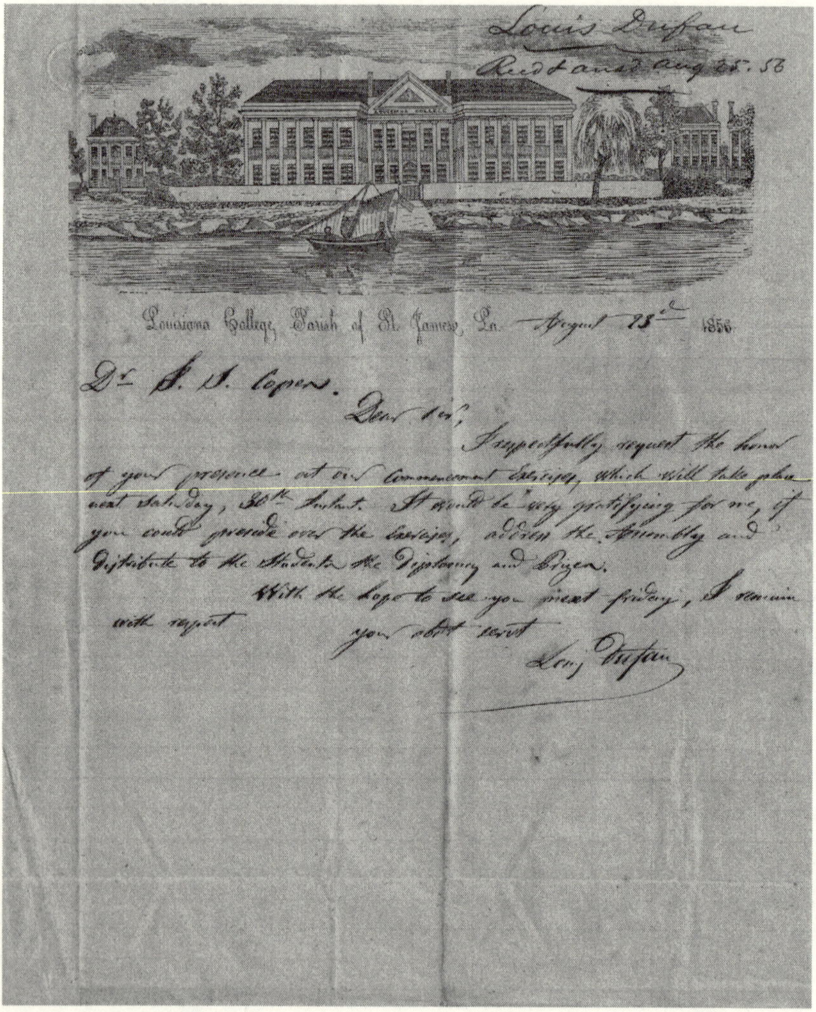

Louis Dufau
Rec'd & ansd Aug 25. 56

Louisiana College, Parish of St. James, La. August 23rd 1856.

Dr. J. S. Copes.

Dear Sir,

I respectfully request the honor of your presence at our Commencement Exercises, which will take place next Saturday, 30th Instant. It would be very gratifying for me, if you could preside over the Exercises, address the Assembly and distribute to the Students the Diplomas and Prizes.

With the hope to see you next friday, I remain with respect

your obdt servt

Louis Dufau

Figure 3: "Louis Dufau to Joseph Copes," August 23, 1856. Joseph Copes Papers, Box 11: Personal and Business Correspondence, 1855 August–1857 March, Folder 13: Personal and Business Correspondence, 1856 August 2–30, Louisiana Research Collection, Tulane University, New Orleans.

Figure 4: Jacques Guillaume Lucien Amans, "Portrait of Francois Gabriel (Valcour) Aime," c. 1838. Image courtesy of the Historic New Orleans Collection, New Orleans, LA.

Figure 5: "Sacred Heart of Jesus Chapel," c. 1924. Sugar Bowl [Year Book], 1924, Volume III. American Province Records, Series 4. Education, Box 3. Image courtesy of the Archives of the Society of Mary, Atlanta, GA.

Figure 6: "Rev. Henri Bellanger, S.M.," undated. *The Catholic Church in Algiers, Louisiana. The Marist Fathers and the Holy Name of Mary Church.* By Rev. J. Francis Georgelin, S.M., S.T.L., edited by Rev. Charles F. Brogley, S.M., M.A. (New Orleans: privately printed by the Society of Mary, 1949), 16. Marist Houses and Communities Records, Holy Name of Mary Parish (Algiers, LA), Box 2. Image courtesy of the Archives of the Society of Mary, Atlanta, GA.

Figure 7: "St. Mary's Hall [with cross]," c. 1905. Catalogue of Jefferson College (St. Mary's), 1905–1906. American Province Records, Series 4. Education, Box 6. Image courtesy of the Archives of the Society of Mary, Atlanta, GA.

Figure 8: "Rev. James Blenk, S.M. [bottom row, center], Rev. Matthew Thouvenin, S.M. [left of Blenk], and Marist Faculty," c. 1890s. General Photograph Collection, Marist Houses and Communities, Jefferson College, Cabinet 189, Drawer 1. Image courtesy of the Archives of the Society of Mary, Atlanta, GA.

Figure 9: "Jefferson College Brass Band," c. 1905–1906. Catalogue of Jefferson College (St. Mary's), 1905–1906. American Province Records, Series 4. Education, Box 6. Image courtesy of the Archives of the Society of Mary, Atlanta, GA.

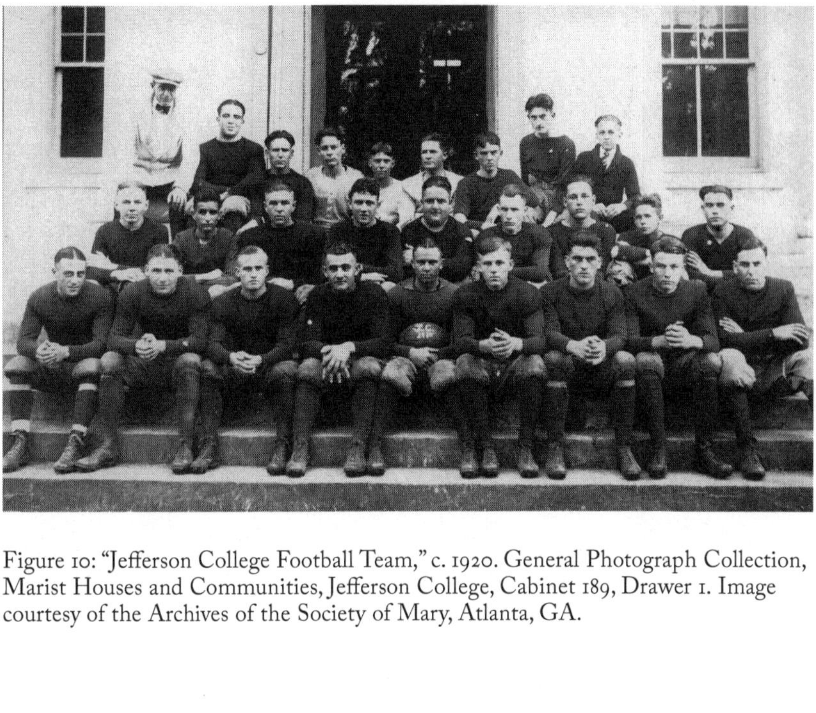

Figure 10: "Jefferson College Football Team," c. 1920. General Photograph Collection, Marist Houses and Communities, Jefferson College, Cabinet 189, Drawer 1. Image courtesy of the Archives of the Society of Mary, Atlanta, GA.

Figure 11: "Blenk Science Hall and Oak Ave, Jefferson College, Convent, La.," c. 1910s. Postcard Collection, Box 1. Image courtesy of the Archives of the Society of Mary, Atlanta, GA.

5
Marists and Americanization

Fiat voluntas Dei!
—Archbishop John William Shaw

When Martha R. Field, better known as Catharine Cole, the late nineteenth-century correspondent for the *Daily Picayune*, visited the gold coast, she was impressed by the decaying old world grandeur of Louisiana's French Creole aristocracy. In the midst of sugarcane fields and sharecroppers' cottages, Cole visited the large, whitewashed buildings of St. Mary's Jefferson College.[1] Recounting her experience Cole wrote,

> St James Parish has always been the home of the aristocrats. Its traditions are all of splendor, brilliance, wealth, and of refined and pleasant society, excellent educational institutions, superb homes, grand families wearing with honor brave and distinguished names. To this day it is famous for its marvelous crops . . . for its Jefferson College and Convent of the Sacred Heart, and for its old plantation homes. It is the land of magnificent mansions, those grand old Doric-columned dwellings, great square edifices like the Parthenon, with porches like cloisters rising one above the other, three and four tiers high on all four sides, and topped by a roof studded with blinking dormer windows, like old faded eyes looking calmly out on an outlined world. Seen from the river, brimming to its banks and giving back its own mirrored idea of sky and cloud, or oak alley and church, and convent and home—how beautiful it is, this parish of St. James.[2]

What Cole did not report was the region's changing demographics and steady Americanization of a college originally established, in large part, to preserve Creole heritage. Surely the observant writer must have noticed the remaining fragments of a bygone class system struggling to survive among a social context that included Acadians, Americans, Germans, Irish, and Italians. By 1870, approximately sixteen percent of Louisiana's inhabitants were either immigrants or born to migrant parents.[3]

These immigrants, mostly poor farmers, lacked formal instruction to match the remaining gentry's educational background. At the start of the 1870s, only twenty-six of the estimated 740-member Acadian population in St. James Parish had received some form of primary schooling.[4] Waking from their wartime slumber, colleges and universities across the American South admitted students from migrant populations to increase tuition revenue.[5] Likewise, sons of Acadian and immigrant families were enrolled at St. Mary's Jefferson College alongside ancienne population students. As a result of the 1873 economic panic and subsequent recession, nearly two thousand men, women, and children left New Orleans in search of work and wages on sugar plantations.[6] The Marist college, with its preparatory department, became a rural instructional mainstay for lower class families. Although sharecropping had swept the South, cheap immigrant labor was welcome.[7] Combined, this inexpensive workforce of foreigners and freedmen bolstered sugar production.[8]

By the late 1870s, the remaining sugar elite, including French Creoles and Anglo-Americans, acknowledged emerging farming technologies, joined forces, and plied for a tangible form of information dissemination. In an attempt to further rebuild south Louisiana's sugar wealth by means of shared knowledge and communal desires to dominate lower-class workers, the Louisiana Sugar Planters Association was formed in 1877.[9] The association, in conjunction with the United States Department of Agriculture, held meetings to propagate salient farming information and published a practical journal titled the *Louisiana Sugar Planter*. Planters connected to the association believed that by combining forces and sharing intelligence, a collective hold on the upper class would be maintained. Mirroring antebellum educational ethos, sugar planters partnered with local colleges and universities. This time, however, planters disregarded liberal arts instruction and invested in agricultural research institutions including the Audubon Park Sugar School, Louisiana State University, and Tulane University.[10] Through research and practical instruction, higher education, it was hoped, would enhance Louisiana's sugar economy. St. Mary's Jefferson College, however, was not chosen to partake in the new association. Despite being surrounded by agriculture, the Society of Mary was more concerned with maintaining liberal arts curricula and missionary work. An agreement dated September 13, 1869, between Father Pére Favre and New Orleans Archbishop Jean-Marie Odin specified that the Marists were responsible for all theological instruction in and around St. Michael's Church. To facilitate the Society of Mary's missionary efforts, Odin provided a pension of 100 dollars.[11] The fathers and brothers of St. Mary's Jefferson College, now under the presidency of Father Henry Gaud, which lasted from 1870 until 1874, welcomed their expanded religious purview and did what they could to impress upon passersby the denominational ascendancy of their college, but did little to illustrate the handful of scientific courses included in the curriculum.[12]

A sense of overt Catholicism undeniably presided at St. Mary's Jefferson College. After the Marists assumed control, they affixed a large cross onto the roof of the main building and placed statues of the Blessed Mother and St. Joseph in front of the campus. As time passed, the statue of Mary increased in import to the college staff. The Marist priest, Phillip H. Dagneau, narrated a legend surrounding this particular effigy: "Tradition states that in former years, during the hurricane that did much havoc for miles around, the College was spared and when the storm subsided, the statue was found turned and facing the direction of the oncoming storm. The Blessed Virgin had protected her consecrated Marist sons."[13] While Catholicism permeated the college's culture, so, too, did the region's geography and climate. For newly arrived Marists, the quasi-tropical weather of south Louisiana was an uncomfortably novel experience. Steeped in humidity, insects, and the ever-present passage of the Mississippi River with its occasional floods and inconsistently protective levees, the campus locale irritated the priestly française étrangers. On hot, muggy days, St. Mary's Jefferson College was riddled with muddy paths while buzzing, biting mosquitoes attacked the cassocked clergy. Brief respites came at night as cotton mesh enclosed students' and instructors' beds in an attempt to ward off the airborne pest.[14]

Mosquitoes, mud, and humidity were not the only threats. The Mississippi River provided its own dangers. Levee crevasses were frequent when the river ran high. Flooding due to levee breaks had the potential to decimate fields of sugarcane, homes, and livelihoods. When breaches occurred, action was swift. As Cole details, "The firing of a gun is the signal of a break. Then the guard rush away crying 'Crevasse! Crevasse!' The plantation's bell rings out its alarm. The people rush out and the valiant fight begins." Parish men, regardless of race or social class, were swift to repair levee breaks since all could suffer as a result of a devastating breach.[15] During a moonlit night in 1892, while occupying a room at the campus hotel first established by Louis Dufau, "a big, cheerful house under great oak trees that stands on the corner of the Jefferson College grounds," Cole reported that "the gallant Marist Fathers of Jefferson College worked at an incipient crevasse on their levee until four in the morning. It was a picturesque spectacle to see those black-gowned shapes, in broad-brimmed hats, spading in the half moonlight, their soft robes beating in the wind as they strenuously worked to save the parish from a crevasse."[16] In the face of such hardships, the Marists had no choice but to acclimate themselves to their surroundings.

After assuming control of the formerly Creole-governed college, the fathers and brothers, ignoring both elevated temperatures and bothersome insects, advertised the benefits of their institution's relatively isolated setting halfway between the capital city of Baton Rouge and the "greatest city of the South," New Orleans. For members of the Society of Mary, the college sat reasonably close to supplies but was well enough removed from the distracting "hum of city life."[17] This type of solitude was preferable to the Marists; however, it was far from student ideals. In 1875, a flustered student wrote to his family in New Orleans and complained that his friends and relatives were simply too distant to visit.[18] Similarly, remoteness proved problematic for family ties and, per college guidelines, if parents lived at a considerable distance from the campus, they were to designate a responsible guardian in either New Orleans or Vicksburg.[19]

In spite of the college's distance from urban centers, there were still celebrations to be had. A yearly Christmas Eve tradition in St. James Parish included the lighting of bonfires along the river levee. These large bonfires not only announced the culmination of the holiday season, they also signaled the start of a new planting period.[20] Although the bonfire tradition has been closely coupled with sugar parish customs, Martha Gaudet speculated that the Marists at St. Mary's Jefferson College started this practice in the 1870s while the college was under the presidential leadership of either Father John Baptist Bigot (from 1874 until 1879) or Father John Joseph Grimes (from 1879 until 1880).[21] The *Daily Advertiser* provides additional details regarding the bonfire tradition and its connection to the Marist college: "Lighting bonfires on Christmas Eve in St. James Parish has been a ritual since [before] the 1880s. Though its origin is debated . . . most likely, they were introduced into the area by French Marist priests who took over Jefferson College . . . in 1864. They built fires on the battures [levees] next to the river on New Year's Eve to entertain their boarding students. The practice caught on and somehow got moved to Christmas Eve—perhaps at the same time the Louisiana French tradition of exchanging gifts on New Year's moved to Christmas."[22] Lending credence to the entertainment theory, a

St. Mary's Jefferson College student described the bonfire festivities to an unrecorded recipient:

Jefferson College
December 27, 1881

My dearest friend,

It is with a great pleasure I am going to write you these few lines to let you know my news. I was sick all last week, but now am well. I am going to tell you how we pass Christmas. We had a great big fire on the batture [levee] we light for Christmas, but I could not go to the fire because I was sick. We had a bonfire, crackers, cotton ball [popcorn and sugar packed to resemble a sphere], and everything. We [also] had a great big cannon. My dearest friend, write to me and tell me how did you pass your Christmas? Dear friend write to me as soon as you can because it is for a long while I did not receive any letters from you. It is all I have to tell you for this time.

Your devoted Friend,
Sidney Montegin[23]

Traditions and student life, though not emphasized during the previous Jefferson and Louisiana College administrations, became an important fixture during the Marist regime, particularly under the second administration of Father Bigot (from 1880 until 1886) and first presidential governance of Father George James Horton Rapier (1886–1887).[24] Not only were extracurricular activities present, they contributed to the overall Americanization of St. Mary's Jefferson College.

By the time Father Thomas Henry became St. Mary's Jefferson College's president in 1887, the sugar industry was feeling the effects of a much-needed monetary boost. In the face of daunting sugar imports from Cuba, Hawaii, the Philippines, and Puerto Rico, Louisiana officials charged planters to increase production of the popular sweetener. Federal authorities had already expanded subsidies to enhance the quality and quantity of US sugar—particularly beet sugar, which was fast outstripping cane sugar production. In response, south Louisiana became a wilderness of cane fields punctuated by hulking refineries riddled with pipes and vats that considerably improved output.[25] In order to ease financial burdens, plantation owners leased their properties to northern corporations who installed modern refineries. In time, these leases gave way to sales. While several planters lost their plantations due to foreclosure, others sold their estates to pay debts.

As the nineteenth century waned, Louisiana's sugar industry yielded greater profits, but, by 1893, the ancienne population no longer controlled gold coast sugar production and commerce.[26]

As the sugar industry continued to expand and modernize, so, too, did St. Mary's Jefferson College. By 1898, the physical campus had grown to include a bathhouse, carpenter's shop, cattle barn, horse stable, infirmary, a new kitchen and storeroom, laundry facility, music rooms, refectory, and servants' quarters.[27] While the Marists occupied rooms on the second floor of the college's main building, lay instructors resided in a two-story wooden structure. Since many of these instructors were Americans and their campus residence was located "south" of the main building, students often referred to their abode as "South America."[28] From 1864 until 1927, approximately eighty lay professors taught at St. Mary's Jefferson College.[29]

In addition to Anglo-American instructors, the Marists employed African Americans to clean, cook, make beds, serve meals, and tend livestock.[30] One such example was Mary Tircuit Jeannine, who worked as a maid at the college for fifty years. In her book *Cabanocey: The History, Customs, and Folklore of St. James Parish*, Lillian C. Bourgeois recounts a conversation between Mary and a college alumni: "Often when old Jeffersonians drove through Convent, they stopped to see Mary. The story is told that once an old Jeffersonian, now a prominent man, stopped to chat with her. He gave his name and asked if she remembered him. Came the quick reply: 'Yas indeed,—yas indeed! Gimme that quarter you owes me fer pressin your pants.' And with much laughter the longstanding debt was paid with interest. . . . She [Mary] talked of old times and remembered many of the old students and Marist Fathers and her last words as she waved good-bye were: 'When you see my old boys, tell em howdy—you hear—tell em howdy.'"[31] Though the clerical administration of St. Mary's Jefferson College employed black servants, they never admitted black students, as Catholic colleges, universities, and schools in Louisiana remained racially segregated until the latter half of the twentieth century.[32]

Racial segregation permeated the American South as a cultural phenomenon tied to pre–Civil War slavery and post-Reconstruction politics. But, while the cultural exclusivity of segregation remained, other ethnic norms, particularly the age-old rivalry between French Creoles and Anglo-Americans, was beginning to slip away. By the 1870s, college and university administrators in south Louisiana began to encourage student activities and residential practices that fostered ethnic cooperation. At Louisiana State University (relocated to Baton Rouge in 1869 after the Pineville campus burned in October of that year),[33] President David French Boyd insisted that Creoles and Americans cohabitate in order to promote gentlemanliness in the stereotypically aggressive and outspoken Americans and entrepreneurship in Creole youth who had been reared by parents largely unwilling to embrace commercial business management and modern mercantilism. Walter L. Fleming, in his 1936 book *Louisiana State University: 1860–1896*, details Boyd's decision to commingle students whose families had long adhered to the social division between Creoles and Anglo-Americans: "It was . . . Boyd's belief that the association of Creole and Anglo-American was good for both. The gentle manners of the former served as a check upon the latter; while the American in turn stimulated the Creole to energy and enterprise. His policy was to mingle the two nationalities in the dormitories as well as in classes and in play."[34] Likewise, French Creole students at St. Mary's Jefferson College not only lived with and studied alongside students from varying ethnic backgrounds, they were exposed to an influx of younger priests and American lay instructors who encouraged administrators to further adapt the St. James Parish college to include extracurricular activities and organizations similar to other US religious and secular institutions. Athletic events, music, as well as advanced military cadet training became commonplace on a campus that, since its antebellum genesis, had been devoid of such activities. These organized pastimes not only encouraged the intermingling of ethnic groups, they further Americanized the Creole campus.

Student Life and Modern Curricula

On January 22, 1869, sixteen brass instruments, one bass drum, three snare drums, a pair of cymbals, and a triangle were delivered to St. Mary's Jefferson College. The instruments, all of which came from France, allowed for the creation of the popular St. Mary's Brass Band. Each horn and drum, while property of the college, was discharged to a select student according to his ability to play said instrument. Care and maintenance of individually assigned instruments were the responsibility of band members with music rehearsals set three times per week.[35] The objective in founding the brass band was "to encourage the study of music to afford the members of that philharmonic society an agreeable and useful pastime during the year and add solemnity and spirit by their performances to the celebration of the Religious and Literary festivals of the College."[36] Despite these lofty aims, the first cadre of student musicians were recorded as "not very skillful but they progressed rapidly having powerful stimulus in the novelty of the thing [band]. They devoted whole recreations to their practicing, in fact so anxious were they all of learning, that the prefects were obliged to appoint a certain time during which it was forbidden to use the instruments."[37]

As the brass band improved, it became a regular fixture in the surrounding parish and was invited to play for carnivals, Catholic feast days, funeral processions, private parties, and promenades. On special occasions, the band was permitted to perform on "floating palace" riverboats. Dressed in black military attire with gold trim and red sashes, student musicians executed overtures and period dance tunes. Marches, however, were the ensemble's preference above all. As band members commented, "Marches are combat a la gloire" (the glory of combat). Eventually, the band exchanged their black attire for livery modeled after West Point fatigues. For more than ten years, priests directed the ensemble, but with the onset of the 1880s they relinquished the band to a lay instructor, Professor F. Green. The band, now under the tutelage of a professional musician, performed for annual Yuletide levee bonfires and welcomed, with pomp and circumstance, political and religious figures. In 1887, Louisiana governor Francis Redding Tillou Nicholls traveled to inspect St. Mary's Jefferson College. When Nicholls arrived, the ensemble struck up a rousing rendition of a song titled "A Brave Warrior: True Patriot that Gave our Country Peace and Prosperity." When New Orleans Archbishop Placide-Louis Chapelle toured the campus on May 2, 1898, the band heralded his visit with a performance of "Red, White, and Blue."[38]

The St. Mary's Brass Band was regularly booked for social events at nearby plantations. In early spring of 1893, the band was invited to perform for a picnic at a local planter's home: "An exquisite table had been set in the Pecan Grove just back of the house. The legs of the table seemed to go down into the ground under the weight of all the meats, cakes, fruits, cream, champagne, etc; which they supported. The [brass band] boys as usual were not without their 'Jeffersonian' appetite. There were no petticoats around to intimidate us. All ate heartily."[39] Aside from the college's brass band, a string and woodwind orchestra also existed. The small orchestra performed every Sunday to the delight of parish residents.[40] To accommodate the orchestra and brass band, the original porters' lodges, fronting the main college lawn, were converted into music rooms.[41]

With well-liked student groups performing under the banner of St. Mary's Jefferson College, the institution became, once more, an appreciated entity. Local newspapers featured student successes and announced annual commencements. One such article reads, "On Monday and Tuesday, [July] 15th and 16th, will take place the exercise of the Annual Commencement at that magnificent institution of education, of which Louisiana can well boast, Jefferson College of our parish."[42] In addition to music ensembles, other student organizations and activities took hold. In 1878, the college's debate team engaged in a rather harsh discourse on "Whether the [Theatric] Stage of the Present is More Immoral than Moral in its Tendency."[43] Other students participated in religious organizations such as the League of the Sacred Heart and Choral Society.[44] If the debate team and choir failed to remedy ennui, students were amused by occasional visits from *Price's Water Queen*, a riverboat touting a celebrated steam calliope that could be heard as the ship rounded College Point.[45]

Although student life evolved to more closely resemble that of secular colleges and universities, the academic course structure at St. Mary's Jefferson College remained in drastic need of revision if the Marists were to compete with increasingly admired regional institutions. In 1891, Father James Hubert Blenk, the newly appointed college president, improved the campus to attract potential students and house expanded curricula. Blenk, a former student and professor of St. Mary's Jefferson College, remained president until 1896.[46] He received his primary and secondary education in New Orleans parish schools and was afterward called to don the cassock of a Redemptorist priest. Blenk enrolled at St. Mary's College (owned and operated by the Catholic Congregation of the Most Holy Redeemer) in Ilchester, Maryland, in 1873, but he became ill and was forced to return to New Orleans. After his recovery, Blenk enrolled at St. Mary's Jefferson College and decided not to become a Redemptorist, but instead chose to become a Marist. After he finished his senior year, the Society of Mary sent Blenk to study philosophy in Belley, France. Later, he entered the Marist novitiate in Lyon and studied mathematics and science at the Catholic University of Ireland. The university's administration was impressed with the young Marist and prompted him to accept the endowed chair of mathematics at St. Mary's College in Dunbalk. In 1885, twenty-nine-year-old Blenk took his final vows of priestly formation. Now ordained, he returned to New Orleans and was assigned a teaching post at St. Mary's Jefferson College. Blenk became an instructor in the college's preparatory division and taught arithmetic, English, and German. Eventually, he administered the commercial branch, served as the campus spiritual director, and held a professor of humanities post. Overall, Blenk was well liked by both peers and pupils. He succeeded at moderating the college's debate team and stayed active in campus life. Blenk even participated in student-led dramas—once as the lead character in *Alonzo de Castile*. In 1888, he was appointment vice president of the college, and in 1891 he was promoted to the presidency.[47]

Under Blenk's administration, science curricula were expanded to include anatomy, botany, physiology, and qualitative analysis. Still, the college president believed that a classical liberal arts education was necessary for any student who wished to procure a legal or medical occupation after graduation. To better house science courses and requisite laboratories, Blenk had Alumni Hall constructed. Named in honor of St. Mary's Jefferson College's graduates, the building contained an auditorium and stage on the first floor and laboratories on the second. Blenk also believed that the sons of native Louisianans needed to be thoroughly acquainted with their state if they were to be knowledgeable citizens. Therefore, he incorporated mandatory classes in Louisiana governance and history for both the preparatory and college divisions. Blenk also established the Jefferson Athletic Club and natatorium to promote student health and physical activity. In sum, according to Mary Bernardine Hill, Blenk "sought to promote the full development of the whole man."[48]

Blenk not only wanted to graduate decent citizens, but he also encouraged alumni to take active roles in Louisiana's schools. To further the latter, he incorporated instructional pedagogy courses for students who were interested in teaching as a vocation. As a result of these changes, the college's enrollment swelled to the point that additional faculty were needed. Not only had Blenk reformed the college's curricular structure, he had also revamped the institution's primary means of advertisement: academic catalogues. College catalogues, formerly printed with little adornment, were improved, filled with lively campus pictures of students engaged in recreation, and included lists of young scholars and athletes who won awards for academic or physical merit. As well, descriptions of the college's litany of courses, including a brief historical account of the campus, were presented in large, clear text.[49]

Though successful as St. Mary's Jefferson College's president, Blenk, like his predecessor Father Bellanger, resigned in 1896 to serve at the Holy Name of Mary Church in Algiers. His resignation brought about a series of promotions that culminated with Blenk assuming leadership of the New Orleans archdiocese. After three years in Algiers, Blenk was chosen to become auditor and secretary for the Apostolic Delegate to Cuba under Archbishop Chapelle. On June 12, 1899, Pope Leo XIII appointed Blenk as Bishop of Puerto Rico. While in Puerto Rico, Blenk established a Catholic college as well as a series of parochial schools. With the death of Chapelle, Blenk was promoted to become the seventh Archbishop of New Orleans in 1906. He remained in this post until his death in 1917.[50] Following Blenk's curricular revisions at St. Mary's Jefferson College, the Marists experienced a renewed sense of support from rural Louisianans. In addition to improved classical, commercial, and preparatory courses, a formal science department was added, which housed courses in chemistry (carbon and its compounds, metals, and nonmetals) and natural philosophy (acoustics, dynamics, electricity, optics, and pneumatics).[51]

Blenk's inclusion of teaching curricula prompted the Marists to petition the Louisiana legislature to confer upon the college the ability to grant teaching certificates. By doing so, students would no longer need to take and pass regional teaching examinations with the exception of tests relevant to educational theory, history, and school administration. The legislature agreed and declared, "By act of the state legislature the graduates of Jefferson College are ipso facto qualified as teachers in the state schools without passing the usual examinations, except in Pedagogy and the methods of teaching. Jefferson College is the only Catholic institution in the state that enjoys this privilege."[52] To ensure students' understanding of pedagogical practices, seniors attended three classes that relied on *Thorndike's Principles of Teaching* and *Wlute's Art of Teaching*.[53] As a result, several graduates became Louisiana teachers, principals, and superintendents.[54]

Although social activities and curricular offerings were adapted to meet student interests and regional wants, external activities, including unsupervised trips to attend community events or holidays, were banned. Students could not leave the college until Christmas. For all holidays, save Christmas, students remained on campus. Staying on campus during Mardi Gras proved particularly difficult as it was (and remains) a favorite holiday for Louisianans. To satiate their want for carnival merriment, students improvised their own Mardi Gras parade. They decorated the college's horse and buggy, "elected a king and a queen (a boy in women's apparel)," and processed for fellow pupils and parish occupants alike.[55] The parade, its royal court, and celebratory rituals resembled those of "The School of Design," otherwise known as the Rex Parade, in New Orleans.[56] An excerpt pertaining to the college's 1905 carnival revelry detailed the associated events with levity:

> The Jefferson students always look forward to the carnival celebrations at the college with most pleasurable anticipations. Though they do not attend the processions in New Orleans, yet the day is passed so enjoyably here that they do not feel so much the privation. The boys always get up a procession of their own, with a judicious mixture of dignity and comedy with king and queen and courtiers, and of course the indispensible jester with a retinue of masked characters. The king this year was Cyprian Foret who composed himself with becoming dignity and gallantry towards his gracious queen, Richard Barker. The role of high chancellor was ably filled by Holliday Dupuy. At 8 a.m. the procession entered the college gates, the brass band leading, and

> proceeded up the picturesque drive till it reached the pretty lawn directly below the President's room. Here the king and queen descended from their chariot, the brass band "did its worst" [to summon the college president], and chancellor Dupuy then read in a ringing voice, the proclamation of the king. From the contents of the royal edict, it was clear that the king had had an interview with Father President previous to the ceremony, for his majesty had assumed royal prerogatives and released from durance vile all those who at the time had the misfortune of being punished. A number of comical local hits in the proclamation called forth laughter and cheers. After marching around the College grounds for awhile, the procession went out into the town, but returned in time to enjoy the good dinner which our good procurator, Father Delahaye, had provided for us. In the afternoon, the boys whiled away a couple of hours at the little annual fair of the Sodality, humorously called the "Grab Bag." . . . At night, the Rex Minstrels gave a fine performance, under the direction of Prof. Beadle. It was quite a success in every way.[57]

In addition to holiday merriment, other student organizations included the French literary and dramatic club, polo team, and tennis team.[58] All of which went a long way to appease student requests for entertainment and attract young men interested in attending an "American" institution inclusive of clubs and sports alike.

Under the college presidency of Father Matthew Thouvenin, Cadet Companies A, B, and C were established.[59] With the instillation of military cadet training, the brass band became linked to the corps and led the college's uniformed students in both drill and parades. The cadets, with their accompanying band playing high-spirited marches, turned out to be very popular. Even James Fortier, a descendent of the Fortier and Aime families, served as the bugler for cadet company C.[60] Cadets, as well as members of the band and orchestra, wore West Point fatigues with matching caps fronted with a laurel wreath that encircled the acronym "JC." The order of cadet instruction was described in academic catalogues as follows: "Military drill has been introduced in this Institution but does not interfere in any way with the course of studies. Each division is required to drill 2 hours per week. A thorough practical instruction is given by an experienced officer."[61]

Although Marists and laypersons were responsible for all academic instruction, Captain Joseph Kantz, 2nd Separate Co., Infantry, oversaw military training.[62] Under Kantz's instruction, the three cadet companies grew in both size and ability. According to the *Jefferson College Record*, "The military organizations of the college, under the very able direction of Capt. Jos.

Kantz are attaining a high degree of proficiency. These companies are not so numerous as last year, as it was thought best to exempt the forty members of the Brass Band from the regular drill. The companies are now more easily handled, however, and Capt. Kantz will be able to get even better results than in the past. The captain is quite proud of the showing the boys make, and expects great things of them before the end of the year."[63] As renown for St. Mary's Jefferson College increased, enrollments rose to 141 and later 175 by the end of the nineteenth century.[64] To publicly exhibit the cadets, the college held a competitive drill between the three corps on May 21, 1905. Captain W. D. Gardiner of the Washington Artillery judged the event. Gardiner indicated that the competition proved particularly difficult to evaluate due to the high quality performance of all military companies.[65] Interestingly enough, historians have labeled the implementation of college cadet training as a means to enhance discipline, but the 1912–1913 St. Mary's Jefferson College catalogue explains that discipline was chiefly enforced through "religious principles and practices . . . gentle reprimands and charitable correction, by frequent appeals to the student's conscience and sense of duty," not military training.[66]

In 1918, the cadet corps was reorganized to become a Student Army Training Corps (SATC) unit. The US involvement in World War I prompted the federal government to fund formalized military training programs at colleges and universities across the country. However, St. Mary's Jefferson College did not join until the new acting president, Father Richard H. Smith, realized that several state institutions of higher education possessed SATC units. Smith worried that the lack of federally supported military curricula might negatively influence public support and thereby hinder student enrollments. The college president acted quickly and contacted the war department. As a result, the Marist college corps was registered as an active SATC battalion. Smith fully backed the program, and US army officers were sent to train cadets. As the program grew, positive reports about the St. James Parish corps reached the war department. The SATC program attracted greater attention, and attendance increased, if only slightly. The Marists certainly desired additional students and put monies that accompanied the military affiliation to good use: "The government recomposed the school well and promptly paid all the outstanding bills. Much prestige accrued to the school in the eyes of the public and the government."[67] Such American, patriotic duties were important for both students and locals, as several sons of Louisiana families had been called away to fight in battles across the Atlantic. Before the "Great War" ended, many young Louisianans had died oversees. One former student of St. Mary's Jefferson College, Henry J. Plaisance, served as a private in the 9th Infantry, 2nd Division.

Sadly, Plaisance, at the age of twenty-three, succumbed to wounds inflicted while in France. He passed away a mere thirty-eight days before the war ended. His body was returned to his hometown of Larose, Louisiana, and entombed in Holy Rosary Cemetery.[68]

In 1921, three years after the signing of the Treaty of Versailles, the US War Department issued a certificate of appreciation for the establishment and maintenance of an SATC unit at St. Mary's Jefferson College. Although the document referred to the institution as "Jefferson College" (a common occurrence throughout the late nineteenth and early twentieth centuries), the document heaped praises on the institution for its nationalistic zeal.

THE UNITED STATES OF AMERICA
To all who shall see these presents, greeting:
This is to certify that
Jefferson College
in a spirit of patriotism and of devotion
to country rendered efficient and loyal
service in connection with The World War
through the establishment and
operation at that institution of a unit of
THE STUDENT ARMY TRAINING CORPS
Given at the War Department, District of Columbia this twenty-second
day of November, one thousand nine hundred and twenty-one.
The Adjutant General's Office.[69]

Student organizations like the brass band, college cadets, and debate team were essential, as the Marist college was far removed from busy, entertaining cities. If anything, these activities countered the adage that "idle hands are the devil's workshop" by remedying negative behaviors attributable to boredom. With the inclusion of clubs and organizations, St. Mary's Jefferson College students soon demanded additional American recreations: football and baseball in particular.

College football was fast becoming a staple on various American colleges and university campuses. To appease students, the Marists followed suit. Donning school colors of blue and gold and referring to team members as the "Blues," fall athletic seasons commenced with a football game against either Louisiana State University or Tulane University.[70] As the 1921–1922 academic year commenced, a wooden grandstand with a seating capacity of 700 was constructed.[71] Fellow students cheered athletes with rousing chants of "boom a-lacka, boom a-lacka, wow, wow, wow; chick-a-lacka, chick-a-lacka, chow, chow, chow; boom a-lacka, chick-a-lacka, who are we? We are the boys of old J.C.!"[72] Disappointingly, the 1921 football season resulted in losses against Loyola University New Orleans, Southwestern Louisiana Industrial Institute (present-day University of Louisiana at Lafayette), Spring Hill College in Mobile, Alabama, and St. Stanislaus College in Bay St. Louis, Mississippi. Nevertheless, school spirit remained high. Ordering their peers to ignore all losses, the student editors of the *Jefferson College Record* announced, "When next season rolls around and all the boys return with some new recruits, things are going to 'hum,' and look out Southwestern, for we promise you one of the strongest teams in the entire South."[73] Recovering from previous defeats, the "Blues" almost triumphed over Louisiana State University, losing by a score of 23–22. The margin was too close for comfort for LSU, however, and as a result, the state university dropped St. Mary's Jefferson College from its yearly football schedule.[74]

The college baseball team was also a student favorite and was more successful than the football squad. Overlooking a poor start to the 1922 season, it was reported that the team "came back with some of the old Jefferson spirit which enabled them to take a majority of the games on the schedule." Baseball skirmishes between Louisiana State University and St. Mary's Jefferson College provoked heated rivalries. Students chided the Baton Rouge institution, indicating that "we have at last succeeded in crossing bats with the University team and found out they weren't 'the whole cheese.' We know some 'mo' better players than they.' Their victory is due to fortuitous circumstance. It was not a matter of outplaying us, as we put up as good an article of ball as they. The umpire's decisions on balls and strikes however worked havoc with us by discouraging the team."[75] On occasion, the Detroit Tigers professional baseball team visited south Louisiana during training seasons and played the Marist college athletes for the entertainment of students and locals.[76] Baseball and football aside, the less popular basketball team surpassed the aforementioned teams in wins but never outstripped either in attendance.[77] No matter what the sport, student reporters from the *Jefferson College Record* attended all games to capture and transform lively events into amusing narratives.

Including athletic displays and cadet performances, the *Record* increased transparency between the college and its surrounding milieu. This magazine was originally implemented to improve students' literary prowess, connect alumni to their college, increase public and financial support, and provide a modicum of entertainment. In time, the *Record* became the campus "voice" by chronicling the daily happenings of professors and pupils alike. The magazine not only included short stories and current events but also featured pictures of the campus, its buildings, athletic events, and general details related to the college itself including but not limited to "moderate" fee and tuition costs. The *Record* contained life advice and humor columns under the heading "Just for Fun." The latter tongue-in-cheek section was solely devoted to comedy with pieces of verse such as "Mary had a little lamp, 'twas filled with kerosene, she blew right down the chimney, and vanished from the scene" and "'No smoking allowed,' the conductor exclaimed to friend Mike as he sat in the car. 'I'm not smoking aloud,' Mike calmly replied; then he silently puffed his cigar."[78] The *Record* also published student fiction and verse. In an ode to his institution, one student recollected days spent at the Catholic college with warm affection:

> Hail venerated pile! Remote from worldly din,
> Undisturbed Wisdom, on the throne of truth,
> There the willing minds of light yet glorious youth
> In thy Grecian cloisters have I sometimes strayed.[79]

Having eulogized the college as a "pile" (a large, stately group of buildings),[80] the author proceeded to recount the institution's religious aspects by featuring the gothic chapel constructed by Valcour Aime:

> My pilgrim steps then guide me to the hallowed shrine—
> The chapel quaint beloved—that Holy House divine,
> Where 'neath the Presence of My Savior-Lord beheld,
> I learnt to pray that I, from sin withheld
> Should faithful serve the best of masters with delight—
> My soul's sweet Lord yclept [*sic*] by him the Life and Light.[81]

The ode goes on to detail rows of large oak trees, sunsets over the Mississippi River, the campus's physical locale once home to Native American tribes, and the affirmative influence the Marist college yielded on its wards.[82]

Not only did the *Record* feature student writings, achievements, college news, and photos, it also displayed supporters who paid for print commercial space. Advertisements included attorneys, athletic supply shops, auctioneers, booksellers, clothiers, confectioneries, dentists, furnishers, grain and rice distributors, grocers, hardware and munitions sales, hatters, metal works, oil refineries, publishers, and plumbers. As well, print space was reserved for the local Academy and Convent of the Sacred Heart, Marist College in Atlanta, Georgia, and the Marist-administered All Hallows College in Salt Lake City, Utah. Other supporters included a range of city, parish, and state businesses including the celebrated Hotel Monteleone, where many

St. Mary's Jefferson College business meetings were conducted. In closing, each edition of the *Record* described the college's curricular structure, its Catholic administration, as well as nearby railroad and river access.[83]

The college needed both rail and river transportation to provide adequate food supplies for a host of hungry boys and young men. In the past, cooking supplies arrived by steamboat alone; however, a new train station had been built three miles from the campus, thus enhancing food delivery by way of the Yazoo and Mississippi Valley Railroad. Riverboats delivered ice, but trains transported beef and other goods such as grain and coffee. Grain deliveries permitted bread to be baked on site while vegetables were still grown in neighboring fields. With a growing number of active students, it became a matter of necessity to enhance both the quality and quantity of comestibles served in the recently constructed Thouvenin Dining Hall. Named in honor of former president, Father Thouvenin, the building was a mirror image of Alumni Hall. The influx of edible provisions diversified campus meals. Typically, college attendees and employees enjoyed a breakfast of biscuits, milk, and syrup. Dinners and suppers consisted of potatoes, roast meat, and vegetables, or beans and rice, a local favorite. In the spring, meals were supplemented with boiled crawfish and shrimp collected from the Mississippi River.[84]

With an enhanced physical plant, improved dining conditions, additional scientific course offerings, and a variety of secular American sports and social organizations, the Marists, concerned about the external perception of their religious institution, published advertisements that described the Catholic principles that undergirded St. Mary's Jefferson College.[85] "The aim of the College is to produce the highest type of Christian manhood and citizenship. . . . It is not enough for any institution to turn out trained minds, but its students should go forth with true principles, with a determined will to do what is right because it is right, and with the fear and love of God in their hearts. This is an all-important consideration at Jefferson. Being Catholic in principle, as well as in tone, the moral atmosphere of the College is admirably well adapted for the formation of true, manly Christian character."[86] These sentiments reflected the diversifying Catholic population of St. James and surrounding parishes as well as the decreasing number of French Creoles. In 1912, there were over 2,000 Creole Catholics under the spiritual care of St. Michael's Church. Italians were listed as "very few" while 200 other persons were listed as non-Catholic. More than half of the Catholic population was reported to "understand" English. In 1914, roughly 1,800 Catholics were listed as "Creoles." By 1915, 2,600 Catholics (without ethnic demarcation) were listed and approximately 2,200 were described as white. The other 400 were listed as Negroes. However, in 1919, greater demographic detail was provided. Reports compiled by St. Michael's Church clergy list roughly 1,600 parishioners. One hundred were described as Italian while 1,500 were labeled as both "American and Creole."[87] The Marists, appealing to their ethnically changing, yet religiously steadfast surroundings, printed the following statement regarding St. Mary's Jefferson College: "Parents who set a proper value on the souls of their children, who want a safe as well as a progressive school for their boys, will make no mistake in sending them to an institution which . . . has held this aim ever before it, and is succeeding beyond expectation in its first mission."[88]

In addition to enhanced religious fervor came augmented tuition. As the twentieth century commenced, the cost of goods, services, and taxes increased. To compensate, tuition at St. Mary's Jefferson College rose to $250 per academic term.[89] Regardless, enrollment continued to climb. By 1904, student numbers had reached 175. This was the largest population since the 1842 fire that decimated antebellum Jefferson College.[90] Enrollments, however, plateaued as students took note of pragmatic job-related curricular offerings at state-funded, research-based institutions and applied elsewhere. In an attempt to invigorate attendance at the Marist college, students were "admitted at any time during the season, but [were] charged for the whole month in which they enter."[91] Of those who enrolled, their ages ranged between fifteen and twenty years. To further bolster student numbers, the Society of Mary transported boys and young men from Mexico and South America. During the Civil War, members of the southern planter elite had traveled to Latin countries in an attempt to escape the ravages of war and Reconstruction. But when racial egalitarianism took hold in South America, planters returned to their once Confederate states now under the rule of Jim Crow segregationist law.[92] Alongside planters, potential students from Mexico and beyond journeyed to the American South in pursuit of formal education—an egress motivated by instruction rather than racial divide. As the Marist college passed through the early twentieth century, enrollments fluctuated between 150 and 200—far from the thousands attending state-supported institutions like Louisiana State University.[93]

Most students at St. Mary's Jefferson College spoke English and varying degrees of French, but with the end of the nineteenth century, Louisiana schools, colleges, and universities were fully vested in the English language.[94] Regardless, students from Mexico and South America entered the college with only their mother tongue. Father Dagneau explained that in order to teach these particular students, Sears Roebuck catalogues were ordered and dispersed under the assumption that Spanish-speaking pupils would associate pictures with English titles. This practice yielded interesting results as some of the Latin students "wrote to the King Bee Tobacco Section for a live queen bee in order to start a colony. Invariably, the Spaniards would politely add a postscript to their orders—such as, 'Please give our best regards to Mrs. Sears and Roebuck.'"[95] At other times, Latin students were paired with Louisiana natives to encourage the use of English among the former. This scheme backfired as English-speaking pupils learned Spanish from their wards faster than Spanish-speaking students garnered English from their Louisiana contemporaries. The Marists, however, did record some victories. Some of which were rather humorous: "One afternoon, a Father passing by heard some strong voices cursing one another in real old American profanity. In looking up into a tree, there were three little Spanish boys having it out. He could not help but remark, 'At least they are learning to talk English.'"[96] Although immigrant students interacted with their peers and were exposed to social and educational opportunities provided by St. Mary's Jefferson College, their presence only slightly contributed to the total student population. Stressed to compete with contemporary colleges and universities, the Marists persistently advertised the affordability and piety of their institution; however, the years of positive school spirit and instructional adaptation were fast coming to an end as an assortment of financial hardships plagued St. Mary's Jefferson College throughout the 1910s and 1920s.[97]

Hardships Abound

Concerned about the pecuniary stability of St. Mary's Jefferson College, the Marists, in an attempt to augment revenue, enhanced the scientific branch beyond Blenk's original adjustment to better match that of other local institutions. Accordingly, the scientific course of study was expanded to last four years and terminated with a bachelor of science degree. As well, a campus powerhouse was constructed to generate electricity by means of diesel engines that animated lights and a motion picture projector, the first one in St. James Parish. Father Henri Georges, the college mechanic, not only maintained the powerhouse and generators; he also used the latter for instructive purposes in his electrical engineering class.[98] Dispensing with kerosene lamps and wax candles, the college marked a technological milestone with the instillation of electric lights in an attempt to embrace modernity. Unfortunately, the powerhouse caught fire not long after its construction and the college went dark. Unperturbed, Father Georges located replacement generators, and the college lights soon glowed once more.[99] With the enlargement of the scientific course of study, academic catalogues informed the public that such curricula would "prepare the student to enter the Technological Division of our universities and qualify him to secure advanced standing in any of the different courses of Engineering. For this purpose, Latin and Greek are eliminated from the program, and the time given to these two languages in the Classical Course, is devoted to the Sciences, Mechanical Drawing and a second modern language, either German or Spanish. Special attention is given to laboratory work. The students are aided by well-equipped Physical and Chemical Laboratories, where they are required to make personal experiments and deductions."[100] Along with expanded technological course offerings and diesel-fueled illumination, the removal of ancient languages was yet another attempt to match job-centric educational trends at secular, state-supported universities.

To house the college's broadened scientific mission, the 1906–1907 academic catalogue detailed plans to construct an expansive science hall. This new building would accommodate the anticipated growth in student numbers and house all labs and associated equipment.[101] In 1910, the Marists accepted architectural plans from the New Orleans firm of Andry and Bendernale.[102] Father Smith had the campus renovated in preparation for the building's construction by improving landscaping and purchasing new furnishings for all facilities. To showcase the campus grounds, Smith had the high brick wall, which had fronted the River Road property since the antebellum era, "lowered by half."[103] As planned, the science hall's cornerstone, engraved with an overlapping "M" and "A" indicative of the Marist order, was laid on March 6, 1911. The building, designed to resemble the college's Greek revival architectural style, was named after the former president, Archbishop Blenk. The Marists hoped that this new building, devoted to empirical exploration and experimentation, would promote the college as a Catholic *university* firmly invested in scientific study.[104] In a 1912 letter, President Smith informed the archbishop that the decision had been made to name the building Blenk Science Hall for several reasons:

> The very idea of the project was yours. You may not remember, but about two years ago, in your home, we were discussing affairs of the College and the necessity for improvement. You finally made the suggestion that I should do as the Jesuits had done at Grand Coteau and the Benedictines at Covington—borrow the money and give a mortgage on the College property. The idea had never come to me and at first I despaired of its realization feeling sure we could never obtain permission to make the loan. It was only because I had you behind me that, after some weeks deliberation, I decided to make the attempt and place the project before the superiors. You know with what difficulty the permission was finally obtained and I know now with absolute certainty that it never would have been

> granted but for your enthusiastic and persistence in promoting it. It is no more than the truth that, if the science hall is today a fact, it is due to your advocacy and your influence. . . . Now, what claims have I to have it called after me? I neither began it nor carried it through; and certainly did not pay for it. It will take at least ten to twelve years to do that and I am likely to be changed at the end of this term and the burden of paying will fall upon another. You showed your unselfishness when you built and paid for the Alumni Hall with money you saved yourself and yet modestly attributed the honor to the Alumni. Yet now you wish me to take to myself an honor I in no way deserve. No, dear Archbishop, not only is the honor rightly yours, but Jefferson owes it to herself to give it to you. For all the past work you did here, for your successful efforts in getting the hall; for the previous financial assistance you have given us towards paying for it; as a requital for your modesty In withholding your name from the Alumni Hall; on account of the fact that your long connection with the college and your . . . services in its behalf have no monument here named to commemorate them; to have all, because of the distinguished honor you have conferred upon Jefferson [College] by the high position you will—Jefferson can do no less than show her appreciation of all this . . . ; and I am sure that all the old boys will hail the name as the most fitting and deserving that can be bestowed upon it. So, dear Archbishop, you must consider the matter settled. The final announcement will be made (with the reasons for it) at the Alumni dinner on June 18th next, and we shall confidently expect you to be present. I hope you are in the best of health.
>
> Affectionately and devotedly yours,
> R. H. Smith, S.M.[105]

Smith regularly conferred with Blenk about the college's affairs and made a point to invite his fellow Marist to all commencement ceremonies. However, the busy archbishop rarely attended. Smith wrote to Blenk so often that he apologized for "trespassing" on the prelate's valuable time.[106]

Smith understood that the archbishop split his attention between St. Mary's Jefferson College and the newly opened Loyola University New Orleans. Founded by the Society of Jesus in 1912, Loyola University furthered a long-standing tradition of Jesuit higher education in Louisiana. In 1837, the Jesuits established St. Charles College in Grand Coteau (closed 1922). Members of the religious order opened the College of the Immaculate Conception in New Orleans's Faubourg St. Marie (1849), and in 1904 they founded Loyola College in the city's uptown Garden District (both institutions closed in 1912).[107] With the creation of Loyola University, Blenk encouraged its founder and first president, Father Albert H. Biever, to advertise the new Jesuit institution as a leading Catholic university for the entire American South, not just Louisiana.[108] As time passed, Blenk's correspondence with the Marist college decreased while his communications with Loyola University steadily increased.

As diocesan support waned, debts grew. In all, the cost to improve the campus was an enormous financial gamble that turned a poor profit. In 1912, the Marists reported that the construction of Blenk Science Hall had incurred considerable debt. Consequently, the fathers and brothers made an appeal to "all who believe in the principles of Catholic education" for assistance.[109] In particular, the Marists directed their pleas to the college's graduates, asking them to give as they were able or to remember their alma mater in last wills and testaments. Alumni were informed that their monetary gifts would ensure the college's longevity and safeguard students from "the dangers encountered in so called non-sectarian institutions."[110] To combat escalating debt, an alumni society was formed to "promote the general welfare of Jefferson College; to insure a more perfect bond of friendship among the present Alumni, and to cultivate the friendly feelings of future graduates."[111]

As debts continued to increase, the institution was slowly stripped of its remaining Creole heritage and took on the form of an American college with an associated high school. The 1918–1919 academic catalogue informed prospective students that in order to gain admittance to the college department's freshman class, they must provide documentation of high school attendance recognized by the Louisiana State Board of Education, a diploma, and a reference of good character—requirements heretofore overlooked, as most students matriculated through the preparatory department.[112] The institution's former preparatory department had, in an attempt to emulate national secondary education trends, been converted into the "high school course of study" and issued diplomas under the name "Jefferson College High School."[113] To foster an unambiguous academic divide between the high school and higher education divisions, the Society of Mary solicited the collegiate department as being fervently based on scientific inquiry while heaped with age and scholastic prosperity. An advertisement printed in the *Morning Star* declares, the college was "one of the oldest Institutions of Higher Education in the South. Delightfully situated on the Mississippi River, 50 miles above New Orleans. Stately Buildings, New, Modern School Furniture, Beautiful Grounds, Finely Equipped Gymnasium, Swimming

Pool and Bathrooms, Healthy Climate, Fine Campus, Classical, Scientific, Commercial and Preparatory Departments. Empowered to grant diplomas and all collegiate degrees. A magnificent Science Hall was recently completed and furnished with splendid modern laboratories for all the Sciences. Terms moderate. Session opens Sept. 5th, 1918. REV. R. H. SMITH, S.M., President."[114]

The preparatory department's conversion into a formal high school reflected changes in state-supported education. By the start of the 1920s, public high schools had become the norm. As a result, student enrollments at the Marist college suffered. Instead of paying to board at the restrictive Catholic college, south Louisiana's youth remained at home with their parents and attended schools that admitted students free of charge.[115] Similarly, Louisiana State University, Loyola University, Southwestern Louisiana Industrial Institute, and Tulane University regularly siphoned applicants from St. Mary's Jefferson College.[116] As the federal government separated higher education from primary and secondary schooling via accreditation measures and associated policies, college life in general became more relaxed with decreased administrative oversight; however, rules remained relatively strict in St. James Parish. After World War I, general unrest arose at the Marist college over the requirement for all enrolled to remain on campus. As Dagneau explains, "Confinement was irksome to the new generation. While Father Smith was president, much of the ancient traditions and customs remained in force." The Marists continued to rely on a mantra of "spare the rod, spoil the child." Ever conscious of the negative spiritual ramifications the outside world might inflict on their wards, priests insisted that students stay on site, attend Mass, and make regular confession. They also censored student letters from friends and family to ensure conversational propriety, as the example below suggests:

> My Dear Mama,
>
> I have just received two letters from my little sisters although I haven't received yours, even though they told me that you also wrote me on the same date. The two letters have come opened by the censor and I believe that the delay with yours is due to that. I hope that it arrives soon. Through the letters from my sisters I know that the weather in Macacona [Costa Rica] has been very good and that you haven't missed being able to enjoy your activities. All of this gives me great pleasure.
>
> Angelica tells me about our niece's misfortune, poor thing! I can imagine the blow they received. Write to them for me, because

> it is absolutely impossible for me to write them now. Like I told you previously, we have retreat (in 20 minutes) in order to finish it Saturday afternoon.
>
> Last week I wrote to you and when I finished my letter to you I immediately went to confession, and the following day I went to communion and I prayed to God for you. Tomorrow, like I also promised you in a previous letter, I will receive communion especially for you. You can imagine the sadness I feel upon finding myself so far from you on March 28th, but tonight after 12 from here I will send you all my soul and all my affection, a kiss that brings you all my best wishes of happiness. I hope that next year I have the good luck of reuniting with you all, although it would be for a short time. Don't forget that I always constantly think about the promise that you all made to come and visit me. Aurora always speaks about this in her letters. About Angelica, she was not sure that she could come, but if not she would have a good reason if it weren't possible. It is necessary not to be blinded by affection. I understand that there are certain things in life. One shouldn't be selfish. A world of happiness I always hope for all of you and Macacona. In communion I will pray on your behalf so that my happy wishes for you will be complete. I hurry to the post office.
>
> On Sunday I will write a longer letter for you.
>
> Again, I hope that you receive all the affection from the son who adores you,
>
> Until next time,
>
> Teddy[117]

Although Father Smith maintained rigorous discipline, his Marist peers and members of the lay faculty commended his fifteen-year administration. When Smith resigned in 1920, students wrote poems praising his tenure as St. Mary's Jefferson College's president. One student, commenting on Smith's dedication to growing the campus, stated, "Father Smith . . . You've built a monument to God."[118]

With Smith's departure, Father Peter Quinn was selected to become the new chief administrator. In a letter to Archbishop Blenk's successor, John William Shaw, Quinn wrote,

> Your grace, this evening I received a letter from Father Provincial informing me that Father General had appointed me President of Jefferson College. Knowing for some time past that my name was mentioned for the position and feeling that the responsibility was

> too great for me, I did all in my power to have someone else named, but all to no effect. The Provincial tells me that I must accept the position in virtue of obedience, so there is nothing for me to do but submit. I trust that the appointment meets with your approval, and I hasten to beg your blessing on my work and on all connected with the college. In the person of your Grace, I feel sure that I shall always have not only a real friend but a kind Father ever ready to assist me and to direct me in the proper fulfillment of my arduous duties. It will be a real pleasure for me to allow the college fathers to assist the priests of the diocese whenever and wherever possible. Assuring you that you shall always be welcome at Jefferson College and begging your prayers and blessing, I am obediently yours,
>
> P. F. Quinn, S.M.[119]

Shaw responded with a short, positive note that acknowledged Quinn's humility and concluded that the Marist "may rest assured that the Good Lord will abundantly provide."[120] With the archbishop's blessing, Father Quinn proceeded to modernize the college's regulatory policies and hired a cadre of young American laymen to gradually shift the college away from outdated disciplinary traditions. Accordingly, Quinn softened institutional rules and allowed students to come and go as they pleased.[121]

Subsequent to Archbishop Shaw's message, in 1921, a group of laymen approached the Marists to request that the college host spiritual retreats. Although the Jesuits offered retreats at St. Charles College, laymen complained that the Grand Coteau campus was difficult to access due to its distance from New Orleans. St. Mary's Jefferson College, in contrast, was much closer. The Marists wrote to Shaw requesting permission to sponsor religious retreats, and in August of 1923, the first retreat was held at the St. James Parish college. To show his support for the Marist-led retreat, Shaw attended and said Mass.[122] These silent retreats would become, in time, the perfunctory purpose of this once secular Creole college turned American Catholic institution.

Beginning of the End

As St. Mary's Jefferson College entered the 1920s, federal funds derived from the SATC program partially addressed the science hall debt; however, the college needed additional monies, and the Marists again pressed the college's alumni for support since several graduates had achieved notable careers with substantial incomes. Included among these alumni were Lynn LaGarde, chief engineer of the Coca-Cola Company, as well as a host of Louisiana bankers, lawyers, legislatures, merchants, and physicians.[123] Still, the college never acquired noteworthy donations. Academic catalogues from the 1920s emphasized the importance of supporting Catholic colleges and universities in a short essay titled "To the Friends of Catholic Education." The plea informed patrons that St. Mary's Jefferson College received no state funds but instead relied completely on private donations and tuition dollars. "Now is the time," the Marists implored, "for our well-to-do Catholics to render a transcendent service to church and country by aiding, according to their means, in promoting the onward march of our Catholic institutions."[124]

The essay concluded by proclaiming that only two sizable donations had ever been given to the college since its Civil War recreation as a sectarian institution. The first was the 1864 shareholding gift of Valcour Aime. The second was a financial endowment from E. H. Miles in 1909, which was used to add a sacristy to the campus chapel. According to the Marists, the college survived by their efforts alone.[125] On October 19, 1926, the *State Item* reported that twenty descendants of Aime had received their education at St. Mary's Jefferson College free of charge—a lingering stipulation of the sugar baron's nineteenth-century donation.[126] Although the newspaper painted the Marists in a positive light, the college's twentieth-century priests and brothers grieved over the loss of twenty tuitions. In Father Dagneau's words, "This promise became . . . an embarrassment and a handicap because as the family spread, the burden of free board, lodging, and tuition became too great a burden and a drain upon the school's finances."[127]

In addition to dwindling funds, the Americanized Marist college's enrollment further shrank in spite of the increased religious diversity of the surrounding parish. In 1923, the clergy of St. Michael's Church listed 1,560 Catholics under their spiritual care; however, the ethnic makeup had notably altered. Italians numbered a scant sixty-eight while "colored non-Catholics" were estimated at over a thousand. No French Creoles were listed in the church's records despite the presence of a few remaining ancienne population families.[128] Diversified ethnic and religious demographics notwithstanding, the public education system of St. James Parish remained racially polarized. In the public sphere, social-class demarcation had taken second place to race as separate schools for black and white children opened in accordance with Jim Crow laws.[129] Accordingly, St. Mary's Jefferson College remained a "whites only" institution, though darker-skinned students from Mexico and South America were admitted. Even though alumni and planters provided little monetary support, they did not want the only Catholic college in the sugar parishes to enroll African Americans.[130]

As the economic and social makeup of the gold coast continued to change, the Marists made several missteps in their attempts to advance the college. Though President Smith had ensured the academy's accreditation with the Catholic Education Association, he "either ignored or failed to notice" the enhanced standards promulgated by the Southern Association of Colleges and Schools (SACS). As a result, the college was never incorporated into the accrediting agency, and its academic standing among peer institutions was called into question.[131] Lacking SACS accreditation, enrollment and tuition revenue decreased yet again. In the mid-1920s, both clerical and lay instructors made suggestions to sustain the college. Proposals included student recruiting trips, motion picture advertisements, and relocating the campus to the New Orleans suburb of Metairie. However, none of these suggestions came to fruition.[132] A ladies' auxiliary club and college fair that featured amateur vaudeville performances, maypole dances, music ensembles, popularity contests, and sporting events were organized but raised little money.[133]

In a final endeavor to garner capital, the Marists decided to drill for oil. Before Father Smith vacated the presidency, he purchased two acres of land in Caddo Parish from the Louisiana Real Estate and Development Company.[134] Smith hoped these acquisitions would lead to the discovery of either oil or natural gas. With the 1901 detection of oil in Jennings, Louisiana, numerous companies and private planters abandoned the production of "white gold" in favor of the extremely valuable "black gold."[135] If oil, gas, or both were discovered, property owners would receive one-eighth of the oil produced and $200 per year for gas.[136] In 1918, however, both plots of land purchased by Smith proved too small to erect a derrick. If the lots had adjoined one another, the construction might have been accomplished, but as they were separated by acreage owned by Caddo Parish residents, drilling could not occur.[137]

In the midst of enrollment and financial decline, Father Quinn stepped down as St. Mary's Jefferson College's president and was replaced by Father Alcime Cyr. Cyr remained in office for one year but is better remembered for later becoming the first American Superior General of the Society of Mary. Following Cyr, Father Joseph Petit was selected as the college's new institutional leader and was immediately saddled with an abysmal set of circumstances. If an ailing budget was not bad enough, conflict erupted between the American and Marist instructors regarding the institution's liberal arts curriculum. The former wanted to substantially alter the academic structure to better match twentieth-century instructional standards, while the latter wished to remain true to aged European teaching and disciplinary practices. In an attempt to quell arguments, Father William Maquire, a graduate of The Catholic University of America and an educational researcher from New York, was tasked to assess the situation and provide suggestions for improvement. Maquire's efforts, however, failed to reconcile the faculty's disparate curricular and regimental ideals. Maquire also discovered that several thousand dollars in unpaid tuition existed. As there was little hope to collect all outstanding tuition debts, both budgetary and curricular issues were declared insurmountable.[138] Tired of instructional arguments and administrative difficulties, Petit resigned in 1927. In February of that same year, the *Times Picayune* announced that Father George Rapier would begin his second administration as president.[139]

As the college unremittingly sank in financial standing and student attendance, a ruinous "disaster threatened the entire countryside."[140] Both Dagneau and historian John M. Barry provide elaborate details concerning the event that ultimately ended the Society of Mary's educational enterprise in south Louisiana: "It was in 1927 and the Mississippi [River] rose to unprecedented heights. Water lapped over . . . levees and forecasted ruin to the entire community. Every able man and boy worked day and night to bolster up the levees. Priests and students of . . . [St. Mary's Jefferson] College filled bags with earth and stacked them up against the encroaching waters."[141] Two weeks prior, numerous crevasses were plugged while "hundreds of people were climbing the levee to see the river. It was angry, wide, high, and fast, swirling in whirlpools, the current sweeping logs, lumber, the bodies of mules and horses past. In some stretches it had risen higher than the levee and was contained by planks backed by thick walls of sandbags."[142]

Rains that fueled the destructive flood began in August of 1926 and continued through early 1927. On Good Friday, April 15, the downpour increased. New Orleans alone experienced fourteen inches of rain, which shut down pumps that kept the below-sea-level metropolis dry. Believing city levees would succumb to river pressure, officials decided to create a series of crevasses south of New Orleans, thus sparing urban upper-class properties but destroying homes and livelihoods of poor, rural Louisianans.[143] On April 29, levees in St. Bernard and Plaquemines parishes were dynamited. Levee destruction continued for ten days using thirty-nine tons of dynamite. The ensuing explosions released 250,000 cubic feet of water per second and displaced over 10,000 people.[144]

Prior to the devastation, Acadian farmers were forced from their homes, leaving many of their belongings behind. Lyle Saxon poignantly illustrates the sad scene: "While their parents loaded the mattresses and tables on wagons, the children gathered their dolls and their pet kittens into their arms and looked back at their homes. There was not much time for sentiment. . . . They had to go."[145] Though some Acadians returned after the waters receded, most permanently vacated the sugar parishes. Those who went back found their homes and fields ruined.[146] The Louisiana legislature promised monetary compensation for personal losses; however, the majority of rural citizens never received payment. Originally, it was estimated that property and real estate claims would not exceed six million dollars. In actuality, the total surpassed thirty-five million dollars. In the end, lower-class planters bereft of capital resented wealthy Louisianans and state politicians alike. To make matters worse, levee blasting proved needless. As Jim Bradshaw explains, "Adding to their [rural Louisianans] anger, a natural breach of the levees subsequently eased pressure on the New Orleans levee; the blasting had been unnecessary."[147] The flood not only decimated the sugar parishes, it forced several New Orleans banks to close. Gambling on a profitable sugar production year, bankers had extended copious loans to cover additional harvest expenditures. Due to large-scale crop loss, many banks and sugar dependent businesses closed.[148] Likewise, personal finances plummeted and, as a direct result, student enrollment at St. Mary's Jefferson College decreased to a dismal level.

Areas surrounding the Marist college were spared from the raging floodwaters, but neighboring parishes were spoiled, and the effect on institutional attendance was catastrophic. All too soon, the Marists realized that St. Mary's Jefferson College could not survive under such circumstances as the student population shrank to ninety-one. Boys and young men who had previously attended the St. James Parish institution enrolled in publicly funded Louisiana universities or were sent to St. Stanislaus College in Bay St. Louis.[149] Of those who returned, most hailed from Louisiana, but one came from Massachusetts, four from Mexico, three from Mississippi, four from Wisconsin, and one from Virginia.[150] On August 8, an advertisement in New Orleans's newspapers described the college's admission requirements and a condensed curricular structure limited to "eighth grade through high school courses, college preparatory—business courses."[151] The institution's final academic catalogue reveals conditions at the dying college. Indeed, the small 1927 booklet contained only twenty-five pages and very few class descriptions. The final page closed with, "Classes Resume September 6, 1927"—an assertion that was never realized.[152] No longer functioning as a legitimate institution of higher education, St. Mary's Jefferson College had become little more than a secondary school with mercantile classes. As south Louisiana slowly recovered from the flood, the Society of Mary was forced to close their wounded academy. In a letter to Archbishop Shaw dated August 15, 1927, Father Rapier conveyed the terminal news: "Your Grace, I bring you bad news: our Superior General finds that Jefferson College should be closed; and our Father Provincial has asked me to announce it. This goes to you in the very first mail after receiving our orders. Begging your blessings, I am, Obediently yours in Christ Jesus, Father Rapier, S.M."[153] At the bottom of the letter, Shaw wrote that he had not been consulted regarding the decision to discontinue but expressed his "deep sympathy on the closing of Jefferson College, one of our oldest education institutions. Fiat voluntas Dei!" (May God's will be done).[154]

Three days later, buried on page sixteen of the *Times Picayune* under the tabloid's "Fact, Fiction, and Fashion" section, a brief obituary for the former Creole college was hemmed in between a "Dorothy Dix's Letter Box" advice column and a "Home Notes" décor editorial. The notice states: "JEFFERSON COLLEGE WILL NOT REOPEN: Institution to Close Doors . . . The Rev. Father George Rapier, president of Jefferson College, has announced that the institution will not open for the session of 1927–28. Due to economic conditions, the Marist fathers say they have no alternative but to suspend operations. July 12, 1864, the Marist fathers took possession of Jefferson College. During these sixty-three years they have sent out graduates who have filled with honor and distinction some of the highest political offices of the state."[155] The once-illustrious college, founded to educate the sons of sugar barons, had closed for a fourth and final time.[156]

Although St. Mary's Jefferson College had officially shut its doors, Father Henri Georges was named acting president of the institution's governing board and Marist corporation on November 21, 1927. This was done so that a priest could represent the religious corporation and oversee the Louisiana property.[157] Less than one year later, on July 26, 1928, Georges was reassigned as pastor of St. Joseph's Church in Paulina, Louisiana, when the Society of Mary resolved that the corporation should be terminated. The Marists reasoned that Georges, "no longer president of Jefferson [College] as it no longer exists, would be able to provide Paulina with some amount of religious governance in the wake of the great Mississippi River flood while serving as a representative for the former college in case a potential buyer emerged."[158] Father Smith, now serving at the Marist province offices in Washington DC, informed Georges, "It remains for me, my dear Father Georges, to thank you for your good work at Jefferson [College] and wish you a happy and successful tenure . . . in your new field of labor."[159]

While several documents recount the final days of St. Mary's Jefferson College, few communicate the depth of compunction felt by the Marists regarding its closure. One such document, a provincial report from 1928, offers a telling statement: "It is no doubt a matter of deep regret to all the Fathers of the Province . . . to see such a venerable institution as Jefferson pass out of existence. None regretted it more than those on whom was placed the responsibility of making the decision which terminated its career. Jefferson had rendered splendid service to the cause of Catholic education in the South. . . . But new conditions in the educational field destroyed its usefulness and there remained no alternative but to bring its existence to a close."[160] Though these melancholic Marist accounts are scarce, alumni and south Louisiana natives voiced their dissent in abundance through letters and telegrams. In one communication E. Pfleger explained to Marist provincial Father N. A. Weber the import of St. Mary's Jefferson College to St. James and adjacent parishes:

> The closing of Jefferson College deprived a certain number of local boys, too poor to study elsewhere, of the higher education and the deeper religious training which would have been an asset to them and an indirect help to the parish. It deprived the district of the distinction which "this great Institution of the South" had reflected on it by its presence and the work in their midst's and likewise ended that uplifting influence, both religious and social, which the College had exercised on a goodly number of workers and parishioners who were in frequent contact with the priests and Catholic boys, without mentioning the money which the College put in circulation in the poor parish. It deprived the Church of the help which the professors would frequently give to their confreres at the Church, of the eloquent sermons they used to preach there, and of the additional pomp and solemnity by which the presence of singing of the student body would enhance the ceremonies on special occasions.[161]

After the Marists shipped all records pertaining to the 429 students who graduated between 1871 and 1927 (not to mention the copious documents pertaining to those who studied at St. Mary's Jefferson College but never completed their degrees of study) to the Marist seminary in Washington, alumni, college and university admission officers, employers, and family members contacted the religious order for academic information such as transcripts and graduation reports.[162]

Many St. Mary's Jefferson College alumni held prominent careers—particularly in the field of education. In 1932, R. P. Lowry, the St. James Parish superintendent of education, indicated that he, as well as the principals of Lutcher High School, Romeville High School, Vacherie Grade School, and White Hall Grade School, were graduates of the former institution. As Lowry stated, "These men [alumni] are doing outstanding work [when compared to] the best that can be shown in the State and Nation. This is not merely a statement formulated under the stimulation of the moment—they are cold facts. To go further and [detail] the achievements of J. C. students . . . in private and public life . . . would be more than the writer could assume at the present writing."[163] Alumni were not alone in their dismay. Catholic authors and clergy were also shocked by the college's closure. Roger Baudier stated, "Jefferson College under the Marist Fathers contributed a substantial share in the task of rearing sterling examples of Catholic manhood and helping establish a body of staunch Catholic laymen who began to wield a beneficent influence in the ranks of business and social life of the state. It is one of the regrettable incidents of the history of the [Catholic] Church in Louisiana that Jefferson College was obliged after more than 60 years of splendid labors, to discontinue its splendid labors on behalf of Catholic education."[164] The recently installed New Orleans Archbishop Joseph Francis Rummel echoed Baudier by describing the supportive roles many St. Mary's Jefferson College graduates played in Louisiana parish and diocesan activities. Rummel, lamenting the Marist college's demise, remarked "that there was a great need and a fertile field for another Jefferson."[165]

Perhaps Rummel did not foresee the forthcoming success of the Society of Jesus and their New Orleans–based Loyola University. Still, the memory of St. Mary's Jefferson College lived on in yet another institution—a fifth life for the old Creole college. Lowry, in a 1932 letter to Father Dagneau, informed the priest, "The old institution of Jefferson College has filled its day splendidly. Why its doors should be closed at the present moment is a puzzle to the writer, but Jefferson has not ceased its spiritual influence. Under the zealous activity of the wonderful organization of the Jesuits that has taken control of the spiritual leadership of our people, only God knows the wonderful work that is being done at the present moment within the walls of Jefferson College. The soul of Jefferson is not dead—its work goes on more effectively than ever."[166] Lowry's letter, though glowing, did not specify Jefferson College's new purpose, the college had reopened as an institution not for higher learning but for spiritual reflection.

In 1931, the Society of Jesus purchased and transformed the college campus into a silent retreat house. The Jesuits dubbed the campus "Manresa" after the Spanish city where St. Ignatius Loyola penned his introspective *Spiritual Exercises*.[167] Even though Father Jean Baptiste Leon Maisounabe declined to purchase the college after the 1842 fire, the Jesuits ultimately acquired the campus in the new century. With the passing of St. Mary's Jefferson College, religious higher education in rural south Louisiana came to a close, but its history, as the only institution built to serve the planter elite yet transformed by fire, war, unstable economics, ethnic rivalries, and religion, does much to illustrate nearly one hundred years of Creole social class decline, the Americanization of student life and instruction, and the resurgence of Catholicism in the sugar parishes. Today, the massive colonnade of white columns remains steadfast amid impressive oak trees and waving fields of sugarcane, while spiritual retreatants walk Manresa's grounds in silent contemplation.

Conclusion

Class and College

> Who could leave this spot [Jefferson College] without carrying away with him a better understanding of the life and culture of the Deep South and how its young men were molded for the responsibilities of preserving its heritage?
>
> —J. Frazer Smith

By the end of the nineteenth century, the prestige once associated with the title Creole had substantially diminished—not without a fight, however. Struggles to uphold the upper French Creole class and its revered social demarcation were characterized by attempts to reinforce the term as one reserved for white individuals of European descent born to Louisiana colonials. Nevertheless, the designation had become confused in the public eye and synonymous with ethnic amalgamation and, to the great disdain of white Creoles, racial mixing. As the ancienne population died away, a series of novels emerged that portrayed Creoles as a blended sect flaunting a crumbling aristocracy. French Creoles fired back in an effort to protect their self-made legacy and reassert Creole as a term reserved for the ancestral white progeny of south Louisiana. Unfortunately, their efforts proved futile, as the formerly esteemed name of Creole no longer carried definitive race or class differentiation.

Louisiana historian Charles Étienne Arthur Gayarré listed that in 1806, 13,500 Creoles, 3,355 free people of color, 3,500 Anglo-Americans, and 5,714 European immigrants resided in south Louisiana. As the twentieth century approached, Creole surnames grew sparse predominantly due to marriages between Creole daughters and Anglo-American sons.[1] As well, the nearing epoch saw Creoles of Color refer to themselves *only* as Creole instead of the class-delineating title *gens de couleur*.[2] Piling confusion on the social signifier, American author George Washington Cable painted white Creoles as ethnically mixed and not of pure French heritage. If that was not sufficient to earn Cable the ire of Louisiana's remaining French Creole aristocracy, the author also supported racial equality.

Historian Lawrence N. Powell explains that Cable's books, particularly *Old Creole Days* (1879) and *The Grandissimes: A Story of Creole Life* (1880), incited agitation among the ancienne population. Born October 12, 1844, in New Orleans, Cable keenly understood white French Creole history and constructed his narratives on upper-class Creole stalwartness, "broken" French dialects long separated from the French of New Orleans's colonial immigrants, and the biological relationship between white Creoles and their mixed-race offspring. Cable's books had, whether intended or not, exposed what white Creoles considered their "private affairs."[3] The author's career blossomed between 1874 and 1880. During this period, he submitted fictional excerpts about Creole life for publication in *Scribner's Monthly*, an illustrated periodical produced in New York but distributed nationwide. After the magazine's editors printed a series of articles that culminated in *The Grandissimes*, Cable's notoriety exploded, and his books cemented the definition of Creole as an individual of mixed parentage.[4] While American-born readers applauded Cable's portrayals of "old Louisiana," Creoles reproached the author for misrepresenting their heritage.[5] Bending to Creole rebuttals, Cable published *The Creoles of Louisiana* in 1884 in an effort to correct etymological confusions and satisfy his ancienne population neighbors. This historic account more closely aligned with white Creole self-perceptions; however, the damage had been done. The novelist's fictions had become so popular that when the 1884 World's Industrial and Cotton Centennial Exposition opened in New Orleans, readers from across the United States flocked to south Louisiana to see Cable's Creole world for themselves.[6] As Thomas Ruys Smith surmised, "Visitors nominally came to the city because of the Exposition, but it was Cable's pre-industrial New Orleans that they truly sought."[7] Enraged, French Creoles defamed Cable in printed essays, newspaper articles, and public lectures.[8]

On April 25, 1885, Gayarré tried to clarify the term Creole and discredit Cable by presenting an essay at the recently renamed Tulane University titled "The Creoles of History and the Creoles of Romance." Gayarré railed against Cable and his fictional accounts of Creole heredity. He began by providing the etymology of Creole including its French and Spanish roots. To wit, Gayarré employed a brief history of the comings, goings, and successes of Louisiana's white Creoles.[9] A noted regional scholar and longstanding president of the Louisiana Historical Society (1860–1888), Gayarré existed as a product of Creole breeding, culture, and education—a model specimen to champion what was left of the ancienne population. He was not only a published author; he was respected throughout the state and was, in his youth, encouraged to pursue a life of politics and public oration. Before dying, the Creole historian's former Collège d'Orléans professor, M. M. Nicolas Rochefort, imparted to his pupil, "Let me kiss, child, those capitolian lips before I am wafted across the Styx by old Charon. *Os magna locuturus* (a mouth which is destined to speak great things), I predict."[10] True to his mentor's divination, Gayarré spoke "great things" but none so heated as his 1885 speech.

Gayarré felt the most egregious mistake of Cable and his readers was to confuse white Creoles with Creoles of Color. The historian insisted that from the colonial period forward, there existed "two elements of population—the Europeans and the Creoles constituting one element (the white); the other, embracing what is known under the general appellation of *black*, or *colored*, [italics in original] people, who had a much inferior social standing, and no political social status whatsoever." The original Creole population, Gayarré stressed, was "of pure white blood" and should not be mistaken to have any connection with native-born black individuals who were referred to as Creole simply due to their Louisiana origins. Gayarré chastised those he saw as too ignorant to understand the meaning of the word Creole when dictionaries were so easily attainable. Although he admitted that occasional sexual relations did occur between Creole plantation owners and their female slaves, he dismissed the import of this fact by briefly describing monetary fines a planter might face if found out. Gayarré did not elaborate on the mixed-race children of these unions, nor did he comment on their genetic claim to the Creole title. (He also failed to mention that, at the age of twenty-one, he had fathered a mixed-race son whose mother was Delphine Le Maitre, a former slave in the Gayarré household.) Skirting the subject of miscegenation, Gayarré condemned Cable for being a native of Louisiana and yet utterly wrong about class and racial distinctions between white French Creoles and parties of mixed blood.[11]

Throughout his speech, Gayarré insisted, "The Creoles, that is to say, the white descendants of Europeans—I cannot repeat it too often" were the dominant class in Louisiana.[12] Gayarré admitted that Creoles sometimes intermarried with Acadians, the "very simple and honest people, of unmixed white blood," which resulted in some lineage confusion; however, this perplexity should not have caused Cable to misconstrue Creole genealogy and African bloodlines. As Gayarré stated, "Thus far it is impossible to imagine by what process of ratiocination any human mind could arrive at the conclusion that the Creole population of Louisiana must be looked upon as being colored and as having their veins tainted with African blood."[13] The Creole historian not only criticized Cable's fictions but also denounced the author's "Freedman's Case in Equity," claiming that the American author intended to advance a cause for interracial marriages. As Gayarré saw it, Cable should have renamed his "Freedman's Case in Equity" to be "Social and conjugal fusion of the blacks and the whites."[14] Even Alcée Fortier, the eminent Louisiana historian and relative of Valcour Aime, admonished Cable for his stance on racial justice.[15]

Gayarré's rebuttal did little to reinforce the lexicographical myth that French Creoles were of noble descent. Like many white Creoles before him, Gayarré did not wish to expose the ethnic heritage nor racially mixed offspring of the ancienne population. To do so would have weakened relationships forged between French Creoles and Anglo-Americans based on white supremacy.[16] As a nationally celebrated author, Cable served on a variety of boards and committees such as the American Association for the Advancement of Science, the New York Charities Aid Association, and the Southern Art Union. He lectured at university commencements, visited asylums, and spoke at various US prisons. As a result, Cable's views and publications had become widely accepted, making it virtually impossible to redirect public opinion.[17] The author's rise in national popularity only enhanced Creole distain. Members of the fading ancienne population made the city of New Orleans so uncomfortable for Cable that in 1885 he chose to relocate his family to Northampton, Massachusetts.[18]

In spite of Gayarré's etymological rebuff, Cable's definition of Creole has lasted well into the modern era with several historians citing his literary representations as fact.[19] The Creole scholar went to his grave refusing to accept the alteration of a social class title he himself proudly donned. In 1895, Gayarré died and was buried alongside his grandfather, Jean Étienne de Boré, in St. Louis Cemetery Number 1, opposite the French Quarter and adjacent to the soon-to-be established "Storyville" red-light district.[20] Regardless of their efforts, French Creoles failed to win back the public perception of their pure heritage.[21] Historian Mary Gehman explains that throughout Louisiana, Mississippi, Alabama, and the rest of the Gulf South, "'Creole' grew to be an obsolete word and 'American' [became] applicable to all."[22] In short, Cable's publications further Americanized the ancienne population by uniting Creoles and Americans under the banner of "white."[23] All the while, descendants of Louisiana's Creoles of Color continued to use the single-word moniker.[24] By the mid-twentieth century, the lingering vestiges of French Creole aristocracy completely died out.[25] Today, Creole has come to mean a person born in America but with European (French/Spanish) or African ancestry.[26]

While French Creoles lost control of their class demarcation, St. Mary's Jefferson College slipped into American trends of higher education. As the once Creole college entered the twentieth century, Creole surnames, once prevalent in academic catalogues, were replaced by those of American, Acadian, German, and Irish families. Seeing that state-supported institutions promoted agricultural and scientific curricula, students who desired pragmatic instruction veered away from the Marist academy with its overwhelmingly liberal arts course load. Similarly, when Loyola University New Orleans opened in 1912 boasting advanced scientific training, Catholic enrollments at the rural St. James Parish college declined. As well, instructional intransigence and faculty disagreements plagued the institution and hampered attendance. In the end, the flood of 1927 left the Society of Mary with little alternative but to cease educational operations. As both state and national higher education evolved, Jefferson College and its various administrations shrunk to a footnote in Louisiana history. Diminished by scholarly neglect, the memory of this defunct Creole college became blurred.

After the institution closed, documents that recount the history of Jefferson College were dispersed throughout the American South. As well, memorials to the college are scant. The greatest and largest testament to Jefferson College is the Manresa House of Retreats. Although several academic buildings have been demolished, the remaining structures easily impress. The campus remains closed to visitors due to silent retreats, but those traveling through Louisiana's sugar parishes can view the facilities as they journey to tour gold coast plantations such as Bocage, Destrehan, Evergreen, Houmas House, Laura, Nottoway, Oak Alley, San Francisco, Whitney, St. Joseph, and its neighbor Félicité. As for the men who influenced Jefferson College, their resting places are scattered with little or no information to connect them with the former educational institution.

St. Michael's Church in Convent, Louisiana—the red brick, Tudor Gothic church that provided religious services for antebellum Jefferson College students, hosted raucous debacles between Louis Dufau and Father Claude Anthony Tholomier, and employed Henri Bellanger, the first Marist president of St. Mary's Jefferson College—stands in front of the St. Michael's Parish cemetery. The cemetery, longer than it is wide, extends away from the Mississippi River. In the center, next to family tombs engraved with old Creole names, rises a tall iron cross. Nearby lies a large cement slab with an elevated plaque. The remains of twenty-one Marist fathers and brothers who died while serving at St. Mary's Jefferson College are buried underneath. At first, their remains rested in a stepped tomb next to the college's Gothic Chapel. After the Jesuits purchased Jefferson College, the deceased Marists were exhumed and placed in a simpler crypt in the parish cemetery. The words "Marist Fathers," inscribed in marble, topped the sepulcher. Over time, the little mausoleum crumbled and was deconstructed. The sacerdotal bones were then removed and later buried beneath the protective slab. The metal plaque, donated by the Acme Marble & Granite Co., Inc., reads, "Fathers and Brothers of the Society of Mary Who Served People of God at Jefferson College and St. Michael Catholic Church, Convent, Louisiana," followed by a short list, which includes Joseph Roman, the grandnephew of A. B. Roman.[27] Separated at a considerable distance from the mass grave of Louisiana Marists is the final resting place of former St. Mary's Jefferson College president and New Orleans archbishop James Hubert Blenk. Two years after a crippling heart attack, Blenk died on April 20, 1917, and was interred under the sanctuary of the St. Louis Cathedral in the Vieux Carré. A marble marker affixed to the church wall commemorates his life next to an altar venerating the Virgin Mary and Christ child.[28]

Across the Mississippi River from the former college stands St. James Catholic Church. Next to the river levee is a large cemetery where graves descend in age. Near the earthen embankment's upward slope is the weatherworn Aime family mausoleum, a large, multi-vaulted crypt adorned with inverted, lit torches carved in stone. The inverted torch represents the end of life, but ablaze it "suggests that the soul (fire) [continues] to exist in the next realm."[29] In 1929, the Fortier family, fearing that the Mississippi River might one day change course and overtake the tomb, relocated the remains of Valcour, his wife, and children to the well-kempt Fortier family crypt in New Orleans's Saint Louis Cemetery Number 3.[30] Not long before the removal of Aime's remains, Le Petite Versailles burned in 1920. The gardens were left unattended and are today overgrown.[31] In 1961, Aime's great-granddaughter, Estelle Fortier, and her son, Louis, placed a marker in front of their ancestor's plantation property.[32] Close to the Aime tomb are the remnants of A. B. Roman's crypt. In years past, his tomb collapsed, and a cement slab comparable to the Marist grave was placed over the governor's remains. The tomb's original marble marker inscribed with Roman's name and the dates of his birth and death was placed atop the slab. Engraved above his name is a small wreath and cross. A decorative, albeit rusted, iron fence encircles the slab. Both the Aime and Roman tombs appear but a shadow of the grandeur these men knew in antebellum Louisiana.

Of all those who taught and toiled at the Creole college, none was so unpopular as Louis Dufau. The French immigrant, despite his desire to construct a lasting educational enterprise, never achieved the glory he so desperately desired. Dufau's life in Louisiana was filled with frustration, disappointment, and defeat. In death, he shares a tomb with in-laws and his two daughters, Camille and Marie. Muddled mid-nineteenth-century New Orleans cemetery records imply that Dufau's body was either interred in St. Louis Cemetery Number 1 *or* 2.[33] His body was, in fact, placed in the Street family vault in St. Louis Cemetery Number 2. Crypts fill this age-old burial ground—some new, others in need of repair. Dufau's mausoleum remains in relatively poor condition; plaster is missing, cracks traverse the brick walls, and vegetation grows on the tomb's roof. The faded tablet that bears Dufau's name and death date possesses no link to his life's work in education. Aside from the few printed documents related to Louisiana College, the only palpable remnant of Dufau excluding his burial chamber is a set of handwritten letters stored in the archives of Howard-Tilton Memorial Library, Tulane University. As no images of Dufau have been uncovered, the few extant examples of his handwriting and legible signature are the only personal relics of this particular member of the française étrangers.

* * *

The multiple lives and deaths of Jefferson College, as with so many closed colleges and universities, serve as reminders of the difficulties many institutions of higher education faced and continue to confront today. As small, private, and/or religious institutions fight to remain viable in modern America, studies focused on institutional closure, survival, and contributing environmental factors provide important lessons regarding the need to enhance adaptability, increase positive communications with external populations, and illustrate how interactions with contributory social groups influence higher education longevity. In essence, institutional histories, such as this book, serve to address John R. Thelin's assertion: "Troubling today is that for those who are seeking to understand and solve the problems now facing colleges and universities, there does not seem much inclination to seek genuine historical perspectives about higher education as a lens."[34] To echo and add to Thelin's comments, studies that examine the histories of living *and* deceased colleges and universities allow for a deeper understanding of factors related to the progression, struggle, and demise of higher education. By critiquing institutional histories, educational researchers and historians peel away incorrect information, seek clarification, and tender analyses that better explain the present and elucidate the future.

The establishment, evolution, and failure of multiple educational administrations on a single college campus serves as a contextual loupe through which to view the reliance of nineteenth- and early twentieth-century Louisiana higher education on local human and financial capital. As well, the examination of Jefferson College permits an enhanced insight into the mindset of wealthy French Creole sugar planters regarding education and gentrification. By investigating a single location in the midst of a social strata swayed by wealth, privilege, resistance, religion, and decline, we can better understand how an ethnic faction of the larger southern planter class impressed their administrative, instructional, and religious ideologies on academic life for good and ill.[35] In essence, the history of Jefferson College provides an opportunity to better understand ancienne population ruination as, over the course of a century, French Creoles watched "a thoroughly French country became thoroughly American."[36] As such, this declining aristocracy saw fit to counter political control, flaunt their wealth, and establish a college that further entrenched their sons in familial life lessons of gentrification.

Creoles fought for political dominance to counter Anglo-American initiatives via the promotion of peers such as A. B. Roman. After the fall of the Collège d'Orléans and the establishment of the Anglo-American College of Louisiana, Creoles readied themselves to shield their social status and fashioned Jefferson College not only to educate young men but also to oppose their political foes. Following the economic crash of 1837 and the college fire of 1842, sugar planters proved ill equipped to maintain their college. Lacking support, Dufau's Louisiana College suffered a similar fate. Although Roman and a handful of gold coast residents backed Dufau at the onset of his St. James Parish enterprise, Creole planters abandoned the French immigrant when his vexing demeanor became apparent. As well, Dufau disinterested the local populous by presenting an institution that more closely resembled a French lycée than a relevant form of higher education. As a consequence, he struggled to garner the financial capital necessary to keep his college afloat. Not long after Dufau's sugar parish departure, Creoles, now financially recovered from the economic crash, were poised to take back the institution.

Due primarily to the financial aid provided by Valcour Aime in 1859, Jefferson College reopened under its original title but with a seminal difference between the initial secular regime and the revived institution. By the mid-1860s, Creoles formerly unwilling to practice their faith on a regular basis attended Mass to quell their postwar, dystopianistic depression. With the steady influx of Catholic German, Irish, and Italian immigrants, regional Catholicism asserted itself as the dominant religious force in spite of French secularism.[37] Though the majority of French Creoles remained steadfast in their simulacra of Parisian secularism, Aime had become an overt Catholic devotee subsequent to the death of several family members. Unsurprisingly, Aime established Catholicism as the denomination of choice at restored Jefferson College. He highlighted this spiritual transformation by constructing an ornate Gothic chapel that was architecturally set apart from the Greek revival columns and moldings of the institution's larger academic buildings. Despite Creole support, the Civil War cut short Jefferson College's third regime and sorely damaged Creole planter affluence and influence.[38] By the 1870s, Creoles had realized their return to political control was principally unattainable. As Irish and German immigrants supported American politicians over Creole or French immigrant legislatures, Anglo-American voting numbers proved far superior.[39]

When Anglo-American political dominance took hold, aspects of traditional French Creole life began to shift. In 1862, English became Louisiana's official instructional language, despite the fact that many New Orleans public and parochial schools had switched to English as early as 1842 following the installment of John Angier Shaw as Faubourg St. Marie's superintendent of public schools. Hailing from Bridgewater, Massachusetts, Shaw launched a string of schools based on the New England educational system stocked with able-minded northern teachers. In 1845, Anglo-Americans secured a majority in the Louisiana legislature. Accordingly, Shaw's plan was expanded and additional monies were allocated to enhance the New Orleans public school system.[40] These schools did much to enhance the Americanization of a traditionally French region even though private tutors retained French as the preferred teaching language.[41] Private instructors and institutions served the linguistic wants of those with requisite funds alone. As Robert Reinders explains, "For the rich who felt the public schools were too vulgar and the parochial schools too narrow, there were numerous private non-denominational schools in New Orleans. . . . If parents wished to send their children out of the city, the rural parts of Louisiana and Mississippi were sprinkled with academies, female institutes, boarding schools, and day schools. Louisiana alone had over 150 private schools in the 1850s."[42] The vast majority of families, however, could not afford the luxury of private tutelage and publicly funded education became an eventual norm.

With the onset of Reconstruction, Creole sugar planters lacked legitimate collateral due to cannon fire, looting soldiers, and plantation property decay. Planters were forced to seek aid from northern or European investors if they were to restore south Louisiana's sugar industry. Nevertheless, many planters simply could not raise the capital needed to keep plantations afloat, so they foreclosed or sold their properties to northern contractors for far less than they were worth.[43] Indeed, war and Reconstruction depleted the preponderance of French Creole assets. Wealthy Americans and European immigrants emerged as the principal landholders while the countryside filled with lower-class Catholic families. With the establishment of the Marist regime, a palpable wave of religion took hold at the renamed St. Mary's Jefferson College. This final administration expanded preparatory education in order to stock the college department with homegrown students. But, like other religious orders in the post–Civil War South, the Marists were slow to adapt their college's upper-level curricula.

As regional universities rivaled the instructional offerings at St. Mary's Jefferson College, enrollments diminished, and a portion of the institution's name was dropped or at least relegated to a secondary titular presence. Academic catalogues, newspapers, print advertisements, and student publications during the late nineteenth and early twentieth centuries referred to St. Mary's Jefferson College as "Jefferson College" and printed "St. Mary's" in smaller print, betwixt parentheses above or below the college's antebellum name. Similarly, letters and local communications referred to the Marist college simply as Jefferson College, Jefferson, or J.C. This colloquial abbreviation reduced the college's cumbersome title to a manageable yet recognizably American sobriquet. Marist fathers and brother observed the preference of alumni and students to refer to St. Mary's Jefferson College as Jefferson College and produced print material accordingly. Regardless of the institution's label, enrollments and financial support fell far below adequate means, and the college, like the remaining ancienne population, dissolved into relative obscurity. According to LeBreton, "With the passing of the great plantations, with the establishment of elementary schools and high schools in the parish, with the increased facilities for transportation to and from the city, Jefferson College has been forced once more to close its doors."[44]

The Society of Mary's administration ended in failure due to curricular and instructional intransigence, meager social support, and a lack of local capital to cover tuition costs; however, from 1864 onward, the college and its priests marked a revival in south Louisiana Catholicism. In a 1906 commencement speech, St. Mary's Jefferson College alumnus John L. Peytavin emphasized the importance of religious education as it applied to the college's graduates: "Christian education, aye, Catholic education, places every man on an equal footing with mankind; it places us on the top rung of the elevated ladder of hope, awaiting the reward of faith and charity. . . . Jeffersonians, your Alma Mater trains your minds and your hearts; she refines and elevates your souls, she uplifts your nature, she teaches you your duty to yourselves, to society, to your state, to your God!"[45] Despite the college's closure, the Society of Mary maintained parish religious and missionary activities throughout the twentieth century. In 2002, after 139 years of service, the Society of Mary exited Louisiana owing to the declining number of Marists. However, the fathers and brothers left an ineffaceable mark on the region. Their influence prompted journalist Elizabeth Barrouquere to eloquently suggest, "The Marists are as much a part of St. James Parish history as sugarcane."[46]

In 1932, Louisiana superintendents of education Thomas H. Harris and Chas F. Trudeau contacted Father Phillip H. Dagneau about the history of

St. Mary's Jefferson College. Both Harris and Trudeau hoped that documenting the college's past might "help perpetuate the memory and achievements of one of the most outstanding institutions of its character ever to exist in Louisiana, perhaps, in the entire Southland."[47] Despite Dagneau's wishes, a complete history of a "college once crowded with sons of Creole planters" never materialized. In spite of this shortcoming, the campus itself remains a seminal historic marker reflective of regional social influence, southern Catholicism, and the evolution of Louisiana higher education.[48] In the words of Charles P. Roland, "Today three landmarks of the ante-bellum sugar plantations remain: the broad, canal-gridded cane fields; occasional sugar houses . . . and the great, proud mansions—ghosts along the waters."[49] Jefferson College, now the Manresa House of Retreats, with its white veneer and Hellenistic structural design, serves as a fourth reminder of a bygone era of French Creole aristocracy. The history of this well-preserved institution, standing tall on Louisiana's Great River Road, offers a strong refute to claims of planter-class disinterest concerning general higher education. Indeed, the campus itself showcases ancienne population attempts to assert social class and political dominance by means of a college intended to educate the sons of sugar.

Hail venerated pile! that nigh the running flood
Of mighty waters doth so gravely stand—
So stately staid, so solemn and so grand!
—Caedmus (excerpt from "Ode to Jefferson College")

Epilogue

Manresa, the Fifth Life of Jefferson College

> The house of silence and sacred sod, where nobody speaks to anybody, yet everybody speaks to God.
>
> —Judge Louis H. Yarrut

Not long after St. Mary's Jefferson College closed, the Society of Mary decided to sell the college buildings and campus for $75,000. The Marists relinquished control of the sale to Ernest A. Carrere's Sons. By 1928, however, potential buyers had yet to surface, so the real estate agency approached the St. James Parish school board to inquire if the district wanted to purchase the defunct college. The board, however, declined. Immediately after the school board's rejection, Louisiana's state superintendent of education, Thomas H. Harris, contacted the firm and inquired about the selling price. Harris worried that the aged State School for the Deaf in Baton Rouge, founded in 1852, had become inadequate and in need of extensive repair. Accordingly, Harris advocated purchasing St Mary's Jefferson College, selling the Baton Rouge facility, and moving the State School for the Deaf to St. James Parish. Governor Huey Pierce Long favored Harris's plan and brought the proposal before the Louisiana legislature. Even with Long's support, Harris's proposition was denied.[1]

The asking price of $75,000 remained constant through 1930. In July of that year, Elinor Judith Gast, sister of famed Algiers jazz musician Emmett Louis Hardy, offered to purchase the college and transform it into a convalescent hospital.[2] Unfortunately, for unknown reasons, this offer fell through. Not long after Gast's offer, a group of diocesan priests expressed interest in the property but were unwilling to pay more than $30,000.[3] The Marists, in response, agreed to reduce the asking price to $60,000.[4] Dissatisfied with the counter proposal, the diocesan priests rescinded their offer. At this point, the Society of Mary was in possession of a college three years closed and already showing signs of wear. Holes had formed in the main building's roof, and foliage had overtaken the once well-manicured grounds. Owing to a deflated economy, the reduced asking price remained well beyond most buyers' budgets.[5] In September of 1930, Jesuits from the New Orleans Province of the Society of Jesus approached the Marists and offered $27,500 for the entire campus. Unable to care for the decaying college, the Marists, on September 5, 1930, agreed to sell the former Creole college to become a spiritual retreat house.[6]

On January 5, 1931, members of the Marist order convened at Paulina, Louisiana, to finalize the college's sale to the Jesuits under the corporation title *Société Catholique d'Education Religieuse et Littéraire* (Catholic Society for Religious and Literary Education).[7] The only caveat was that the Marist "corporation [would] reserve to itself a one undivided one-half interest in and to all oil and mineral rights on the above described property, which reservation is to be exercised and used by this corporation, its assigns and lessees, in cooperation with the purchaser of said property or its successors in title, and all drilling, mining or other operations shall be done at such places and under such requirements and restrictions as may be determined by the purchaser or subsequent owner of said property."[8] Agreeing to this condition, the fathers and brothers at Jesuit High School New Orleans cobbled together $27,500. Through the sale of physical properties, liberty bonds, and retreat houses in New Orleans and south Mississippi, the Jesuits were able to raise $19,500. To complete the asking price, $8,000 was taken from a 1915 donation meant to promote Jesuit-led retreats. Lee Boarman Keith, the Kansas City, Missouri, nurse turned philanthropist, had inherited the financial estate of her husband, Richard Henry Keith (former president of the Central Coal and Coke Company), when he died in 1905. Throughout the remainder of her life, she sponsored various Catholic-based initiatives including hospitals, orphanages, and retreat centers. Offering the full amount to the Marists, St. Mary's Jefferson College was purchased and reestablished as the Manresa House of Retreats under the supervision of the Jesuit High School clerics.[9]

Father John M. Salter, provincial for the New Orleans Jesuits, finalized the purchase on January 9, 1931, and directed its reorganization as a retreat house.[10] Michael Kenny recorded that "Father Salter . . . acquired Jefferson College and grounds at Convent, Louisiana to supply ampler facilitates for lay retreats, and this model retreat house [Manresa] has since been numerously attended through the weeks and week-ends of the year by laymen and students of city and state."[11] With the sale complete, Manresa hosted the first retreat under Jesuit leadership in February of 1931. Fifteen men, in addition to their spiritual director, Father Robert Bryant, experienced an aged physical plant poorly heated and in need of refurbishment.[12] The former college, even if decayed, was excellently placed for silent retreats. The rural location might have impeded a college competing against urban universities but it benefited a silent, religious haven. Prior to the campus purchase, the Jesuits had conducted retreats at Covington, Grand Coteau, and New Orleans, Louisiana. They also held retreats at Villa Gonzaga Retreat House in Long Beach, Mississippi.[13] By the end of 1931, 200 retreatants had graced Manresa's halls.[14]

Once an educational institution, the campus became a haven for theological meditation in both name and function. "Manresa" refers to "the small Spanish town of Manresa on the Cardoner River where St. Ignatius of Loyola, the founder of the Jesuits, spent ten months in a cave making the first Jesuit long retreat. From this experience came his book, the *Spiritual Exercises*, the format for all [Jesuit spiritual] retreats given."[15] Though renamed, retreatants continued to refer to the institution, in part, as Jefferson College. Retreat logs reveal that attendants often referred to the campus as "Manresa House, Jefferson College." This practice continued well into 1943. Moving forward, logbooks reference the campus as "Manresa House" alone. Today, registration accounts style the campus by its full name, Manresa House of Retreats. The Jesuits not only renamed the campus, but also renamed several of the college's buildings—mostly after saints who, in life, belonged to the Society of Jesus. The main building was dubbed St. Mary's Hall in honor of the Marists who previously served in the surrounding parish. Blenk Science Hall became Xavier Hall (after St. Francis Xavier), and the president's home, once used as the college hotel, became St. Ignatius House of Prayer.[16]

Other buildings, too costly or too damaged to restore, were torn down. These included Alumni Hall, the college gymnasium and pool, the lay professor residence (referred to by students as "South America"), student recreation rooms, Thouvenin Dining Hall, the wooden grandstand, and portions of the surrounding antebellum brick wall. Though demolished, building materials were reused in other St. James Parish construction projects. In 1940, St. Michael's Parochial School, next to St. Michael's Church, was "built with materials from Jefferson College school buildings which were torn down and purchased by the parish through a donation made by Mr. E. J. Claire, Sr. of Edgard [Louisiana]."[17] In 1944, another change occurred that established Manresa as an entirely Jesuit-owned property. In August of that year, the law offices of Martin and Himel in Lutcher, Louisiana, informed both the Jesuits and Marists that as a result of a failure to use oil and mineral rights for ten years, any claims the Society of Mary had in regards to said minerals were null and void.[18]

Despite Manresa's progress toward the end of the 1940s, Jesuit High School New Orleans was in need of physical expansion. As well, the buildings of Manresa required additional renovations. Unfortunately, both projects could not be undertaken due to Jesuit High School's limited finances. In need of funds, the Jesuits at their New Orleans school put Manresa up for sale. Oliver P. Carriere recorded that the Society of Jesus and Louisiana governor Richard Webster Leche discussed selling the campus to Louisiana State University, but the purchase never occurred and retreats continued.[19] In time, the two Jesuit entities needed to separate if Manresa was to subsist. In a letter dated July 6, 1949, the Jesuit Superior General in Rome, Jean-Baptiste Janssens, raised Manresa to the status of "independent house," which reestablished the retreat as an autonomous cooperation with its own rector and board of supervisors.[20] In order to facilitate the separation, Manresa administrators repaid Jesuit High School the original purchase price of $27,500 and allowed the high school to receive a half share in all mineral rights from the gold coast property.[21] This act of legal separation was concluded on March 14, 1950.[22] After Manresa's severance from Jesuit High School and the Catholic Society of Religious and Literary Instruction, the Society of Jesus established a new corporation titled "Manresa House of Retreats." The Jesuits of Manresa drafted a constitution and formulated a board of directors, which included the provincial of the New Orleans Province, Manresa's director, and sixteen retreat captains and/or assistant captains "who [had] served for a period of at least three years."[23]

The new constitution dictated that "retreats for laymen at Manresa House [would] be closed retreats and conducted according to the method based on the *Spiritual Exercises* of St. Ignatius Loyola." Members of the first governing board, in addition to the New Orleans Jesuit provincial, provincial treasurer, Manresa retreat director, and assistant retreat director, included seven laymen: Harold R. Ainsworth, Oliver P. Carrier, Durbin J. Gibbens, Walter Leaumont, Sr., Caye A. Melson, Sam A. Micelie Sr., and Irvin P. Plaisance Sr.[24] After Carrier stepped down from Manresa's board, he wrote and published a chronology of Jefferson College and Manresa. This small volume remains the genesis attempt to include *all* relevant historic figures connected to the Creole college prior to the signing of its 1831 charter through the 1931 Jesuit purchase and beyond.[25]

With the Manresa House of Retreats recognized as a separate entity, the governing board focused on repairing St. Mary's Hall.[26] After the Jesuit Superior General gave his approval, the Manresa corporation acquired a loan of $175,000.[27] These funds allowed for the 1950 renovation, which increased the number of private rooms to fifty-five with hot and cold running water, larger bathrooms, and a sprinkler system in case of fire.[28] With repairs complete, retreat attendance increased. In 1980 alone, 4,555 men retreated at the former college campus.[29] To solidify Manresa's permanence as both a spiritual and historically significant site, the campus was placed on the National Register of Historic Places on January 31, 1985.[30] Although attendance remained high, the "Grand Dame," as laymen were referring to the old college, showed an increasing need for preservation if it was to remain a functional historic site. In 2002, Trahan Architects, the Manresa Board of Director's Building Committee, and Father Anthony Ostini filed a report that detailed Manresa's physical status and the imperative need for extensive renovations. The aged buildings suffered from a musty smell (resulting from water damage known as "rising damp"), peeling paint, and cracked brickwork. As such, the building committee proposed a full renovation of St. Mary's Hall. The overhauled structure would contain a minimum of eighty-eight rooms, private bathing and toilet facilities, a single bed and simple furnishings, a new interior staircase, and an elevator. Planners took special care to preserve the building's external architectural features.[31] Laymen who patronized the retreat provided financial assistance for this and other building projects.[32] More recently, Xavier Hall was renovated, and a new building dedicated to Our Lady of Good Counsel was constructed to provide private space for spiritual consolation and the Sacrament of Reconciliation.[33] According to Manresa's current director, Tim Murphy, over $20 million has been raised since 2000. All donations are dedicated to Manresa's maintenance and further development.[34]

Since 1931, retreatants have been privy to St. Ignatius's *Spiritual Exercises*. This guided, meditative process focuses on the annunciation, birth, life, death, resurrection, and ascension of Christ. In brief, the personal examination is "a time to let Jesus convince [retreatants] that true religion and true spirituality is a matter of the heart as well as of the lips, and to learn to relate all the commandments and all religious practices to an authentic love of self and neighbor and God."[35] All of this is done while laymen and clergy observe "the rule of silence." The rule, practiced by those on retreat, requires both "interior and exterior" silence if a person is "to receive the graces of the days of the *Exercises*."[36] As explained in Manresa's retreat manual:

> The rule is cherished by the men of Manresa because it brings a welcome, soothing relief from the stresses and strains of public life. Its therapeutic value gradually reveals itself in deep tranquility where I may hear God's word. Silence is not a penance nor is it something I do for God, but it is something I can do for myself so that I may be

> present to God who is always present to me. Silence is the servant of the retreat. It enables us to make contact with our innermost self and thus satisfy our anxious quest for self-identity. It also enables us to explore more deeply the purpose of life and examine more closely our fundamental orientations and to tap the deep resources of human power, which each one has, but seldom uses. . . . Most important of all, silence is necessary for prayer. It provides the climate for interior stillness in which God's voice can be heard. Abstaining from dialogue with others unleashes the easy, spontaneous communication with God which is prayer. Small wonder that at Manresa, "Silence is the Golden Rule."[37]

Retreats begin on Thursday evenings and close on Sunday after lunch. Each day, retreatants listen to instructional sessions that focus on portions of the *Spiritual Exercises*. In between lectures, attendees are encouraged to wander the grounds and meditate, pray in the Gothic chapel of the Sacred Heart of Jesus, participate in the Stations of the Cross, pray the Rosary, and read the Bible. Before breakfast, lunch, and dinner, all gather in the chapel for Mass, communion, and recitation of the Angelus.[38]

Throughout each retreat, individuals concentrate on their relationship with God the Father, God the Son, and God the Holy Spirit. By reflecting on the ministry of Jesus and the presence of "God in all things" (an important contemplative phrase germane to the Society of Jesus), laymen attempt to grow in their faith. Like Jefferson College of old, Manresa does not exclude retreatants based on religion. A moving account of this fact is the booklet written and published by Louis H. Yarrut titled *The Reflections of a Jew at a Catholic Retreat*. Yarrut recounts years of personal retreats at Manresa and his spiritual experiences while immersed in meditative silence. To Yarrut, a New Orleans judge, Manresa served as an escape from the commotion of daily life, a place of stillness and spiritual observation "where nobody speaks to anybody, yet everybody speaks to God."[39]

From an academic institution governed by secularism and later Catholicism, to a spiritual refuge from the distracting turmoil of life and labor, the Manresa House of Retreats, formerly Jefferson College, remains a striking example of French Creole sugar planter influence on regional higher education. The four governances of this Creole college, as well as the fifth, solely religious life of Manresa, provide a focused illustration of differing eras of social class conflict, sugar planter prosperity, political turbulence, and regional Catholic resurgence, as well as the evolving cultural climate of a once French territory Americanized by time and indoctrination.

Appendix A

Institutional Presidents at the Jefferson College Site

Jefferson College

Bernard Granet, 1834
Claudius Crozet, 1834–1836
Thomas R. Ingalls, 1837–1841
Alexander Hill Everett, 1842–1844
F. Chevet, 1844–1848

Louisiana College

Louis Dufau, 1853–1856

Jefferson College (Valcour Aime period)

Charles Oscar Dogué, 1859–1862

St. Mary's Jefferson College

Henri Bellanger, 1864–1865
Stephen Chaurain, 1865–1870
Henry Gaud, 1870–1874
John Baptist Bigot, 1874–1879 (first administration)
John Joseph Grimes, 1879–1880
John Baptist Bigot, 1880–1886 (second administration)
George James Horton Rapier, 1886–1887 (first administration)

Thomas Henry, 1887–1890
James Hubert Blenk, 1891–1896
Matthew Thouvenin, 1896–1905
Richard H. Smith, 1905–1920
Peter F. Quinn, 1920–1923
Alcime Cyr, 1923–1924
Joseph Petit, 1924–1927
George James Horton Rapier, 1927 (second administration)

Post–St. Mary's Jefferson College Closure

Henri Georges, 1927–1928 (acting president representing Marist corporation)

Appendix B

Acts to Incorporate and Support Jefferson College

An Act to Incorporate the College of Jefferson, in the Parish of St. James, Approved February 28, 1831.

24. Sec. I. That Étienne Mazureau, D. F. Burth, J. H. Shepherd, and others, who have subscribed, or who may become subscribers to the College of Jefferson, in the parish of St. James, be, and they are hereby constituted a body corporate and politic, under the name of the "Directors of the college of Jefferson," and by that name, they and their successors in office, shall have perpetual succession, and power to sue and be sued, defend and be defended, in all courts and places whatsoever, and may have a common seal, which they may change at their pleasure; and they shall also be able to take by purchase, gift, grant, or in any other legal matter, and to hold any real or personal, moveable or immovable estate.

25. Sec. II. That the directors for the said college, shall be chosen from amongst the subscribers, in such numbers, and according to such formalities, and at such times and places, as are prescribed in the constitution of the said college, adopted by the subscribers, in the parish of St. James, on the first day of September, 1830, or may be prescribed in any other constitution the said subscribers may adopt.

26. Sec. III. That the subscribers shall have full power to amend or change the constitution of said college in the manner therein pointed out.

27. Sec. IV. That the said directors shall have full power to prescribe the discipline to be observed in said college, and establish plans for education, appoint preceptors, masters and tutors, and all other necessary officers and assistants; and make all by-laws which shall be thought proper for carrying into effect said

college: *Provided*, that such by-laws shall not make the religious tenets of any person a condition of admission as a teacher or student, into said college, nor be inconsistent with the constitution and laws of the United States or of this State.

28. Sec. V. That the professors and tutors shall have the power of granting such literary honors and degrees as are usually granted by any seminary of learning in the United States, in such mode as may be pointed out by the directors, and in testimony whereof, may deliver suitable diplomas.

29. Sec. VI. That the subscribers to said college, a majority of the whole consenting, may, at any time previous to the erection of the buildings fix the location of said college, in any other parish in this State.

An Act Relative to the College of Jefferson, Approved, March 18, 1833.

30. Sec. I. That the president and directors of the college of Jefferson be, and they are hereby authorized and empowered, on mortgage of the property of the college or otherwise, to borrow such sums as to them may seem expedient.

31. Sec. II. That the treasurer be, and he is hereby authorized to pay out of any funds in the treasury, not otherwise appropriated, on the warrant of the president of the board of directors of the college of Jefferson, the sum of ten thousand dollars being the amount of the two installments of five thousand dollars each, which were to be paid in the years one thousand eight hundred and thirty-three, and one thousand eight hundred and thirty-four, conformably to the act entitled "an act in favor of the Louisiana, Jefferson, and Franklin Colleges."

32. Sec. III. That the sum of twenty-thousand dollars be, and the same is hereby appropriated for the benefit of the college of Jefferson, to be paid in the following manner, to wit: five thousand on the first day of June, in eighteen hundred and thirty-four, and five thousand dollars on the first day of June of each successive year thereafter, till the whole sum is paid. . . .

An Act for the Relief of the Colleges of Jefferson, Louisiana, and Franklin, Approved March 31, 1835.

36. Sec. I. That the sum of twenty-one thousand seven hundred and seventy-five dollars, be, and the same is hereby appropriated for the relief of the college of Jefferson, in the State of Louisiana, to pay the mortgage debt due by the said college of Jefferson to the Union Bank of Louisiana, and that the treasurer of the State be, and he is hereby authorized and required to pay the same to the aforesaid Union Bank of Louisiana, annually, as the said mortgage obligations of the aforesaid college of Jefferson may become due to the aforesaid Union Bank, out of the appropriation.

37. Sec. II. That the sum of twenty-seven thousand dollars be, and the same is hereby appropriated for the relief of the said college of Jefferson, to pay the balance due by the aforesaid college, for obligations due for the buildings, and to the banks for the promissory notes of the board of trustees of the aforesaid

college, and that the treasurer of the State be, and he is hereby authorized to pay the aforesaid sum of money, out of any monies in the treasury not otherwise appropriated, to the treasurer of the board of trustees of the aforesaid college of Jefferson, upon the warrant of the president, countersigned by the secretary of the aforesaid board of trustees.

38. Sec. III. That the sum of fifteen thousand dollars be annually appropriated to each of the colleges of Jefferson and Louisiana, situated in the parishes of St. James and East Feliciana, for the space of ten years, for the purpose of enabling the aforesaid colleges to pay the salaries of their professors, so as to enable them to lower the rates of boarding and tuition, and that the treasurer of the State be authorized and required to pay the aforesaid appropriation to the treasurers of the respective colleges, upon the warrants of the presidents of the boards of trustees of the aforesaid colleges, countersigned by their secretaries, and that this appropriation be paid quarter annually, out of any monies in the treasury not otherwise appropriated; that a sum of fifteen thousand dollars shall be paid by the State treasurer out of any funds in the treasury, not otherwise appropriated, to the directors of Franklin college, on their warrant, as soon as a majority of the said directors certify, that the said college has been commenced: *Provided*, that the same be done within the twelve months which will follow the passage of this act.

39. Sec. IV. That the affairs and condition of these institutions shall be examined into every year by a joint special committee of the legislature of this State.

Appendix C

Sale of Jefferson College to Louis Dufau

In the city of New Orleans, State of Louisiana on the 5th day of September, 1853, and the 78th year of the Independence of the United States of America.

Before me Adolphe Boudousquie, Notary Public duly commissioned and sworn in and for said city and parish, here residing and in the presence of the undersigned witnesses.

Personally appeared Mr. James D. Denegre, residing in this city, President of the Citizens Bank of Louisiana, acting by virtue of a resolution adopted by the Administrators of the said Bank at their session on the 26th day of July, 1853, as will appear from a certified copy of the said resolution hereto annexed and made part thereof.

Which the said appearer declares in his capacity as above mentioned, sells, cedes, transfers and abandons with a full guarantee of rights, to Mr. Louis Dufau, Professor, domiciled in the Parish of Saint James, this State, who is represented by Mr. Gustave Leroy, by this effect also present, accepting in the name and as Special General Agent of the said Mr. Louis Dufau, by virtue of a procuration given him by act of the record of Alfred Roman, Notary Public in and for the said Parish of Saint James on the 29th day of August, 1853, as per a certified copy of said procuration hereto annexed and made part thereof.

The Jefferson College, the land upon which is constructed the said College and its dependencies, as well as all books, the Physics and Chemistry cabinets, the furniture and all movables belonging to said College, without exception; the said land comprising two portions of land as follows:

A lot of land situated in the Parish of Saint James, on the left bank of the Mississippi River, at about 60 miles above the City of New Orleans, measuring three arpents on the river, by 29 arpents, more or less, in depth between lines closing in the rear, more or less, together with all the buildings thereon.

Another lot of land preceding the first and bounded in the following manner; on the upper side by the line forming the lower boundary of the land on which the Jefferson College is built as above described, below by a straight line on the outside of the foundation walls of the yard of said College and extending on that side of the Mississippi River and the other, that is to say, to the depth up to the junction of the said line with the other forming the lower line of the lot above described, together with the tools, the latrines and the ditch which drains the same, and touching said wall, as well as the employment and use of the lot upon which is constructed the said mentioned latrines and the ditch which drains the same and of the lot upon which the barn is erected, and said lot extending to the said straight line by the outside foundation of said wall, and extending beyond the limits of the lot of the land above described, and for the use of the said lots as presently designated and employed, and it shall not be changed by the Vendee or the future Vendees.

The two above lots described with their dependencies belonging to the said Citizens Bank were acquired in the following manner, to wit:

The first above described was acquired at the public sale made January 1, 1848, by Eugene Cry, Sheriff of the Parish of Saint James by virtue of a write of Fieri facias, addressed by the Honorable Court of the Judicial District of this State in the case of the Citizens Bank of Louisiana against the Directors of Jefferson College, and the lot as above secondly described, the barn, the latrines, as well as that portion of land upon which presently is erected said barn and latrines by mean of the acquisition made to the said Bank, from Mr. Edmond Druilhet, as per an act of record of Alfred Roman, Notary Public in and for the said Parish of Saint James on August 12, 1853, as per the certificate of the parish recorder in and for the said Parish of Saint James, attached to the copy of the Adjudication Act to the said Citizens Bank, for the portion of land secondly described above, which said copy remained in the possession of said Bank and shows therein that Jefferson College and the two tracts of land as above described are free from mortgage and in the same name to the said same Bank.

By which the said Mr. Louis Dufau will make and dispose of the said Jefferson College and the said pieces of land presently sold, and which the said agent recognizes that he has been, will and truly put in possession of same, and the said Mr. Denegre, in the name of the said Citizens Bank, declare by these presents, appointed and gave the said claimant in and to all rights of property and actions of what so ever sort which the said Bank has or might have to exercise against all Vendors or former owners of the said College and all other improvements attached or situated thereon.

The present sale is made and accepted for the price and sum of $20,000 in liquidation of which the said Gustave Leroy, as present by furnished notes of the sum of $5,000 each, subscribed and endorsed by him in name and as agent of the said Louis Dufau, dated August 15, 1853 and made payable on the 1st of October of each year 1854, 1855, 1856, 1857, at the domicile of the said Citizens Bank of Louisiana in New Orleans and bearing interest at the rate of 8% from the date until final payment; which notes after having been signed Ne Vatietur, by me the undersigned Notary, for identification with present act, was returned to Mr. James D. Denegre in his capacity as above mentioned.

And for the assurance of the payment of the said notes in capital and interest, the said Mr. Gustave Leroy, in the name of the said present acquirer, affects and specially hypothecates and mortgages, privileges in favor of the Citizens Bank of Louisiana, or all future holders of the said notes, the property presently sold with the moveable effects attached hereto, and obliges his constituent not to sell, alienate or mortgage same to the prejudice of this act.

And here the said Mr. Gustave Leroy, in his capacity as above mentioned, expresses, and acting in the name of Special Agent of Cecelia Street, wife of age of the said Louis Dufau, by virtue of the procuration which was given to him with the assistance and authorization of her spouse following the act of record of the said Alfred Roman, Notary Public, of the date of August 9, 1853, declares by these presents, that for the purpose of assuring in a more efficacious manner, the payments in principle and interest of the notes above mentioned for the sum of $5,000 payable October 1, 1854; he affects and specially mortgages same in the name of the said Louis Dufau and the said lady, his spouse, consenting and declaring, in favor of said Citizens Bank of Louisiana and all future holders of the said notes of $5,000, which is accepted by the said James D. Denegre, in his capacity, the property, as described as follows, to wit:

A lot of land situated in this City in the street bounded by Dauphine, Ursuline and Hospital, having 58 feet front on said Dauphine Street by 140 feet in depth, together with the buildings and improvements thereon and also all rights, actions and privileges which is dependent thereon.

And another portion of land situated back of the property above described, measuring 29 feet and 8 inches on the rear line of the above mentioned lot; 29 feet 3 inches on the line which divides the property of Mrs. Widow Guerin; 17 feet on a line running towards the street called Hospital; 20 feet on a line running towards Bourgoyne Street which is parallel to Ursuline Street; 12 feet on a line running on Hospital Street, parallel to Dauphine Street and 40 feet 3 inches on a line running on Dauphine Street, parallel to Ursuline Street, together with all buildings and improvements thereon, and with all rights, actions, privileges dependent thereon.

The said Mr. Leroy, promising in the name of his constituents, not to sell, alienate or mortgage this said described property to the prejudice of the present mortgage and consenting in fact that the present act imparts a confession of judgment and execution in favor of the said Citizens Bank and to all portions of the said note of $5,000.

The property above mortgaged, belongs to the said Mr. and Mrs. Dufau by virtue of the acquisition made from Mrs. Celestine Genois, wife of Mr. Joseph Soniat Dufossant, following an act of record of D. Doricourt, Notary Public of this City on March 9, 1852.

The said Mr. Leroy declares by these facts that he consents in the name of his constituents that in case of the nonpayment of the said note of $5,000 issued October 1, 1854, the Bank will have the right to have seizure made and to sell together or separately, as per their judgment, the said Jefferson College and the property situated on Dauphine Street or that one alone by these effects, and he renounces in his capacity, to all privileges and discussions.

In short, the said Mr. Leroy, declares that he obliges his constituents by these presents to keep Jefferson College and the property on Dauphine Street constantly insured against the danger of fire until final payment of the four notes of $5,000 each as above described and to transfer the Insurance Policy to the said Bank as collateral for the payment of the said notes in principle and interest.

As per the Certificate of the Recorder of the Mortgages for this City, bearing a date of September 2, 1853, it appears that the said property situated on Dauphine Street is free from mortgages in the name of the said Mr. and Mrs. Dufau, with the exception of the following, to wit:

That which was consented hereto by act passed before D. Doricourt, Notary Public on March 9, 1852, in favor of Mrs. Celestine Genois, wife of Joseph Soniat Dufossant, their Vendor, for surety of $4,000 plus interest stipulated in said act.

That which was consented hereto by the act passed before A. Ducatel, Notary Public on April 29, 1852, in favor of Jean Avengo, have concurred to guarantee the payment up to the sum of $2,000 on a property given as bail by these parties to the said Mr. and Mrs. Dufau by this said act.

And that which was consented hereto by this act passed before the same Notary on August 34, 1852, in favor of Thomas Harington, contractor for surety of the payment of the sum of $3,280 which was the amount of the price of certain buildings upon the land which these last named persons have promised to construct upon the property above described together with accrued interest.

Said act made and passed at the office in New Orleans, the day month and year mentioned above, in the presence of Mr. Duplantier and Mr. Eugene Bonny, lawful and legal witnesses, domiciled in this City, who have signed with the appear [*sic*] and me, the Notary, after due reading of the whole.

Signed, Gustave Leroy
James D. Denegre
E. Bonny
G. Duplantier
Adolphe Boudousquie, Notary.

Certified copy given at New Orleans September 9, 1853.
Signed, Adolphe Boudousquie, Notary.

Recorded at the Parish of Saint James on September 22, 1853.
Signed, B. Ferchand, Recorder.

Appendix D

Property Deed: Valcour Aime to the Jefferson College Board of Directors

Having acquired the old Jefferson College and having finished repairing it and for the purpose of restoring it to its original purpose, the undersigned cedes this property whole and entire to his four sons-in-law,

Messrs. Florent Fortier
Alexis Ferry
Septime Fortier
Alfred Roman

So they in turn can hand over this college in part to those who will join them for the purpose of forming a company of share holders and of selling shares at one hundred dollars each until the sum of ten thousand dollars is arrived at, the sum considered necessary to reopen the college for public instruction, the college having cost twenty thousand dollars. Messrs. Fortier, Ferry, Fortier, and Roman will be, after the shares have been sold, share holders, each having fifty shares, representing one-fourth of the actual value of the college, and amounting to the total sum of twenty thousand dollars.

Those who will join them will become co-owners of the college for the sum of their respective share. The shareholders will form a corporation by an act signed by a notary or by a special act of the legislature.

The business of the corporation will be administered by eight directors, also by the president of the college who will be ex officio a member and president of the board of directors. Five directors will constitute a quorum. Vacancies on the board caused either by death or dismissal or for whatever reason will be filled by the directors themselves. Those who fill these vacancies will become shareholders elected either for a year or for that time until their successors can be named. Elections will take place on the annual awards day of the college or at whatever time the directors will designate. The shareholders have a right to one vote for each share, but whatever the number of shares they have, no share holder will have more than twenty votes and they will vote either personally or by proxy.

The directors will have the full right to adopt the rules and the discipline which will be observed at the college, and to establish the whole educational system, to name the president, and following the advice of the president, to name definitely the teachers and the Catholic chaplain who will become part of the institution. Besides that, they will be the ones responsible for the appointment of all the employees, making the particular rules that they will think necessary for the good administration of the college, and until the board decides otherwise, the presidency of the college will be given over to Mr. Oscar Dugué of Jefferson Parish.

An account of the expenses and receipts of the institution will be made annually by the board. Any profits will be divided among the shareholders according to the proportion of their shares. The undersigned, renouncing for himself and for his legal heirs in this particular business, has made no profit at all based on the property itself of the college, and the dividends of the shares will each year be allocated and given over to the president of the college either at the annual awards day of the college or until the time that their successors will be named by the shareholders. The board of directors will be composed of the following:

[*Space provided for the names of board members and college president*]

They will exercise all the rights delegated and besides that they will be in charge of the annual subscriptions, and they should also see to it that the necessary steps are taken for the incorporation of the college. It is understood that the subscription of the shares will be paid half in cash and the rest by bill or note.

By order of the President and the Board of Directors of Jefferson College, payable at New Orleans by the month of March, 1860. St. James [Parish], the 10th of August, 1859.

Valcour Aime

Appendix E

Acts of Transfer: The Jefferson College Board of Directors and Valcour Aime to the Society of Mary

State of Louisiana
City of New Orleans

By Adolph Boudousquie, Notary Public, duly commissioned by this city and this parish of Orleans and in the presence of witnesses whose names are herein signed, and also Mr. Alexie Robert of this City, lawyer representing the Congregation of the Marists, who, by these presents hand it over to the Notary whose name is undersigned, the documents here attached. These documents are word for word as follows:

We, the undersigned, share holders of Jefferson College, incorporated by an act of Legislature of the State of Louisiana, approved on the 5th of March, 1861, declare to have ceded, deeded, and transferred over for the stock value according to the conditions stipulated later on, to the Reverend Mr. Bellanger, acting in the name of and representing the Congregation of the Marists, all the rights, titles and interests which we have acquired and could have acquired by the shares or other possessions of each of the share holders in the form of capital of this college.

The number and value of the shares are indicated after our respective names:

Florent Fortier	50 shares	$5,000.00
Alexis Ferry	50 shares	$5,000.00
Septime Fortier	50 shares	$5,000.00
Alfred Roman	50 shares	$5,000.00

The conditions of transferring these shares and accepted by the said Reverend Mr. Bellanger of the Congregation of the Marists are:

That the said Marists, until the sum of the shares is completely transferred, assume all the rights and that they will engage in fulfilling the obligations of the share holders of the college according to the dispositions of the act of incorporation mentioned below.

That the said Marists will oblige themselves after the expiration of the date of the limitation of the act of this incorporation to consecrate and to sustain in the same place a college of the first class [including] all of the properties to which they have right and which are being transferred by these shareholders. . . .

The directors will give over to Mr. Bellanger the possession of the said college and all that which it owns without exception or reservation and that they bind themselves not to trouble or bother in the use that they will make of it until such time as they can become shareholders with legal title of the property, or until they themselves will obtain the necessary means to open and sustain the cost of the said college.

The transmission of the college will take place according to these following conditions. The Marists will engage themselves on the pain of losing their rights to this transfer to establish as soon as possible and to maintain always a collegiate institution of the first class for the classical and religious education of the young men of Louisiana. They will take in charge all of the costs, they will assume all of the expenses for repairs and the other things that might become necessary during the time of possession, and will become responsible for the debts of the college in the manner that will be here explained. The institution will be placed and always maintained under the protection of the governments of the Reverend Archbishop of Louisiana.

According to these conditions the directors will oblige themselves to transfer over freely to the Reverend Mr. Bellanger within fifteen days following the signing of this contract, at least $25,000 up to $35,000 which represents the actual cost of all the shares of the college. They promise besides to use their influence to help increase the number of these transferred shares, and it is understood that the directors cannot nor will they interfere with the share holders who want to keep their shares as co-owners of the college, who could continue to exercise as long as their shares are in their own name according to the dispositions of the title. . . .

The debts of the college are the following:

It is owed to Mr. Valcour Aime	$6,433.21
It is owed to Mr. Florent Fortier	$822.00
It is owed to Mr. Perilloux	$190.00
to Mr. Emile Jacob	$86.61
to Mr. Petiville	$10.00
to Mr. Labauve	$390.00
to Mr. Genti	$206.00
to Mr. Plaisance . . .	$246.00 . . .

Mr. Valcour Aime will hand over what is owed to Mr. Bellanger freely. Mr. Fortier consents to await payment of what is owed to him without claiming interest for the period of ten years. And Mr. Perilloux and Mr. Jacob will wait one year with the same conditions. For the balance of the sum of $862.16, Mr. Bellanger will take the means to pay in cash.

Acted upon in St. James Parish
6th of May, 1864
Signed: Florent Fortier
Accepted: Bellanger

The Reverend Mr. Bellanger is authorized by the directors of Jefferson College to take possession immediately of the said college according to the clauses of the contract signed the 6th of this month between Mr. Bellanger and the directors represented by Mr. Florent Fortier. The said contract is approved and ratified by the following signatures.

Parish of St. James
8th of May, 1864

Signed: Florent Fortier
Septime Forter
A. Ferry
Joseph Bourgeois
J. Gaudet
John Easley [*sic*] . . .

Appendix F

An Act to Incorporate the Society of "The Fathers of the Society of Mary"

We the undersigned and our successors, desiring to associate ourselves for a literary, scientific, religious and missionary purpose within the state of Louisiana, and to enjoy the powers and immunities of a Corporation or body politic, so that such purposes may be the more efficiently accomplished, do hereby by virtue and in pursuance of section six hundred and seventy seven of the Revised statutes, adopted by the general Assembly of the state of Louisiana, in the year eighteen hundred and seventy, create, organize and establish in the Parish of St. James, in the state of Louisiana a Corporation or body politic under the name and style of "The Fathers of the society of Mary," for the purpose of diffusing knowledge, promoting learning and morality, instructing youths in the liberal arts and sciences, and in missionary works in the towns and Parishes of the state of Louisiana.

We do declare and ordain as our Charter and organic law, the following articles, viz:

Article I.

The name and style of the Corporation, hereby created and established, is "The Fathers of the society of Mary," and as such shall have perpetual successors in law, sue and be sued, plead and be impleaded in any courts of law or equity and before any judges or justice of the Peace in all manner of suits, actions, pleas and proceedings in a full and effectual manner as any other person or persons, body corporate or politic within this state may or can do, and shall be competent at law or in equity to take, receive and hold all and all manner of lands, tenements, rents, and hereditaments or any state or interest therein, and all property real or personal whatsoever, all sums of money goods and chattels, rights and credits, for any estate or interest therein, by gift, bequest, devise, donation, legacy or succession shall be accepted by, nor shall any purchase be made for this Corporation in violation of the said Revised statutes of the state of Louisiana, prohibiting corporations organized under its provision from acquiring property exceeding in value the amount of three hundred thousand dollars.

And this Corporation shall have full power to grant, bargain, sell, transfer, assign, lease, hire, exchange, donate, pledge, mortgage, and exercise all the rights of complete and absolute ownership, over any and all its property, corporeal, moveable and immovable, and in the management of its affairs make and enter into all and all manner of conduct and obligations not prohibited by law.

Article II.

The President of this Corporation shall be the party representing it, in all its business and affairs, and upon whom process of petition and citation must be served, and who shall appear before or in all court or courts of this state, in person by proxy or counsel.

Article III.

The seat and domicile of this Corporation is hereby fixed and established at Jefferson College, in the Parish of St. James, in the state of Louisiana.

Article IV.

That the capital of this Corporation is hereby fixed at one hundred and eighty thousand dollars, but said capital may in future be increased so as to raise it to the amount fixed by the laws under which it is now organized.

Article V.

That the Board of Trustees shall have power to adopt a common seal with such device and emblems as the said trustees shall deem proper as the seal of this Corporation, and alter the same at pleasure; and a design of said seal together with a copy of this Charter duly approved by the District Attorney of the Parish of St. James shall be deposited and recorded in the office of the Recorder of St. James Parish, and all deeds, acts and legal instruments of this Corporation, shall be attested by the signature of the President of the Board of Trustees, and the affixing of the seal of the Corporation thereto.

Article VI.

This Corporation shall be managed and administrated by a Board of Trustees composed of eight persons, and the first Board of Trustees organized under this act shall consist of the members undersigned, one of whom, viz: Henry Bellanger [*sic*] is hereby constituted and appointed President of this Corporation, and in case of his death, resignation, removal or otherwise, the Board shall by and with the consent of the superior of the order of the Marist Fathers appoint his successor or successors, to be chosen among the members of this Corporation.

Article VII.

That when a vacancy shall happen in the said Board of Trustees hereby constituted, by death, resignation, removal, insanity or otherwise, the remaining Trustees with the President shall have full power to fill such vacancy, and the said Board of Trustees shall at its discretion increase the number of its members, when ever the exigencies of the Corporation may require such increase, and the President with the approbation in writing of three members of the Board, may in his discretion remove any member of the Board.

Article VIII.

That the Board of Trustees shall have full power to adopt such by laws, rules and regulations not repugnant to the constitution and laws of the United States, nor of this state, as to it shall also have the power of creating such offices as it shall deem necessary, prescribe their functions and duties appertaining to such offices, and remove the said officers at discretion, and such officers may be selected from among the members of the Corporation or not, as said Board may deem expedient. And the appointment of a member of the Board to any office shall not disqualify him from acting as a member of said Board.

Article IX.

The Board of Trustees shall meet as often as they may be notified to do so by the President, and in case of sickness or prolonged absence of the President, the Board of Trustees shall appoint one of its members Vice President with the approbation of the President, who shall have and exercise all the powers vested in the President by this Charter.

The presence of five members the President included, shall be sufficient to constitute a quorum. The majority of votes, including the President, shall suffice to decide finally on all questions presented to said Board. But if the majority does not include the President, two thirds of the votes shall be required to make a decision final.

Article X.

No trustee nor any member of this Corporation shall ever be personally responsible for the debts or liabilities of this Corporation. No private acts or contracts of any of the members of this Corporation shall, in any way bind the Corporation. Nor shall any member of this Corporation be ever entitled to any claims for personal services or otherwise, past or present, nor shall any member be entitled to or pretend to possess in his own and individual right any part or portion whatsoever of the property and capital of this Corporation.

Article XI.

That this constitution may be amended at any time by the Board of Trustees, in pursuance of and in conformity with the laws regulating the creation of Corporations to be found in section six hundred and seventy nine of the Revised statutes of eighteen hundred and seventy.

Article XII.

That the duration of this Corporation is hereby fixed at twenty five years from the date hereof.

Article XIII.

That no misnomers of this Corporation shall defeat or annul any donation, gift, grant, devise bequest or conveyance to or from said corporation.

Thus done, signed and accepted in the Parish of St. James, in the state of Louisiana, on this 10th day of April 1875.

President Reverend H. Bellanger, S.M.
Vice President Reverend G. B. Paigot, S.M.
Secretary Reverend O. Renaudier, S.M.
Treasurer Reverend U. Durel, S.M.

Appendix G

Sale of St. Mary's Jefferson College to the Society of Jesus

Seventy-third Meeting of the Corporation of: The Fathers of the Society of Mary

On this day January, the fifth, 1931, the Corporation of "The Fathers of the Society of Mary" held its meeting at Paulina, La. The following members constituting a quorum were present: Very Rev. Father Butin, President, Rev. H. Georges, Rev. J. Collins, Rev. J. Carroll, Rev. R. Hebert. . . .

On mention of Rev. H. Georges seconded by Father Hebert, the following resolution was unanimously passed. Resolved that this Corporation sell to SOCIETE CATHOLIQUE D'EDUCATION RELIGIEUSE ET LITTERAIRE, for the price and sum of $27,000 and $500 cash: the following described property, to wit:

> (1) A certain tract of land situated in the Parish of St. James on the left bank of the Mississippi River, at about 60 miles above the city of New Orleans, measuring three arpents and twenty feet, more or less, in front of said river by a depth of twenty-nine arpents, between lines closing in the rear, so as to meet at said point, bounded above by lands formerly belonging to, J. N. Dufont, now to Rev. F. Chauvin, and below by lands of Widow Pierre Millet, together with the College building, the chapel building and all other buildings and improvements thereunto belonging, situated and all the privileges, rights, servitudes and accessories thereto belonging under the act of donation by Valcour Aime passed before J. B. Ferchaud, recorder of the Parish of St. James on January 3, 1860.
>
> (2) A tract of land or plantation, situated in the Parish of St. James on the left bank of the Mississippi River, at about 60 miles above the City of New

> Orleans, measuring two arpents front on said river, by a depth of twenty-six arpents, bounded above by the property of Henry Baude, and below by the property of St. Mary's Jefferson College, with all buildings and improvements thereon.
>
> (3) A certain portion or lot of ground, situated in the Parish of St. James, on the left bank of the Mississippi River, at a distance of 60 miles above the City of New Orleans, measuring one arpent, more or less on said river by a depth of about twenty-nine arpents, bounded above by the property of Jefferson College and below by lands formerly belonging to Edmond Druilhet and after to Edgard Webre, together with all the buildings and improvements thereon and thereunto belonging.
>
> (4) A certain tract of land situated in the Parish of St. James, on the left bank of the Mississippi River at about 60 miles above the City of New Orleans, measuring 74 feet front on said river, more or less, bounded above by the property belonging to "the Fathers of the Society of Mary" and below by that of Widow Antoine Ockman, together with all the buildings and improvements thereon and thereunto belonging. . . .

The act of sale is also to include all of the buildings, appurtenances, fixtures, machinery, etc., presently upon the property.

This Corporation to reserve to itself a one undivided one-half interest in and to all oil and mineral rights on the above described property, which reservation is to be exercised and used by this Corporation, its assigns and lessees, in cooperation with the purchaser of said property or its successors in title, and all drilling, mining or other operations shall be done at such places and under such requirements and restrictions as may be determined by the purchaser, or subsequent owner of the said property.

This Corporation to obligate itself, in the exercise of said reserved rights, either by itself, or its assigns and lessees to do nothing or to permit nothing to be done in prospecting for oil and minerals or thereafter in the operations incidental to the removing and handling of oils and minerals from the soil, that would impair the free use of the purchaser and its successors in title of and to the property sold to said purchaser, or do damage to any standing timber or to in anywise impair or affect the insurance upon any improvements now or hereafter to be erected on the herein described property or to cause an increase in the rates of such insurance.

Resolved further that Rev. Romain Butin, the [Corporation] President, and Rev. H. Georges, the secretary of this Corporation, be and are therefore authorized, instructed and empowered to sign the act of sale and transfer of said property, to receive the purchase price thereof and to grant full acquittance and discharge thereof, for and in the name of the Corporation.

Further giving and granting unto said Rev. Romain Butin, the President and Rev. H. Georges, the secretary of this Corporation full power and authority to sign any and all acts and do to all things necessary and proper in the premises.

Resolved further that this resolution and authority to sell the foregoing property at the price stipulated and the authority to receive the purchase price and grant acquittal and discharge thereof is vested by this board under the terms of this resolution unto the President and Secretary of this Corporation, acting for and on behalf thereof, is consistent with the charter and by-laws of this Corporation and is nowise inconsistent with any stipulation or provision thereof. . . .

Signed: Rev. R. Butin, S.M. President
Rev. H. Georges, Secretary and Treasurer

Notes

Introduction

1. Craig A. Bauer, *Creole Genesis: The Bringier Family and Antebellum Plantation Life in Louisiana* (Lafayette: University of Louisiana at Lafayette Press, 2011), ix.

2. James G. Dauphine, "Catholic Education in North Louisiana: The Pre-World War One Foundations of Catholic Life in a Southern Diocese," in *Cross, Crozier, and Crucible: A Volume Celebrating the Bicentennial of a Catholic Diocese in Louisiana*, ed. Glenn R. Conrad (New Orleans: Archdiocese of New Orleans in cooperation with the Center for Louisiana Studies, 1993), 289.

3. Craig A. Bauer, *A Leader among Peers: The Life and Times of Duncan Farrar Kenner* (Lafayette: University of Louisiana at Lafayette Press, 1989); Lillian C. Bourgeois, *Cabanocey: The History, Customs, and Folklore of St. James Parish* (Gretna, LA: Pelican, 1957); Richard Follett, *The Sugar Masters: Planters and Slaves in Louisiana's Cane World, 1820–1860* (Baton Rouge: Louisiana State University Press, 2005); Harnett T. Kane, *Plantation Parade: The Grand Manner in Louisiana* (New York: William Morrow, 1945); Donna McGee Oneband, *The House That Sugarcane Built: The Louisiana Burguières* (Jackson: University Press of Mississippi, 2014); J. Carlyle Sitterson, *Sugar Country: The Cane Sugar Industry in the South, 1753–1950* (Westport, CT: Greenwood, 1973); Mary Ann Sternberg, *River Road Rambler: A Curious Traveler along Louisiana's Historic Byway* (Baton Rouge: Louisiana State University Press, 2013.

4. Clare D'artois Leeper, *Louisiana Place Names: Popular, Unusual, and Forgotten Stories of Towns, Cities, Plantations, Bayous, and Even Some Cemeteries* (Baton Rouge: Louisiana State University Press, 2012), 220–221.

5. "Manresa House of Retreats," *Louisiana Historical Marker Program*, Historical Marker, River Road, Convent, LA, n.d.

6. Leroy E. Willie, *A Look at Louisiana's First Century, 1804–1903* (Baton Rouge: SAR Books, 2002), 27.

7. Cline, *Pioneer Leaders and Early Institutions in Louisiana Education*, 259.

8. Edwin Whitfield Fay, *The History of Education in Louisiana* (Washington, DC: Government Printing Office, 1898), 52.

9. Oliver P. Carriere, *A Sketch of the History of Jefferson College and Manresa House of Retreats, Convent, Louisiana: 1830–1973* (Convent, LA: Manresa House of Retreats, 1974), 9.

10. Bourgeois, *Cabanocey*, 26; Rodney Cline, *Pioneer Leaders and Early Institutions in Louisiana Education* (Baton Rouge: Claitor's Publishing Division, 1969), 262.

11. Roger Baudier, *The Catholic Church in Louisiana* (New Orleans: A. W. Hyatt Stationary Mfg. Co. Ltd., 1939), 418; John M. Barry, *Rising Tide: The Great Mississippi Flood of 1927 and How It Changed America* (New York: Simon and Schuster, 1997), 409; P. H. Dagneau, *Memories of Seventy-Five Years of the Centenary of the Marist Fathers in America* (Atlanta: Keystone Press, 1963), 49–50.

12. Bourgeois, *Cabanocey: The History, Customs and Folklore of St. James Parish*, 81.

13. Baudier, *The Catholic Church in Louisiana*, 418; Barry, *Rising Tide*, 409; "About Us," Manresa House of Retreats: Convent, LA, http://manresala.org/About-Us.aspx, accessed July 19, 2012.

14. Earl F. Niehaus, "Jefferson College: The Early Years," *Louisiana Historical Quarterly* 38, no. 4 (October 1955): 88; Joseph Maselli and Dominic Candeloro, *Italians in New Orleans: Images of America* (Mount Pleasant, SC: Arcadia Publishing, 2004), 14; Mary Bernardine Hill, "The Influence of James Hubert Blenk on Catholic Education in the Archdiocese of New Orleans, 1885–1917" (EdD diss., Louisiana State University, 1964), 82; *Catalogue of the Officers and Students of Jefferson College (St. Mary's): Distribution of Premiums for the Academic Year 1875–76, St. James Parish* (New Orleans: Morning Star Office, No. 116 Poydras Street, 1876), 15, in Louisiana State University Special Collections, Hill Memorial Library, Baton Rouge; *The Citizens' League: A History of the Great Reform Movement in New Orleans, April 21st 1896, with Biographical Sketches of Those Who Took Prominent Part in the Movement* (New Orleans: S. W. Taylor, 1897), 56, in Louisiana Research Collection, Tulane University; Alicia P. Long, *The Great Southern Babylon: Sex, Race, and Respectability in New Orleans, 1865–1920* (Baton Rouge: Louisiana State University Press, 2005), 111.

15. Roger L. Geiger, "Introduction: New Themes in the History of Nineteenth-Century College," in *The American College in the Nineteenth Century*, ed. Roger L. Geiger (Nashville: Vanderbilt University Press, 2000), 20; Thomas D. Snyder, ed., *120 Years of American Education: A Statistical Report* (Washington, DC: National Center for Education Statistics, US Department of Education, Office of Educational Research and Improvement, 1993), 6.

16. Fredrick Rudolph, *The American College and University: A History* (Athens: University of Georgia Press, 1990), 280.

17. Ibid., 241–244; John R. Thelin, *A History of American Higher Education*, 2nd ed. (Baltimore: Johns Hopkins University Press, 2011), 129; Snyder, *120 Years of American Education: A Statistical Report*, 6.

18. Darryl L. Peterkin, "'Within These Walls': Reading and Writing Institutional Histories," in *The History of U.S. Higher Education: Methods for Understanding the Past*, ed. Marybeth Gasman (New York: Routledge, 2010), 9.

19. Dagmar Renshaw LeBreton, *Valcour Aime: A Biography* (n.p., 1948), 150, in Louisiana and Special Collections Department, Earl K. Long Library, University of New Orleans.

20. Nathan A. Buman, "Two Histories, One Future: Louisiana Sugar Planters, Their

Slaves, and the Anglo-Creole Schism, 1815–1865" (PhD diss., Louisiana State University, 2013); Rien Fertel, *Imagining the Creole City: The Rise of Literary Culture in Nineteenth-Century New Orleans* (Baton Rouge: Louisiana State University Press, 2014); Gwendolyn Midlo Hall, *Africans in Colonial Louisiana: The Development of Afro-Creole Culture in the Eighteenth Century* (Baton Rouge: Louisiana State University Press, 1995); Alfred N. Hunt, *Haiti's Influence on Antebellum America: Slumbering Volcano in the Caribbean* (Baton Rouge: Louisiana State University Press, 1988).

21. Baudier, *The Catholic Church in Louisiana*; Glenn R. Conrad, ed., *Cross, Crozier, and Crucible: A Volume Celebrating the Bicentennial of a Catholic Diocese in Louisiana* (New Orleans: Archdiocese of New Orleans in Cooperation with the Center for Louisiana Studies, 1993); Randall Miller and John Wakelyn, eds., *Catholics in the Old South* (Macon, GA: Mercer University Press, 1999).

22. Carriere, *A Sketch of the History of Jefferson College and Manressa House of Retreats Convent, Louisiana*, 6.

23. Fay, *The History of Education in Louisiana*, 49.

24. Follett, *The Sugar Masters*, 41; Willie, *A Look at Louisiana's First Century, 1804–1903*, 36.

25. Dagneau, *Memoirs of Seventy-five Years for the Centenary of the Marist Fathers in America*, 48; J. Frazer Smith, *White Pillars: The Architecture of the South* (New York: Bramhall Publishing, 1970), 197.

26. John B. Rehder, *Delta Sugar: Louisiana's Vanishing Plantation Landscape* (Baltimore: Johns Hopkins University Press, 1999), 172; Robert Mark Rudd, "Apocryphal Grandeur: Belle Grove Plantation in Iberville Parish, Louisiana" (MA thesis, University of Delaware, 2002), 22.

27. Donald G. Tewksbury, *The Founding of American Colleges and Universities before the Civil War* (1932; reprint, Mansfield Centre, CT: Martino, 2011), 25–27, 193.

28. Thelin, *A History of American Higher Education*, xii–xiii.

29. Ibid., xxvii.

30. Roger Wallace Shugg, "Survival of the Plantation System in Louisiana," in *The Louisiana Bicentennial Series in Louisiana History: Volume VI, Reconstructing Louisiana* (Lafayette: Center for Louisiana Studies, University of Louisiana at Lafayette, 2001), 469, 473–474.

31. Raymond H. Schmandt, "An Overview of Institutional Establishments in the Antebellum Southern Church," in *Catholics in the Old South*, ed. Randall Miller and John Wakelyn (Macon, GA: Mercer University Press, 1999), 72.

32. Cline, *Pioneer Leaders and Early Institutions in Louisiana Education*, 268.

33. Elizabeth Barrouquere, "Last of the Marists Catholic order leaving Convent after arriving in 1862," *The Advocate (Baton Rouge, La.)* (July 13, 2002): 1F.

34. Alcée Fortier, *Louisiana Studies: Literature, Customs and Dialects, History and Education* (New Orleans: F. F. Hansell and Bro., 1894), 253, 302–304.

35. Alcée Fortier, ed., *Louisiana: Comprising Sketches of Parishes, Towns, Events, Institutions, and Persons, Arranged in Cyclopedic Form* Volume I (Century Historical Association, 1914), 579–582.

36. Otto S. Varnado, "A History of the Early Institutions of Higher Learning in Louisiana" (MA Louisiana State University, 1927), 41, 47.

37. "Finds Traces of State College that Vanished," *Times Picayune*, September 5, 1949, n.p.; Vernon Elmer Kappel, "Louis Dufau's Louisiana College" (MA thesis, Tulane University, 1949), 56.

38. Edward John Power, *A History of Catholic Higher Education in the United States* (Milwaukee: Bruce Publishing Company, 1958), 263.

39. Ibid.

40. Dagneau, *Memories of Seventy-Five Years of the Centenary of the Marist Fathers in America*, 47–73.

41. Hill, "The Influence of James Hubert Blenk on Catholic Education in the Archdiocese of New Orleans, 1885–1917," 46–73.

42. Carriere, *A Sketch of the History of Jefferson College and Manressa House of Retreats Convent, Louisiana*, 4–14.

43. William Reeves, *Manresa on the Mississippi: For the Greater Glory of God* (Convent, LA: Manresa House of Retreats, 1996), 26–27.

44. Carriere, *A Sketch of the History of Jefferson College and Manressa House of Retreats Convent, Louisiana*, 33–34.

45. "Manresa Iconography," Manresa House of Retreats, Convent, LA: Society of Jesus USA, http://manresala.org/Prayers/Iconography.aspx, accessed September 7, 2014.

46. Gary McCulloch and William Richardson, *Historical Research in Educational Settings* (Buckingham: Open University Press, 2000), 125.

Chapter 1

1. Rodney Cline, "Early Colleges and Schools in Louisiana," in *The Louisiana Bicentennial Series in Louisiana History: Volume XVII, Education in Louisiana*, ed. Michael G. Wade (Lafayette, Louisiana: Center for Louisiana Studies, University of Louisiana at Lafayette, 1999), 230.

2. Richard Follett, *The Sugar Masters: Planters and Slaves in Louisiana's Cane World, 1820–1860* (Baton Rouge: Louisiana State University Press, 2005) 41.

3. Dagmar Renshaw LeBreton, *Valcour Aime: A Biography* (n.p., 1948), 60–61, Louisiana and Special Collections Department, Earl K. Long Library, University of New Orleans.

4. Michael O'Brien, *Intellectual Life and the American South, 1810–1860* (Chapel Hill: University of North Carolina Press, 2010), 97.

5. Earl F. Niehaus, "Jefferson College: The Early Years," *Louisiana Historical Quarterly* 38, no. 4 (October 1955): 64.

6. "Proceedings and Debates of the Convention of Louisiana, 1844," as cited in *A Documentary of Education in the South before 1860 in Five Volumes: Volume V, Educational Theories and Practices*, ed. Edgar W. Knight (Chapel Hill: University of North Carolina Press, 1953), 285–286.

7. Ned Sublette, *The World That Made New Orleans: From Spanish Silver to Congo Square* (Chicago: Lawrence Hill Books, 2008), 79; Gwendolyn Midlo Hall, "The Formation of Afro-Creole Culture," in *Creole New Orleans: Race and Americanization*, ed. Arnold R. Hirsch and Joseph Logsdon (Baton Rouge: Louisiana State University Press, 1992), 60; Joseph G. Tregle Jr., "Creoles and Americans," in *Creole New Orleans: Race and Americanization*, ed. Arnold R. Hirsch and Joseph Logsdon (Baton Rouge: Louisiana State University Press, 1992), 133.

8. Sublette, *The World That Made New Orleans*, 79.

9. Tregle, "Creoles and Americans," 137; Philomena Hauck, *Bienville: Father of Louisiana, Louisiana Life Series, No. 10* (Lafayette: Center for Louisiana Studies, University of Southwestern Louisiana, 1998), 143–145.

10. José de Pedro, *The Spanish in New Orleans and Louisiana*, trans. Richard E. Chandler (Gretna, LA: Pelican, 2000), 17, 65, 68.

11. Dianne Guenin-Lelle, *The Story of French New Orleans: History of a Creole City* (Jackson: University Press of Mississippi, 2016), 88.

12. Kimberly S. Hanger, *Bounded Lives, Bounded Places: Free Black Society in Colonial New Orleans* (Durham, NC: Duke University Press, 1997), 163–164; Emily Epstein Landau, *Spectacular Wickedness: Sex. Race, and Memory in Storyville, New Orleans* (Baton Rouge: Louisiana State University Press, 2013), 67.

13. Landau, *Spectacular Wickedness*, 79–80; Jocelyn Hazelwood Donlon, "Louisiana's Three Fold Regions," http://www.louisianafolklife.org/LT/Articles_Essays/la_3_folk_reg.html, accessed February 26, 2013.

14. Sublette, *The World That Made New Orleans*, 79–80; Donlon, "Louisiana's Three Fold Regions."

15. Emily Clark, *The Strange History of the American Quadroon: Free Women of Color in the Revolutionary Atlantic World* (Chapel Hill: University of North Carolina Press, 2013), 66.

16. Powell, *The Accidental City*, 108.

17. Claude F. Jacobs and Andrew J. Kaslow, *The Spiritual Churches of New Orleans: Origins, Beliefs, and Rituals of an African-American Religion* (Knoxville: University of Tennessee Press, 1991), 21.

18. Landau, *Spectacular Wickedness*, 85; Tregle, "Creoles and Americans," 131–185; Jennifer M. Spear, *Race, Sex, and Social Order in Early New Orleans* (Baltimore: Johns Hopkins University Press, 2009), 142–143.

19. Justin A. Nystrom, *New Orleans after the Civil War: Race, Politics, and the Birth of Freedom* (Baltimore: Johns Hopkins University Press, 2010), 155.

20. George W. Cable, *The Creoles of Louisiana* (1884; reprint, Gretna, LA: Pelican, 2005), 41, 42.

21. Virginia R. Dominguez, *White by Definition: Social Classification in Creole Louisiana* (New Brunswick, NJ: Rutgers University Press, 1986), 263.

22. Powell, *The Accidental City*, 214.

23. Charles Gayarré, *The Creoles of History and the Creoles of Romance: A Lecture Delivered in the Hall of the Tulane University by Charles Gayarré, on the 25th of April, 1885* (New Orleans: C. E. Hopkins, 1885), 1.

24. George Dargo, *Jefferson's Louisiana: Politics and the Clash of Legal Traditions*, rev. ed. (Clark: Lawbook Exchange, 2009), 13.

25. Tregle, "Creoles and Americans," 140.

26. Landau, *Spectacular Wickedness*, 67.

27. Powell, *The Accidental City*, 328; Paul F. Lachance, "The Foreign French," in Hirsch and Logsdon, *Creole New Orleans*, 101–130 (Baton Rouge: Louisiana State University Press, 1992), 102.

28. Albert E. Fossier, *New Orleans the Glamour Period, 1800–1840: A History of the Conflicts of Nationalities, Languages, Religion, Morals, Cultures, Laws, Politics and Economics during the Formative Period of New Orleans* (Gretna, LA: Firebird Press, Pelican, 1998), 274.

29. Tregle, "Creoles and Americans," 140.

30. John Malcolm Forbes Ludlow, *A Sketch of the United States from Independence to Secession* (Cambridge, MA: Macmillan and Co., 1862), 316.

31. Powell, *The Accidental City*, 328; Lachance, "The Foreign French," 101, 102, 121.

32. Henry Augustine Washington, "The Social System of Virginia," *SLM* 14 (February 1848): 65–81, as cited in Michael O'Brien, ed., *All Clever Men, Who Make Their Way: Critical Discourse in the Old South* (Athens: University of Georgia Press, 1992), 242, 243.

33. Cable, *The Creoles of Louisiana*. 260.

34. Carl A. Brasseaux, *Acadian to Cajun: Transformation of a People, 1803–1877* (Jackson: University Press of Mississippi, 1992), 4.

35. Glenn R. Conrad, "The Faces of French Louisiana," in *Cross, Crozier, and Crucible: A Volume Celebrating the Bicentennial of a Catholic Diocese in Louisiana*, ed. Glenn R. Conrad, 5–15 (New Orleans: Archdiocese of New Orleans in cooperation with the Center for Louisiana Studies, 1993), 13.

36. Nathan A. Buman, "Two Histories, One Future: Louisiana Sugar Planters, Their Slaves, and the Anglo-Creole Schism, 1815–1865" (PhD diss., Louisiana State University, 2013), 9, 90, 174.

37. J. Carlyle Sitterson, *Sugar Country: The Cane Sugar Industry in the South, 1753–1950* (Westport, CT: Greenwood, 1973), 68.

38. Powell, *The Accidental City*, 109.

39. Carl A. Brasseaux, "Acadian Education: From Cultural Isolation to Mainstream America," in *The Louisiana Bicentennial Series in Louisiana History: Volume XVII, Education in Louisiana*, ed. Michael G. Wade, 236–242 (Lafayette, Louisiana: Center for Louisiana Studies, University of Louisiana at Lafayette, 1999), 236.

40. Conrad, "The Faces of French Louisiana," 14.

41. Lachance, "The Foreign French," 115.

42. Lillian C. Bourgeois, *Cabanocey: The History, Customs and Folklore of St. James Parish* (Gretna, LA: Pelican, 1957), 79.

43. Jon L. Wakelyn, "Catholic Elites in the Slaveholding South," in *Catholics in the Old South*, ed. Randall Miller and John Wakelyn (Macon, GA: Mercer University Press, 1999), 230.

44. Roger W. Shugg, *Origins of Class Struggle in Louisiana: A Social History of White Farmers and Laborers during Slavery and After, 1840–1875* (1939; reprint, Baton Rouge: Louisiana State University Press, 1966), 27–28.

45. Charles P. Roland, *Louisiana Sugar Plantations during the Civil War* (Baton Rouge: Louisiana State University Press, 1997), 6.

46. Ibid., 142.

47. Ibid., 144, 175, 182.

48. John C. Rodrigue, *Reconstruction in the Cane Fields: From Slavery to Free Labor in Louisiana's Sugar Parishes, 1862–1880* (Baton Rouge: Louisiana State University Press, 2001), 24.

49. Shugg, *Origins of Class Struggle in Louisiana*, 31–33, 86, 127.

50. Sitterson, *Sugar Country*, 72.

51. Jennifer R. Green, *Military Education and the Emerging Middle Class in the Old South* (Cambridge: Cambridge University Press, 2008), 153.

52. Ibid.

53. Shugg, *Origins of Class Struggle in Louisiana*, 313.

54. Follett, *The Sugar Masters*, 85.

55. Shugg, *Origins of Class Struggle in Louisiana*, 29, 30.

56. Sitterson, *Sugar Country*, 88.

57. Henry O. Robertson, *"In the Habit of Acting Together": The Emergence of the Whig*

Party in Louisiana, 1838–1840 (Lafayette: Center for Louisiana Studies, University of Louisiana at Lafayette, 2007), 6–8, 24.

58. Shugg, *Origins of Class Struggle in Louisiana*, 149.

59. Freddi Williams Evans, *Congo Square: African Roots in New Orleans* (Lafayette: University of Louisiana at Lafayette Press, 2011), 15; Albert H. Biever, *The Jesuits in New Orleans and the Mississippi Valley: Jubilee Memorial* (New Orleans: Hauser Printing Company, 1924), 34.

60. John B. Rehder, *Delta Sugar: Louisiana's Vanishing Plantation Landscape* (Baltimore: Johns Hopkins University Press, 1999), 42–43; William Stubbs, *The Cultivation of Sugar Cane* (Savannah, GA: D. G. Purse, 1900), 7.

61. Cable, *The Creoles of Louisiana*, 111–113.

62. Herbert Asbury, *The French Quarter: An Informal History of the New Orleans Underworld* (New York: Thunder's Mouth Press, 1939), 33; Leonard Huber, *New Orleans: A Pictorial History, From the Earliest Times to the Present Day* (New York: Crown Publishers, 1971), 5.

63. Stubbs, *The Cultivation of Sugar Cane*, 7–8.

64. Ibid., 8; Follett, *The Sugar Masters*, 17–18; Buman, "Two Histories, One Future," 28; Rodrigue, *Reconstruction in the Cane Fields*, 11–12.

65. Ibid., 19.

66. Follett, *The Sugar Masters*, 33–34; Solomon Northup, *Twelve Years a Slave*, ed. Sue Eakin and Joseph Logsdon (1853; reprint, Baton Rouge: Louisiana State University Press, 1996), 161–162; Richard Follett, "The Rise and Fall of American Sugar," in *Plantation Kingdom: The American South and its Global Commodities*, ed. Richard Follett, Sven Beckert, Peter Coclanis, and Barbara Hahn (Baltimore: Johns Hopkins University Press, 2016), 64–65, 68.

67. Follett, *The Sugar Masters*, 24, 25, 92–94, 104–105; Buman, "Two Histories, One Future," 19.

68. Follett, *The Sugar Masters*, 47.

69. Rodrigue, *Reconstruction in the Cane Fields*, 75–76.

70. Joe Gray Taylor, *Negro Slavery in Louisiana* (Baton Rouge: Louisiana State University Press, 1963), 179; Eugene D. Genovese, *Roll, Jordan, Roll: The World the Slaves Made* (New York: Random House, 1976), 321.

71. Henry Marie Brackenridge, *Views of Louisiana, Together with a Journal of a Voyage up the Mississippi River in 1811* (1814; reprint, Chicago: Quadrangle Books, 1962), 173.

72. John David Smith, introduction to *Louisiana Sugar Plantations during the Civil War* by Charles P. Roland (Baton Rouge: Louisiana State University Press, 1997), xii.

73. Rodrigue, *Reconstruction in the Cane Fields*, 11.

74. Edwin Hergesheimer, *Map Showing the Distribution of the Slave Population of the Southern States of the United States. Compiled from the Census of 1860* (Washington, DC: Henry S. Graham, 1861), Library of Congress, Geography and Map Division; Susan Schulten, *Mapping the Nation: History and Cartography in Nineteenth-Century America* (Chicago: University of Chicago Press, 2012), 119–121.

75. Sitterson, *Sugar Country*, 88; Smith, introduction to *Louisiana Sugar Plantations during the Civil War*, xii.

76. O'Brien, *Intellectual Life and the American South, 1810–1860*, 73.

77. Rodrigue, *Reconstruction in the Cane Fields*, 26.

78. Roland, *Louisiana Sugar Plantations during the Civil War*, 5.

79. Powell, *The Accidental City*, 261.

80. Buman, "Two Histories, One Future," 34–35, 97; Rehder, *Delta Sugar*, 19–21; Ned Sublette and Constance Sublette, *The American Slave Coast: A History of the Slave-Breeding Industry* (Chicago: Lawrence Hill Books, 2016), 12.

81. Tregle, "Creoles and Americans," 134.

82. Dargo, *Jefferson's Louisiana*, 9.

83. Rodrigue, *Reconstruction in the Cane Fields*, 12.

84. Sitterson, *Sugar Country*, 69.

85. Dargo, *Jefferson's Louisiana*, 15, 19.

86. Nystrom, *New Orleans after the Civil War*, 167; Randall M. Miller, "A Church in Cultural Captivity: Some Speculations on Catholic Identity in the Old South," in *Catholics in the Old South*, ed. Randall Miller and Jon Wakelyn, 11–52 (Macon, GA: Mercer University Press, 1999), 34.

87. Lawrence N. Powell, introduction to *The New Orleans of George Washington Cable: The 1887 Census Office Report* by George Washington Cable (Baton Rouge: Louisiana State University Press, 2008), 11.

88. Patricia L. Duncan, "From Creole College to Religious Retreat: The Story of a Louisiana Landmark," *Preservation in Print* (August 9, 1994): 1; Nathalie Dessens, *Creole City: A Chronicle of Early American New Orleans* (Gainesville: University Press of Florida, 2015), 171–172.

89. Margaret Humphreys, *Yellow Fever and the South* (New Brunswick, NJ: Rutgers University Press, 1992), 48.

90. Ibid., 108, 113–114; *François Drouilly, 15 Days of Prayer with Jean-Claude Colin* (Hyde Park: New City Press, 2012), 17–18.

91. Humphreys, *Yellow Fever and the South*, 116, 119; Richard Campanella, *Bourbon Street: A History* (Baton Rouge: Louisiana State University Press, 2014), 26.

92. Duncan, "From Creole College to Religious Retreat: The Story of a Louisiana Landmark," 1.

93. Bourgeois, *Cabanocey*, 78.

94. Ibid., 78, 13.

95. Dawdy, *Building the Devil's Empire*, 170.

96. Niehaus, "Jefferson College: The Early Years," 63.

97. Curtis A. Manning, *The History of Higher Education in Louisiana* (Bloomington, IN: Xlibris Corporation, 2006), 14–15.

98. Powell, *The Accidental City: Improvising New Orleans*, 321.

99. John Charles Dawson, *Lakanal, The Regicide: A Biographical and Historical Study of the Career of Joseph Lakanal* (Tuscaloosa: University of Alabama Press, 1946), 125.

100. Dargo, *Jefferson's Louisiana*, 130.

101. Ibid., 89–92.

102. Ibid.

103. Powell, *The Accidental City*, 340–342.

104. Ibid., 342–343.

105. Dargo, *Jefferson's Louisiana*, 102–103, 131, 269, 271.

106. Powell, *The Accidental City*, 343.

107. Ibid., 343; Sacher, *A Perfect War of Politics*, 14; Glen Jeansonne and David Luhrssen, "William C. C. Claiborne," in *KnowLA: Encyclopedia of Louisiana*, ed. David Johnson (Louisiana Endowment for the Humanities, 2010–, 2013), http://www.knowla.org/entry/921/; Governor Claiborn to Thomas Jefferson, Monticello, Virginia, January 12,

1810, cited in Clarence Edwin Carter, ed., *The Territorial Papers of the United States: Volume IX, The Territory of Orleans, 1803–1812* (Washington, DC: US Government Printing Office, 1940), 864–865; John M. Sacher, *A Perfect War of Politics: Parties, Politicians, and Democracy in Louisiana, 1824–1861* (Baton Rouge: Louisiana State University Press, 2003), 14; Daniel Rasmussen, *American Uprising: The Untold Story of American's Largest Slave Revolt* (New York: HarperCollins, 2011), 119; Follett, "The Rise and Fall of American Sugar," 66.

108. Lee Morgan, *Centenary College of Louisiana, 1825–2000: The Biography of an American Academy* (Shreveport: Centenary College of Louisiana Press, 2008), 3.

109. Dawson, *Lakanal, The Regicide*, 127.

110. Morgan, *Centenary College of Louisiana, 1825–2000*, 2, 3.

111. Sublette, *The World That Made New Orleans*, 209.

112. Order of the Board of Trustees, *The Charts and Statues of Jefferson College, Washington, Mississippi, as Revised and Amended: Together with a Historical Sketch of the Institution from its Establishment to the Present Time: To Which Is Prefixed a List of the Trustees, Officers and Faculty, The Acts of Congress and of the Legislature Relating to the Institution, and a Catalogue of Its Library, Apparatus, &c.* (Natchez, MS: Book and Job Office, Main Street, Near the Post Office, 1840), 19, 31; Memorial to Congress by the Trustees of Jefferson College, Mississippi Territory, January 3, 1803, in *The Territorial Papers of the United States: Volume V, The Territory of Mississippi, 1798–1817*, ed. Clarence Edwin Carter (Washington, DC: US Government Printing Office, 1937), 181–182; Memorial to Congress by the Territorial Legislature, Mississippi Territory, December 22, 1812, in *The Territorial Papers of the United States: Volume VI, The Territory of Mississippi, 1809–1817 Continued*, ed. Clarence Edwin Carter (Washington, DC: US Government Printing Office, 1937), 344–345.

113. William T. Blain, *Education in the Old Southwest: A History of Jefferson College Washington, Mississippi* (Natchez, MS: Judson Printing Company, 1976), 13, 25; Edward Turner to John C. Breckinridge, Washington, Mississippi, November 2, 1803, in *The Territorial Papers of the United States: Volume V, The Territory of Mississippi, 1798–1817*, ed. Clarence Edwin Carter (Washington, DC: US Government Printing Office, 1937), 268; Memorial to Congress by the Trustees of Jefferson College, Natchez, Mississippi, October 5, 1804, in *The Territorial Papers of the United States: Volume V, The Territory of Mississippi, 1798–1817*, ed. Clarence Edwin Carter (Washington, DC: US Government Printing Office, 1937), 333–335.

114. Order of the Board of Trustees, *The Charts and Statues of Jefferson College, Washington, Mississippi, as Revised and Amended*, 3, 7.

115. Blain, *Education in the Old Southwest*, 132.

116. "Jefferson College Today," MDAH: Historic Jefferson College, http://mdah.state.ms.us/hprop/hjccurrent.html, accessed August 3, 2014; "Jefferson Military College Foundation," MDAH: Historic Jefferson College, http://mdah.state.ms.us/hprop/jmcf.html, accessed August 3, 2014.

117. Edwin Whitfield Fay, *The History of Education in Louisiana* (Washington, DC: Government Printing Office, 1898), 31, 67; Donald E. DeVore and Joseph Logsdon, *Crescent City Schools: Public Education in New Orleans, 1841–1991* (Lafayette: Center for Louisiana Studies, University of Southwestern Louisiana, 1991), 7.

118. Adam Otis Hebert Jr., "History of Education in Colonial Louisiana," in *The Louisiana Purchase Bicentennial Series in Louisiana History, Volume XVIII: Education in Louisiana* (Lafayette: Center for Louisiana Studies, University of Louisiana at Lafayette,

1999), 30–31; Emily Clark, *Voices from an Early American Convent: Marie Madeleine and the Development of a New Society, 1724–1834* (Baton Rouge: Louisiana State University Press, 2007), 15–16; Trent Angers, *Grand Coteau: The Holy Land of the South* (Lafayette, LA: Acadian House Publishing, 2005), 32–33; *The Yearbook of Education for 1879. Being the Second Annual Supplement to the Cyclopedia of Education: Dictionary of Information for the Use of Teachers, School Officers, Parents, and Others* (New York: E. Stieger, 1879), 211.

119. Claude L. Vogel, *The Capuchins in French Louisiana (1722–1766)* (New York: Joseph F. Wagner, 1928), 69–101; R. Eric Platt, *Sacrifice and Survival: Identity, Mission, and Jesuit Higher Education in the American South* (Tuscaloosa: University of Alabama Press, 2014), 25–27; Marcel Giraud, *A History of French Louisiana, Volume 5: The Company of the Indies, 1723–1731*, trans. Brian Pearce (Baton Rouge: Louisiana State University Press, 1991), 78, 299–303.

120. Sitterson, *Sugar Country*, 85; Sarah L. Hyde, "Punctuated Progress: Education Developments in the Antebellum Gulf South," in *The Enigmatic South: Toward Civil War and Its Legacies*, ed. Samuel C. Hyde Jr. (Baton Rouge: Louisiana State University Press, 2014), 24, 25–26.

121. Shugg, *Origins of Class Struggle in Louisiana*, 68.

122. Sitterson, *Sugar Country*, 84.

123. Shugg, *Origins of Class Struggle in Louisiana*, 35.

124. Lorri Glover, *Southern Sons: Becoming Men in the New Nation* (Baltimore: Johns Hopkins University Press, 2007), 51; Dawson, *Lakanal, The Regicide*, 124.

125. Robert C. Reinders, *End of an Era: New Orleans, 1850–1860* (1964; reprint, Gretna, LA: Pelican, 1998), 143; Liliane Crété, *Daily Life in Louisiana, 1815–1830*, trans. Patrick Gregory (Baton Rouge: Louisiana State University Press, 1981), 120.

126. Dawdy, *Building the Devil's Empire*, 56; Sarah L. Hyde, *Schooling in the Antebellum South: The Rise of Public and Private Education in Louisiana, Mississippi, and Alabama* (Baton Rouge: Louisiana State University Press, 2016), 69–71; Sitterson, *Sugar Country*, 85; Shugg, *Origins of Class Struggle in Louisiana*, 68.

127. Dan R. Frost, *Thinking Confederates: Academia and the Idea of Progress in the New South* (Knoxville: University of Tennessee Press, 2000), 18–19.

128. Stuart Grayson Noble, "Governor Claiborne and the Public School System of the Territorial Government of Louisiana," in *The Louisiana Bicentennial Series in Louisiana History: Volume XVII, Education in Louisiana*, ed. Michael G. Wade, 51–64 (Lafayette: Center for Louisiana Studies, University of Louisiana at Lafayette, 1999), 60.

129. Dawson, *Lakanal, The Regicide*, 122–123; Crété, *Daily Life in Louisiana, 1815–1830*, 120, 126–127.

130. William Hamilton Nelson, *A Burning Touch and a Flaming Fire: The Story of Centenary College of Louisiana* (Nashville: Methodist Publishing House, 1931), 64.

131. Fred Daspit, *Louisiana Architecture, 1714–1820* (Lafayette: Center for Louisiana Studies, University of Louisiana at Lafayette, 2004), 213; Roulhac Toledand and Mary Louise Christovich, *New Orleans Architecture, Volume VI: Faubourg Tremé and the Bayou Road, North Rampart Street to North Broad Street, Canal Street to St. Bernard Avenue* (Gretna, LA: Pelican, 2003), 195.

132. Henry C. Castellanos, *New Orleans As It Was: Episodes of Louisiana Life* (New Orleans: L. Graham & Son, 1895), 317.

133. Fay, *The History of Education in Louisiana*, 31; Leonard V. Huber, *The Battle of New Orleans: New Orleans As It Was in 1814–1815* (Gretna, LA: Pelican, 2011), 42.

134. Joseph A. Breaux, "Some Early Colleges and Schools of Louisiana," in *Publications of the Louisiana Historical Society: New Orleans, Louisiana, Volume VII—1913–14*, 136–142 (New Orleans: Louisiana Historical Society, 1915), 137; Dawson, *Lakanal, The Regicide*, 128–129; Memorial to Congress from the Regents of the University [*sic*] of Orleans, New Orleans, Louisiana, April 20, 1812, as cited in Clarence Edwin Carter, ed., *The Territorial Papers of the United States: Volume IX, The Territory of Orleans, 1803–1812* (Washington, DC: US Government Printing Office, 1940), 1014–1016.

135. Nathalie Dessens, *From Saint-Domingue to New Orleans, Migration and Influences* (Gainesville: University Press of Florida, 2007), 88; William D. Reeves, *Historic Louisiana: An Illustrated History* (San Antonio: Historical Publishing Network, 2003), 17.

136. Dawson, *Lakanal, The Regicide*, 128–129; Memorial to Congress from the Regents of the University [*sic*] of Orleans, in Carter, *The Territorial Papers of the United States: Volume IX*, 1015; Edward Larocque Tinker, "Charles Gayarré 1805–96," in *The Papers of the Bibliographical Society of America* 27, no. 1 (1933): 27.

137. Charles Gayarré, *Fernando de Lemos: Truth and Fiction, a Novel* (New York: G. W. Carleton & Co., 1872), 13.

138. Ibid., 24, 25.

139. Ibid., 19–29; Crété, *Daily Life in Louisiana, 1815–1830*, 121–124.

140. Breaux, "Some Early Colleges and Schools of Louisiana," 129.

141. Alfred N. Hunt, *Haiti's Influence on Antebellum America: Slumbering Volcano in the Caribbean* (Baton Rouge: Louisiana State University Press, 1988), 57.

142. Dawson, *Lakanal, The Regicide*, 130.

143. Ibid., 3–5; Simone de la Souchére Deléry, *Napoleon's Soldiers in America* (1950; reprint, Gretna, LA: Pelican, 1998), 101.

144. Deléry, *Napoleon's Soldiers in America*, 101–103.

145. Dawson, *Lakanal, The Regicide*, 10–11.

146. Ibid., 91–96; Thomas L. Purvis, "The Seven Years War and Its Political Legacy, in *A Companion to the American Revolution* (Malden, MA: Blackwell, 2004), 112–117; Deléry, *Napoleon's Soldiers in America*, 78–79.

147. Dawson, *Lakanal, The Regicide*, 98–99, 101.

148. Ibid., 118.

149. Deléry, *Napoleon's Soldiers in America*, 78–79.

150. Dawson, *Lakanal, The Regicide*, 112–113, 117.

151. Ibid.

152. Mikko Macchione, *Napoleon House* (New Orleans: Vissi d'Arte Books, 2006), 8–45.

153. Dawson, *Lakanal, The Regicide*, 117.

154. Ibid., 101–103.

155. Dawson, *Lakanal, The Regicide*, 134.

156. Deléry, *Napoleon's Soldiers in America*, 102–105.

157. Ibid., 104–105.

158. Ibid., 106.

159. Hunt, *Haiti's Influence on Antebellum America*, 57.

160. Fossier, *New Orleans the Glamour Period, 1800–1840*, 233–232.

161. Dawson, *Lakanal, The Regicide*, 136–140; Deléry, *Napoleon's Soldiers in America*, 104, 105; Hyde, *Schooling in the Antebellum South*, 70.

162. Dawson, *Lakanal, The Regicide*, 104, 105, 140, 163, 178; "Joseph Lakanal," *Find*

a Grave, http://www.findagrave.com/cgi-bin/fg.cgi?page=gr&GSln=lakanal&GSfn=joseph+&GSbyrel=all&GSdyrel=all&GSob=n&GRid=7754&df=all&, accessed December 20, 2016.

163. Ibid., 136–140.

164. Toledand and Christovich, *New Orleans Architecture, Volume VI*, 194, 195.

165. Castellanos, *New Orleans As It Was*, 317–318; Daspit, *Louisiana Architecture, 1714–1820*, 213.

166. Annemarie Kasteel, *Francis Janssens, 1843–1897: A Dutch-American Prelate* (Lafayette: Center for Louisiana Studies, University of Southwestern Louisiana, 1992), 283–304.

167. Grace Elizabeth King, *Creole Families of New Orleans* (New Orleans: Cornerstone, 2010), 273–274; Crété, *Daily Life in Louisiana, 1815–1830*, 121.

168. Gayarré, *Fernando de Lemos: Truth and Fiction, a Novel*, 18.

169. Rien Fertel, *Imagining the Creole City: The Rise of Literary Culture in Nineteenth-Century New Orleans* (Baton Rouge: Louisiana State University Press, 2014), 23; Hunt, *Haiti's Influence on Antebellum America*, 55–57; Dessens, *From Saint-Domingue to New Orleans, Migration and Influences*, 142.

170. Niehaus, "Jefferson College: The Early Years," 64.

171. Stanley Clisby Arthur and George Campbell Huchet de Kernion, *Old Families of Louisiana* (Gretna, LA: Firebird Press, Pelican, 1998), 270–271; Par L. Moreau Lislet, *Digeste Général Des Actes de la Législature de la Louisiane, Passés Depuis L'Anée 1804 Jusqu'en 1827, Volume 1* (New Orleans: Imprimé Par Benjamin Levy, 1828), 250.

172. Tregle, "Creoles and Americans," 149.

173. Bourgeois, *Cabanocey*, 62.

174. John P. Dyer, *Tulane: The Biography of a University, 1834–1965* (New York: Harper & Row, 1966), 18.

175. Niehaus, "Jefferson College: The Early Years," 64; Nelson, *A Burning Touch and a Flaming Fire: The Story of Centenary College of Louisiana*, 64.

176. Arthur and de Kernion, *Old Families of Louisiana*, 270–271.

177. Tregle, "Democracy Triumphant," 115.

178. Ibid., 128.

179. Sacher, *A Perfect War of Politics*, 44–46.

180. Ibid.; Walter Greaves Cowan and Jack B. McGuire, *Louisiana Governors: Rules, Rascals, and Reformers* (Jackson: University Press of Mississippi, 2008), 71.

181. Arthur and de Kernion, *Old Families of Louisiana*, 270–271.

182. Shugg, *Origins of Class Struggle in Louisiana*, 68, 69.

Chapter 2

1. Dan R. Frost, *Thinking Confederates: Academia and the Idea of Progress in the New South* (Knoxville: University of Tennessee Press, 2000), 18.

2. Roger W. Shugg, *Origins of Class Struggle in Louisiana: A Social History of White Farmers and Laborers during Slavery and After, 1840–1875* (1966; reprint, Baton Rouge: Louisiana State University Press, 1939), 475; Albert E. Fossier, *New Orleans the Glamour Period, 1800–1840: A History of the Conflicts of Nationalities, Languages, Religion, Morals, Cultures, Laws, Politics and Economics during the Formative Period of New Orleans* (Gretna, LA: Firebird Press, Pelican, 1998), 237.

3. Christina Vella, *Intimate Enemies: The Two Worlds of the Baroness De Pontalba* (Baton Rouge: Louisiana State University Press, 1997), 38.

4. Lorri Glover, *Southern Sons: Becoming Men in the New Nation* (Baltimore: Johns Hopkins University Press, 2007), 54.

5. Vella, *Intimate Enemies*, 38.

6. Lee Morgan, *Centenary College of Louisiana, 1825–2000: The Biography of an American Academy* (Shreveport, Centenary College of Louisiana Press, 2008), 3, 4; Walter L. Fleming, *Louisiana State University: 1860–1896* (Baton Rouge: Louisiana State University Press, 1936), 14–15.

7. Edwin Whitfield Fay, *The History of Education in Louisiana* (Washington, DC: Government Printing Office, 1898), 49.

8. Earl F. Niehaus, "Jefferson College: The Early Years," *Louisiana Historical Quarterly* 38, no. 4 (October 1955): 65.

9. Mary Bernardine Hill, "The Influence of James Hubert Blenk on Catholic Education in the Archdiocese of New Orleans, 1885–1917" (EdD diss., Louisiana State University, 1964), 51.

10. Morgan, *Centenary College of Louisiana, 1825–2000*, 5.

11. Stanley Clisby Arthur and George Campbell Huchet de Kernion, *Old Families of Louisiana* (Gretna, LA: Firebird Press, Pelican, 1998), 270–271; Estelle M. Fortier Cochran, *The Fortier Family and Allied Families* (San Antonio: E. M. F. Cochran, 1963), 76.

12. "Death of Ex-Gov. A. B. Roman," *Daily Picayune*, January 30, 1866, n.p.

13. Cochran, *The Fortier Family and Allied Families*, 76; Walter Greaves Cowan and Jack B. McGuire, *Louisiana Governors: Rules, Rascals, and Reformers* (Jackson: University Press of Mississippi, 2008), 71; Anna Blanche McGill, *The Sisters of Charity of Nazareth, Kentucky* (New York: Encyclopedia Press, 1917), 133.

14. Miriam G. Reeves, *The Governors of Louisiana* (Gretna, LA: Pelican, 1998), 51, 53.

15. Morgan, *Centenary College of Louisiana, 1825–2000*, 3, 4; William Hamilton Nelson, *A Burning Touch and a Flaming Fire: The Story of Centenary College of Louisiana* (Nashville: Methodist Publishing House, 1931), 67.

16. Niehaus, "Jefferson College: The Early Years," 66.

17. *Biographical and Historical Memoirs of Louisiana: Embracing an Authentic and Comprehensive Account of the Chief Events in the History of the State, a Special Sketch of Every Parish and a Record of the Lives of Many of the Most Worthy and Illustrious Families and Individuals, Vol. 2* (Chicago: Goodspeed Publishing Company, 1892), 347.

18. Samuel Wilson Jr., "Manresa—Jefferson College, Convent, Louisiana," *Louisiana Architect* 9, no. 2 (February 1970): 10.

19. Robert F. Hunter and Edwin L. Dooley Jr., *Claudius Crozet: French Engineer in America, 1790–1864* (Charlottesville: University Press of Virginia, 1989), 98; Fay, *The History of Education in Louisiana*, 51; Reeves, *The Governors of Louisiana*, 51.

20. Niehaus, "Jefferson College: The Early Years," 67.

21. Oliver P. Carriere, *A Sketch of the History of Jefferson College and Manresa House of Retreats Convent, Louisiana: 1830–1973* (Convent, LA: Manresa House of Retreats, 1974), 4, 8; J. D. B. De Bow, ed., "A. B. Roman, of Louisiana, Agriculturalist," in *De Bow's Review of the Southern and Western States, Volume 11* (New Orleans: Office, 22 Exchange Place, 1851), 437; Henry A. Bullard and Thomas Curry, *A New Digest of the Statutes and Laws of the State of Louisiana, from the Change of Government to the year 1841, Inclusive, Vol. I* (New Orleans: E. Johns & Co. Stationers' Hall, 1842), 300.

22. Rodney Cline, *Pioneer Leaders and Early Institutions in Louisiana Education* (Baton Rouge: Claitor Publishing Division, 1969), 259; Earl Francis Niehaus, "Jefferson College in St. James Parish, Louisiana: 1830–1875" (MA thesis, Tulane University, 1954), 18n.

23. Fay, *The History of Education in Louisiana*, 47, 50.

24. Niehaus, "Jefferson College: The Early Years," 68.

25. Niehaus, "Jefferson College in St. James Parish, Louisiana: 1830–1875," 9–10.

26. "Historic Convent in St. James Is of French Provincial Design," *Morning Advocate, Baton Rouge* (November 1, 1936): 2B.

27. Karen Kingsley, *Buildings of Louisiana* (New York: Oxford University Press, 2003), 182.

28. Robert Mark Rudd, "Apocryphal Grandeur: Belle Grove Plantation in Iberville Parish, Louisiana" (MA thesis, University of Delaware, 2002), 19.

29. P. H. Dagneau, *Memories of Seventy-five Years of the Centenary of the Marist Fathers in America* (Atlanta: Keystone Press, 1963), 47.

30. Hunter & Dooley, *Claudius Crozet*, 98; Wilson, "Manresa—Jefferson College, Convent, Louisiana," 10.

31. Niehaus, "Jefferson College: The Early Years," 71.

32. Frank Schneider, "Even Oaks Find Peace at Manresa," *Times Picayune*, May 7, 1987; Lillian C. Bourgeois, *Cabanocey: The History, Customs, and Folklore of St. James Parish* (Gretna, LA: Pelican, 1957), 40.

33. Simine Short, *Locomotive to Aeromotive: Octave Chanute and the Transportation Revolution* (Urbana: University of Illinois Press, 2011), 3.

34. Debra Goldschmidt, "Colleges Come to Terms with Slave-owning Pasts," *CNN*, May 23, 2011, http://www.cnn.com/2011/US/05/23/university.slavery/; Craig Steven Wilder, *Ebony and Ivey: Race, Slavery, and the Troubling History of America's Universities* (New York: Bloomsbury Press, 2013), 17.

35. Jessie Poesch and Barbara Sorelle Bacot, *Louisiana Buildings 1720–1940: The Historic American Building Survey* (Baton Rouge: Louisiana State University Press, 1997), 89–94.

36. Carriere, *A Sketch of the History of Jefferson College and Manresa House of Retreats Convent, Louisiana*, 56.

37. Richard Anthony Lewis, *Robert W. Tebbs, Photographer to Architects: Louisiana Plantations in 1926* (Baton Rouge: Louisiana State University Press, 2011), x; Wilson, "Manresa—Jefferson College, Convent, Louisiana," 11.

38. R. P. Martin, *The Plantation Mansion and Estate of Valcour Aime St. James Parish or A Brief Discussion of the Historical and Architectural Aspects of the Plantation Known as the St. James Refinery* (n.p., 1968), 3, in Louisiana Research Collection, Special Collections, Howard-Tilton Memorial Library, Jones Hall, Tulane University, New Orleans.

39. Wilson, "Manresa—Jefferson College, Convent, Louisiana," 10–11.

40. Hunter and Dooley, *Claudius Crozet*, 98.

41. Fay, *The History of Education in Louisiana*, 51.

42. Carriere, *A Sketch of the History of Jefferson College and Manresa House of Retreats Convent*, Louisiana, 8.

43. Patricia L. Duncan, "From Creole College to Religious Retreat: The Story of a Louisiana Landmark," *Preservation in Print* (August 9, 1994): 1.

44. Wilson, "Manresa—Jefferson College, Convent, Louisiana," 11, 12.

45. John L. Peytavin, "Jefferson College and Education: Addressed Delivered on

June 20th, 1906," *Jefferson College Record* (July 1906), 138, Box: American Province Jefferson College 3, Marist School, Atlanta, Georgia, Provincial Archives of the Society of Mary.

46. Hunter and Dooley, *Claudius Crozet*, 98; "A. B. Roman, of Louisiana, Agriculturalist," 437.

47. *Statutes, Plan of Education, and Rules, Adopted in Jefferson College, Louisiana* (New Orleans: J.C. de St. Romanes, 18—?), 9, Louisiana State University Special Collections, Hill Memorial Library, Baton Rouge.

48. Fay, *The History of Education in Louisiana*, 51.

49. Hunter and Dooley, *Claudius Crozet*, 98; Fossier, *New Orleans the Glamour Period, 1800–1840*, 235.

50. Fay, *The History of Education in Louisiana*, 51.

51. Michael O'Brien, *Intellectual Life and the American South, 1810–1860* (Chapel Hill: University of North Carolina Press, 2010), 33–34, 91.

52. De Sennegy, *Une Paroisse Louisianaise*, 49.

53. Niehaus, "Jefferson College in St. James Parish, Louisiana: 1830–1875," 19.

54. O'Brien, *Intellectual Life and the American South, 1810–1860*, 12.

55. Charles P. Roland, *Louisiana Sugar Plantations during the Civil War* (Baton Rouge: Louisiana State University Press, 1997), 6.

56. "Historic Convent in St. James Is of French Provincial Design," *Morning Advocate, Baton Rouge* (November 1, 1936): 2B; Short, *Locomotive to* Aeromotive, 3.

57. Fleming, *Louisiana State University*, 15.

58. Niehaus, "Jefferson College in St. James Parish, Louisiana: 1830–1875," 25.

59. De Sennegy, *Une Paroisse Louisianaise*, 49.

60. Dagneau, *Memoirs of Seventy-five Years for the Centenary of the Marist Fathers in America*, 49.

61. James B. Scott, *Outline of the Rise and Progress of Freemasonry in Louisiana from Its Introduction to the Reorganization of the Grand Lodge in 1850* (New Orleans: Order of the Grand Lodge, 1923), 6–7.

62. Frank M. Uter, *Stones beside a River: A History of the Catholic Church on the East Bank of St. James Parish, 1809–2009* (Baton Rouge: Catholic Diocese of Baton Rouge, 2009), 36.

63. Cline, *Pioneer Leaders and Early Institutions in Louisiana Education*, 265–266.

64. Niehaus, "Jefferson College: The Early Years," 81.

65. Dagmar Renshaw LeBreton, *Valcour Aime: A Biography* (n.p., 1948), 32, in Louisiana and Special Collections Department, Earl K. Long Library, University of New Orleans; Arthur and de Kernion, *Old Families of Louisiana*, 47, 139; *Biographical and Historical Memoirs of Louisiana*, 341, 348, 438.

66. J. A. C. Chandler, *The South in the Building of the Nation: A History of Southern States Designed to Record the South's Part in the Making of the American Nation; to Portray the Character and Genius, to Chronicle the Achievements and Progress and to Illustrate the Life and Traditions, Volume 12* (Richmond: Southern Historical Publication Society, 1909), 357–358; Grace Elizabeth King, *Creole Families of New Orleans* (New Orleans: Cornerstone Book Publishers, 2010), 457.

67. John Howard Brown, ed., *The Cyclopedia Of American Biography: Comprising the Men and Women of the United States Who Have Been Identified with the Growth of the Nation, Volume 3* (Boston: James H. Lamb Company, 1900), 6.

68. *Statutes, Plan of Education, and Rules, Adopted in Jefferson College, Louisiana*, 10.

69. Rex Laney, "Andre Roman, Pioneer in Education, Industry," *Sunday Advocate, Baton Rouge* (October 29, 1961): 3.

70. Bourgeois, *Cabanocey: The History, Customs and Folklore of St. James Parish*, 79.

71. Carl A. Brasseaux, "Acadian Education: From Cultural Isolation to Mainstream America," in *The Louisiana Bicentennial Series in Louisiana History: Volume XVII, Education in Louisiana* (Lafayette: Center for Louisiana Studies, University of Louisiana at Lafayette, 1999), 238.

72. Charles Gayarré, *Fernando de Lemos: Truth and Fiction, a Novel* (New York: G. W. Carleton & Co., Publishers, 1872), 30, 36–38.

73. Robert M. Lusher, "History of Early State System of Education in Louisiana" (n.p., n.d.), in Robert M. Lusher Papers, folder: History of Early System of Education in Louisiana, Louisiana State University Special Collections, Hill Memorial Library, Baton Rouge.

74. "Jefferson College: Beginnings, 1831–1842," 1935, Binder, Manresa: Jefferson College, Loyola University, Monroe Library, Archives of the New Orleans Province, Society of Jesus, New Orleans.

75. *Statutes, Plan of Education, and Rules, Adopted in Jefferson College, Louisiana*, 17.

76. Ibid., 21.

77. Robert F. Pace, *Halls of Honor: College Men in the Old South* (Baton Rouge: Louisiana State University Press, 2004), 48–49.

78. Michael David Cohen, *Reconstructing the Campus: Higher Education and the American Civil War* (Charlottesville: University of Virginia Press, 2012), 6–7, 22–23.

79. Bourgeois, *Cabanocey: The History, Customs and Folklore of St. James Parish*, 149.

80. Rex Bowman and Carlos Santos, *Rot, Riot, and Rebellion: Mr. Jefferson's Struggle to Save the University That Changed America* (Charlottesville: University of Virginia Press, 2013), 1–3.

81. Ibid., 19, 20.

82. Ibid., 21–22.

83. Alexander Hill Everett, *Address Delivered at Jefferson College, St. James Parish, La. June 30, 1841, On Assuming the Functions of President of that Institution* (New Orleans: J. C. de St. Romes, 1841), 12–13, Louisiana State University Special Collections, Hill Memorial Library, Baton Rouge.

84. *Statutes, Plan of Education, and Rules, Adopted in Jefferson College, Louisiana*, 22.

85. Ibid., 24, 17–18.

86. Niehaus, "Jefferson College in St. James Parish, Louisiana: 1830–1875," 30; Frost, *Thinking Confederates*, 2.

87. LeBreton, *Valcour Aime: A Biography*, 35.

88. Tregle, "Creoles and Americans," 157, 160.

89. O'Brien, *Intellectual Life and the American South, 1810–1860*, 80.

90. *Statutes, Plan of Education, and Rules, Adopted in Jefferson College, Louisiana*, 7.

91. Ibid., 3.

92. Ibid., 4–7.

93. Ibid.

94. Ibid., 7–9.

95. Roger L. Geiger, "Introduction: New Themes in the History of Nineteenth-Century College," in *The American College in the Nineteenth Century*, ed. Roger L. Geiger (Nashville: Vanderbilt University Press, 2000), 2; Jennifer R. Green, *Military Education and the Emerging Middle Class in the Old South* (Cambridge: Cambridge University Press, 2008), 63.

96. Niehaus, "Jefferson College: The Early Years," 75, 33–36, 40.

97. Niehaus, "Jefferson College in St. James Parish, Louisiana: 1830–1875," 33–36, 40; Hunter and Dooley, *Claudius Crozet*, 99.

98. Niehaus, "Jefferson College in St. James Parish, Louisiana: 1830–1875," 33–36, 40.

99. Niehaus, "Jefferson College: The Early Years," 75–76.

100. Simone de la Souchére Deléry, *Napoleon's Soldiers in America* (1950; reprint, Gretna, LA: Pelican, 1998), 141.

101. Hunter and Dooley, *Claudius Crozet*, 86–101, 103; Deléry, *Napoleon's Soldiers in America*, 140.

102. Deléry, *Napoleon's Soldiers in America*, 142–143.

103. Ibid., 142.

104. Ibid.

105. Hunter and Dooley, *Claudius Crozet*, 100.

106. Deléry, *Napoleon's Soldiers in America*, 142–143; Hugues Jules de La Vergne, "Jefferson College Ledger and Letter Book," July 15, 1835–March 1, 1837, De La Vergne Papers, Tulane Manuscripts Department, Special Collections, Howard-Tilton Memorial Library, Jones Hall, Tulane University.

107. Niehaus, "Jefferson College in St. James Parish, Louisiana: 1830–1875," 33–36, 40.

108. Hunter and Dooley, *Claudius Crozet*, 1, 9–11, 13–15, 31, 86–91, 103.

109. Vella, *Intimate Enemies*, 262–263.

110. Edward Gaylord Bourne, *History of the Surplus Revenue of 1837 Being an Account of Its Origin, Its Distribution among the State, and the Uses to Which It Was Applied* (New York: G. P. Putnam's Sons, 1885), 68–69.

111. Fossier, *New Orleans the Glamour Period, 1800–1840*, 53.

112. Vella, *Intimate Enemies*, 266.

113. Fossier, *New Orleans the Glamour Period, 1800–1840*, 70–71.

114. Ibid., 40–41; Fay, *The History of Education in Louisiana*, 50; John Duffy, *The Tulane Medical Center: One Hundred and Fifty Years of Medical Education* (Baton Rouge: Louisiana State University Press, 1984), 8, 11.

115. Lusher, "History of Early State System of Education in Louisiana."

116. Niehaus, "Jefferson College: The Early Years," 79, 80.

117. Thaddeus Rogers Kilpatrick, "A Conservation Study of the Decorative Paintings at Whitney Plantation, St. John the Baptist Parish, Louisiana" (MS thesis, University of Pennsylvania, 1992), 8–17.

118. Niehaus, "Jefferson College: The Early Years," 80.

119. Lusher, "History of Early State System of Education in Louisiana."

120. Cline, *Pioneer Leaders and Early Institutions in Louisiana Education*, 261.

121. Ibid., 17.

122. Uter, *Stones beside a River*, 36; Roger Baudier, *The Catholic Church in Louisiana* (New Orleans: A. W. Hyatt Stationary Mfg. Co. Ltd., 1939), 356.

123. Ibid., 116, 119.

124. Lusher, "History of Early State System of Education in Louisiana."

125. Ibid.

126. Niehaus, "Jefferson College in St. James Parish, Louisiana: 1830–1875," 45–46.

127. Neil J. Savage, *Extraordinary Tenure: Massachusetts and the Making of a Nation, from President Adams to Speaker O'Neill* (Worcester, MA: Ambassador Books, 2004), 71.

128. Elizabeth Evans, "Alexander Hill Everett: Man of Letters" (PhD diss., University of North Carolina, 1970), 42, 48–49.

129. Ibid., 1; Earl F. Niehaus, "Jefferson College: The Early Years," *Louisiana His-*

torical Quarterly 38, no. 4 (October 1955): 83; Robert L. Gale, *A Henry Wadsworth Longfellow Companion* (Westport, CT: Greenwood, 2003), 75; Carl A. Brasseaux, *In Search of Evangeline: Birth and Evolution of the Evangeline Myth* (Thibodaux, LA: Blue Heron Press, 1988), 7; Henry Wadsworth Longfellow, *Evangeline: A Tale of Acadie* (reprint, Boston: Houghton, Mifflin, and Company, 1866).

130. Everett, *Address Delivered at Jefferson College, St. James Parish, La. June 30, 1841, On Assuming the Functions of President of That Institution*, 3–4, 18.

131. Ibid., 4.

132. Ibid., 15–16.

133. Niehaus, "Jefferson College: The Early Years," 84.

134. Ibid., 10–11.

135. Ibid., 4.

136. *Statutes, Plan of Education, and Rules, Adopted in Jefferson College, Louisiana*, 9.

137. Everett, *Address Delivered at Jefferson College, St. James Parish, La., June 30, 1841, On Assuming the Functions of President of That Institution*, 27, 29, 31.

138. Ibid., 33–34.

139. Niehaus, "Jefferson College: The Early Years," 86.

140. Carriere, *A Sketch of the History of Jefferson College and Manresa House of Retreats, Convent, Louisiana*, 8.

141. Edward Larocque Tinker, *Les Ecrits De Langue Francaise: En Louisiane Au XIXe Siecle* (1923; reprint, Nendeln, Liechtenstein: Kraus Reprint, 1970), 135; Brasseaux, *In Search of Evangeline: Birth and Evolution of the Evangeline Myth*, 20–21.

142. René de Sennegy, *Une Paroisse Louisianaise: St. Michel du Comté d'Acadie* (New Orleans: Mr. Capo, 1877), 53.

143. "Jefferson College Burnt," *The Courier*, March 7, 1842.

144. "Jefferson College," *Times Picayune*, March 9, 1842, 2.

145. "Notice," *Times Picayune*, March 13, 1842, 2.

146. Niehaus, "Jefferson College in St. James Parish, Louisiana: 1830–1875," 62–63.

147. William D. Reeves, *Manresa on the Mississippi: For the Greater Glory of God* (Convent, LA: Manresa House of Retreats, 1996), 74; Wilson, "Manresa—Jefferson College, Convent, Louisiana," 12; Niehaus, "Jefferson College in St. James Parish, Louisiana: 1830–1875," 64.

148. Craig A. Bauer, *Creole Genesis: The Bringier Family and Antebellum Plantation Life in Louisiana* (Lafayette: University of Louisiana at Lafayette Press, 2011), 43.

149. John B. Rehder, *Delta Sugar: Louisiana's Vanishing Plantation Landscape* (Baltimore: Johns Hopkins University Press, 1999), 86; Nathalie Dessens, *Creole City: A Chronicle of Early American New Orleans* (Gainesville: University Press of Florida, 2015), 86–87.

150. J. Carlyle Sitterson, *Sugar Country: The Cane Sugar Industry in the South, 1753–1950* (Westport, CT: Greenwood, 1973), 72, 73.

151. Wilson, "Manresa—Jefferson College, Convent, Louisiana," 12.

152. James Gallier, *Autobiography of James Gallier, Architect* (New York: Da Capo Press, 1973), 6, 9, 11, 17, 21.

153. Samuel Wilson Jr., "Introduction," in James Gallier, *Autobiography of James Gallier, Architect* (New York: Da Capo Press, 1973), vii; Gallier, *Autobiography of James Gallier, Architect*, 36–38.

154. Wilson, "Introduction," in James Gallier, *Autobiography of James Gallier, Architect*, vii.

155. Arthur Scully Jr., *James Dakin, Architect: His Career in New York and the South* (Baton Rouge: Louisiana State University Press, 1973), 98–99.

156. Bauer, *Creole Genesis*, 34; Reeves, *Manresa on the Mississippi*, 74.

157. Scully, *James Dakin, Architect*, 9–10.

158. "The Marine Pavilion at Rockaway, Long Island," *NOAA Photo Library*, www.photolib.noaa.gov/brs/clind50.htm, accessed August 3, 2014; John C. Ferguson, "Charles and James Dakin," in *KnowLA: Encyclopedia of Louisiana*, ed. David Johnson, http://www.knowla.org/entry/572/, accessed November 9, 2016; Julie Dickason, "James Harrison Dakin," *Find a Grave*, http://www.findagrave.com/cgi-bin/fg.cgi/page/gr/fg.cgi?page=gr&Grid=142282254, accessed November 9, 2016; Scully, *James Dakin, Architect*, 188–189; Leonard V. Huber, "New Orleans Cemeteries: A Brief History," in *New Orleans Architecture, Volume III: The Cemeteries*, ed. Mary Louise Christovich, 3–62 (Gretna, LA: Pelican Publishing Company, 1974), 21.

159. Deléry, *Napoleon's Soldiers in America*, 141.

160. Cline, *Pioneer Leaders and Early Institutions in Louisiana Education*, 261; Roulhac B. Toledano, "Louisiana's Golden Age: Valcour Aime in St. James Parish," *Louisiana History: The Journal of the Louisiana Historical Association*, 10, no. 3 (Summer 1969): 218.

161. Vella, *Intimate Enemies*, 266; Richard Follett, "The Rise and Fall of American Sugar," in *Plantation Kingdom: The American South and Its Global Commodities*, ed. Richard Follett, Sven Beckert, Peter Coclanis, and Barbara Hahn (Baltimore: Johns Hopkins University Press, 2016), 71, 76.

162. Vella, *Intimate Enemies*, 268.

163. LeBreton, *Valcour Aime: A Biography*, 82; Gabriel Aime, *Diary of Gabriel Aime*, Vol. 1, entry of July 20, 1856, Alcée Fortier Collection, Louisiana State Museum Historical Center, New Orleans, Louisiana.

164. F. Chevet, "To the President of the Directors of Jefferson College," *Journal of the Senate. Second Session of the Sixteenth Legislature of Louisiana* (1844): viii–xiiii [*sic*].

165. Carriere, *A Sketch of the History of Jefferson College and Manresa House of Retreats, Convent, Louisiana*, 9.

166. Niehaus, "Jefferson College in St. James Parish, Louisiana: 1830–1875," 64; F. Chevet, "Experiments on Bichlorure [*sic*] of Sulphur [*sic*] and Certain Carbures [*sic*] of Hydrogen, made in the Laboratory of Jefferson College (Louisiana)," *American Journal of Science and Arts* 42 (April 1842): 71–74; F. Chevet, "On Bichlorure [*sic*] Sulphur [*sic*]," *American Journal of Science and Arts* and *Annals of Electricity, Magnetism, and Chemistry;* and *Guardian of Experimental Science* 8 (January 1842): 493–495; Chevet, "To the President of the Directors of Jefferson College," viii–xiiii [*sic*].

167. Carriere, *A Sketch of the History of Jefferson College and Manresa House of Retreats, Convent, Louisiana*, 9.

168. Niehaus, "Jefferson College in St. James Parish, Louisiana: 1830–1875," 66.

169. Morgan, *Centenary College of Louisiana, 1825–2000*, 20.

170. Ibid., 21, 23, 26, 27.

171. Ibid., 26, 27.

172. Ibid., 20, 71, 81; Nelson, *A Burning Touch and a Flaming Fire: The Story of Centenary College of Louisiana*, 93–94; "Centenary State Historic Site," Louisiana State Parks, accessed November 10, 2016, www.crt.state.la.us/louisiana-state-parks/historic-sites/centenary-state-historic-site/index; Miriam G. Reeves, *The Felicianas of Louisiana* (Baton Rouge: Claitor's Book Store, 1967), 54.

173. Fay, *The History of Education in Louisiana*, 48, 52, 54; Lusher, "History of Early

State System of Education in Louisiana"; Niehaus, "Jefferson College in St. James Parish, Louisiana: 1830–1875," 70–72; Dan R. Frost and Kou K. Nelson, *The LSU College of Engineering, Volume I: Origins and Establishment, 1860–1908* (Baton Rouge: Louisiana State University College of Engineering, 1995), 23.

174. Lusher, "History of Early State System of Education in Louisiana."

175. Niehaus, "Jefferson College in St. James Parish, Louisiana: 1830–1875," 67–68.

176. Michael Kenny, 1946, *Jesuits in Our Southland 1566–1946: Origin and Growth of New Orleans Province*, Kenny Papers, Archives of the New Orleans Province, Society of Jesus, Loyola University, Monroe Library, New Orleans, 46.

177. Jean Baptiste Maisounabe, *Diary 1847*, entry of August 22–24, 1847, Maisounabe Papers, Loyola University, Monroe Library, Archives of the New Orleans Province of the Society of Jesus, New Orleans; Jerome Neyrey and Thomas Clancy, *Southern Jesuit Biographies: Pastors and Preachers, Builders and Teachers of the New Orleans Province* (Lafayette, LA: Acadian House, 2015), 28.

178. Maisounabe, *Diary 1847*, entry of September 12, 1847.

179. Kenny, *Jesuits in our Southland 1566–1946*, 42, 46.

180. Maisounabe, *Diary 1847*, entry of September 29, 1847.

181. Ibid., entry of October 15, 1847.

182. Kenny, *Jesuits in Our Southland 1566–1946*, 46.

183. John P. Dyer, *Tulane: The Biography of a University, 1834–1965* (New York: Harper & Row, 1966), 26; Niehaus, "Jefferson College in St. James Parish, Louisiana: 1830–1875," 73.

184. John La Tourrette, *Tourrette's Reference Map of the State of Louisiana: From the Original Surveys of the United States, Which Show the Townships, Sections, or Mile Squares, Spanish Grants, Settlement Rights & C., Also the Plantations with the Owners Names Engraved Thereon* (New Orleans: John La Tourette, 1848), Library of Congress, Geography and Map Division.

185. Reeves, *The Governors of Louisiana*, 52.

Chapter 3

1. Lillian C. Bourgeois, *Cabanocey: The History, Customs and Folklore of St. James Parish* (GRetna, LA: Pelican, 1957), 129.

2. Louis Dufau, "Louisiana College Now Situated in the Parish of St. James, in the Extensive and Commodious Buildings, Formerly Known as the Jefferson College," 1853, Tulane Manuscripts Department, Special Collections, Howard-Tilton Memorial Library, Jones Hall, Tulane University, New Orleans.

3. Dufau, "Louisiana College Now Situated in the Parish of St. James, in the Extensive and Commodious Buildings, Formerly Known as the Jefferson College."

4. "Copy of Act of Sale from Edmond Drurlheh to Louis Dufau, 1881," Box: American Province Jefferson College 1, Folder: Jefferson College Property Deeds, 1872–1881, Marist School, Atlanta, Georgia, Provincial Archives of the Society of Mary.

5. Edward Larocque Tinker, *Les Ecrits De Langue Francaise: En Louisiane Au XIXe Siecle* (1923; reprint, Nendeln, Liechtenstein: Kraus Reprint, 1970), 148, 246; Caryn Cossé Bell, *Revolution, Romanticism, and the Afro-Creole Protest Tradition in Louisiana, 1718–1868* (Baton Rouge: Louisiana State University Press, 1997), 179.

6. "A Manual of the Regular and Irregular Verbs of the French Language," *Daily Picayune*, May 31, 1851, 2.

7. Louis Dufau, Louisiana College (Louisiana) to Archbishop (Anthony Blanc):

New Orleans, February 28, 1856, Catholic Church. Archdiocese of New Orleans Collection, University of Notre Dame Archives, Notre Dame; Bell, *Revolution, Romanticism, and the Afro-Creole Protest Tradition in Louisiana, 1718–1868*, 179.

8. Robert Hague, "Cecile *Street* Dufau," *Find a Grave*, http://www.findagrave.com/cgi-bin/fg.cgi?page=gr&GRid=79127533&ref=acom, accessed August 19, 2014; Seventh Census of the United States, 1850; Records of the Bureau of the Census, Record Group 29; National Archives, Washington, DC, roll M432_232, p. 95A, image, 194.

9. Seventh Census of the United States, 1850, p. 95A, image, 194.

10. Dinstel, "Louis Dufau," *Find a Grave*, http://www.findagrave.com/cgi-bin/fg.cgi?page=gr&GRid=148604074, accessed November 29, 2015.

11. Minutes, Board of Administrators, University of Louisiana, February 6, 1849, University Archives, Tulane University, New Orleans; Ruby Van Allen Caulfeild, *French Literature of Louisiana* (Gretna, LA: Pelican, 1998), 44.

12. "University of Louisiana," *Daily Picayune*, February 13, 1849, 2; "Orleans High School," *Daily Picayune*, September 6, 1849, 2; "Cabinet de Lecture (Reading Room), July 2, 1849, 3; John Howard Brown, ed., *The Cyclopedia of American Biography: Comprising The Men and Women of the United States Who Have Been Identified with the Growth of the Nation*, vol. 3 (Boston: James H. Lamb Company, 1900), 17–18.

13. "First Municipal Hall," *Daily Picayune*, May 1, 1849, 2.

14. "A Manuel of the Regular and Irregular Verbs of the French Language," *Daily Picayune*, May 31, 1851, 2.

15. Louis Dufau, *Manuel des Verbes Français, Réguliers et Irréguliers* (New Orleans: Chez Leliévre, libraire, rue Royle, No 210, 1851), n.p.

16. Sally Kittredge Reeves, introduction to *Jacques-Felix Leliévre's New Louisiana Gardner* by Jacques-Felix Leliévre (Baton Rouge: Louisiana State University Press, 2001), 40.

17. Dufau, *Manuel des Verbes Français, Réguliers et Irréguliers*, n.p.

18. Charles Testut, *Portraits Litteraires: Nouvelle-Orleans* (Nouvelle-Orleans: Imprimerie des Veillees Louisianaises, 1850), 65–67.

19. "By reference to the advertisement in another column, it will be seen that Mr. Dufau has opened a new seminary," *Daily Picayune*, September 12, 1851, 2.

20. Tinker, *Les Ecrits De Langue Francaise*, 148, 246; John P. Dyer, *Tulane: The Biography of a University, 1834–1965* (New York: Harper & Row, 1966), 336; "1125–1127 Dauphine St.: Chain of Title," Collins C. Diboll Vieux Carré Digital Survey, http://www.hnoc.org/vcs/property_info.php?lot=23065-01, accessed October 29, 2015.

21. Rodney Cline, *Pioneer Leaders and Early Institutions in Louisiana Education* (Baton Rouge: Claitor Publishing Division, 1969), 262.

22. Sarah L. Hyde, "'Teach Us Incessantly': Lessons and Learning in the Antebellum South" (PhD diss., Louisiana State University, 2010), 282; Lachance, "The Foreign French," 123–124.

23. Bell, *Revolution, Romanticism, and the Afro-Creole Protest Tradition in Louisiana, 1718–1868*, 179–180.

24. Vernon Elmer Kappel, "Louis Dufau's Louisiana College" (MA thesis, Tulane University, 1949), 22.

25. Sarah L. Hyde, *Schooling in the Antebellum South: The Rise of Public and Private Education in Louisiana, Mississippi, and Alabama* (Baton Rouge: Louisiana State University Press, 2016), 108; John P. Dyer, "Education in New Orleans," in *The Past as Prelude: New Orleans, 1718–1968*, ed. Hodding Carter (Gretna, LA: Firebird Press, 1968), 128.

26. "College of Louisiana," *Daily Picayune*, December 10, 1851, 2; "Louisiana College.

Dauphine Street, between Hospital and Ursuline Streets. Director: Louis Dufau," *Louisiana Courier*, February 26, 1852, 4; "Louisiana College. Dauphine Street, between Hospital and Ursuline Streets. Director: Louis Dufau," *Louisiana Courier*, February 27, 1852, 4; "Louisiana College. Dauphine Street, between Hospital and Ursuline Streets. Director: Louis Dufau," *Louisiana Courier*, February 28, 1852, 4; "Louisiana College. Dauphine Street, between Hospital and Ursuline Streets. Director: Louis Dufau," *Louisiana Courier*, March 3, 1852, 1; W. P. Riddell, *A Genealogical Sketch of the Riddell Family, Including a List of the Descendants of the Three Brothers, Hugh, Gawn, and Robert, Who Came to America in 1737* (New Orleans: John F. Trow, 1852), 37.

27. "College of Louisiana," 2.

28. Dinstel, "Louis Dufau"; 1852 Orleans Parish Death Index, vol. 13, p. 248, Louisiana Archives, Baton Rouge, retrieved from http://files.usgwarchives.net/la/orleans/vitals/deaths/index/1852di.txt.

29. "Miss S. S. Hull's Young Ladies' Institution," *Mississippi Free Trader*, September 25, 1852, 3.

30. Glenn R. Conrad, "The Faces of French Louisiana," in *Cross, Crozier, and Crucible: A Volume Celebrating the Bicentennial of a Catholic Diocese in Louisiana*, ed. Glenn R. Conrad, 5–15 (New Orleans: Archdiocese of New Orleans in cooperation with the Center for Louisiana Studies, 1993), 13, 14.

31. "Catalogue of the Officers and Students of Louisiana College, for the Year Ending July, 1853," New Orleans, 1853, Tulane Manuscripts Department, Special Collections, Howard-Tilton Memorial Library, Jones Hall, Tulane University, New Orleans.

32. Oliver P. Carriere, *A Sketch of the History of Jefferson College and Manresa House of Retreats, Convent, Louisiana: 1830–1973* (Convent, LA: Manresa House of Retreats, 1974), 9.

33. Benjamin H. Trask, *Fearful Ravages: Yellow Fever in New Orleans, 1796–1905* (Lafayette: Center for Louisiana Studies, University of Louisiana at Lafayette, 2005), 5, 7; Margaret Humphreys, *Yellow Fever and the South* (New Brunswick, NJ: Rutgers University Press, 1992), 4–5.

34. Humphreys, *Yellow Fever and the South*, 46.

35. Trask, *Fearful Ravages*, 38–39, 43.

36. Dagmar Renshaw LeBreton, *Valcour Aime: A Biography* (n.p., 1948), 115, in Louisiana and Special Collections Department, Earl K. Long Library, University of New Orleans.

37. Kappel, "Louis Dufau's Louisiana College," 23.

38. Estelle M. Fortier Cochran, *The Fortier Family and Allied Families* (San Antonia: E. M. F. Cochran, 1963), 76.

39. "Death of Ex-Gov. A. B. Roman," *Daily Picayune*, January 30, 1866, n.p.

40. Ibid.

41. "Catalogue of the Officers and Students of Louisiana College. Statutes, Plan of Education, Regulations and Discipline. Act of Incorporation, and Contract between President and the Stockholders. Academic Year 1855–1856," Parish of St. James (La.), 1855, Tulane Manuscripts Department, Special Collections, Howard-Tilton Memorial Library, Jones Hall, Tulane University, New Orleans.

42. Joseph G. Tregle Jr., "Creoles and Americans," in *Creole New Orleans: Race and Americanization*, ed. Arnold R. Hirsch and Joseph Logsdon (Baton Rouge: Louisiana State University Press, 1992), 153.

43. *Acts Passed by the First Legislature of the State of Louisiana, Held and Begun in*

the Town of Baton Rouge, on the 17th January, 1853 (New Orleans: Emile La Sere, State Printer, 1853), 210–211; "New Works," *Daily Picayune*, March 13, 1853, 4.

44. John R. Ficklen, "History of Education in New Orleans," in *Standard History of New Orleans: Giving a Description of the Natural Advantages, Natural History in Regard to the Flora and Birds, Settlement, Indians, Creoles, Municipal and Military History, Mercantile and Commercial Interests, Banking, Transportation, Struggles Against High Water, the Press, Educational, Literature and Art, the Churches, Old Burying Grounds, Bench and Bar, Medical, Public and Charitable Institutions, the Carnival, Amusements, Clubs, Societies, Associations, Etc.*, ed. Henry Rightor (Chicago: Lewis Publishing Company, 1900), 239.

45. "Catalogue of the Officers and Students of Louisiana College."

46. "A Transfer of College Sites," *Daily Picayune*, August 5, 1853, 1.

47. Alcée Fortier, *Louisiana Studies: Literature, Customs and Dialects, History and Education* (New Orleans: F. F. Hansell and Bro., 1894), 303.

48. Kappel, "Louis Dufau's Louisiana College," 16.

49. Dufau, "Louisiana College Now Situated in the Parish of St. James."

50. "Louisiana College, Located in the Buildings formerly occupied by Jefferson College, Parish of St. James," *Daily Picayune*, July 27, 1856, 3.

51. "Catalogue of the Officers and Students of Louisiana College, For the Year Ending July, 1853"; "Catalogue of the Officers and Students, Regulations and Discipline of Louisiana College for the Academic Year of 1854–1855. Parish of St. James," 1854, Tulane Manuscripts Department, Special Collections," Howard-Tilton Memorial Library, Jones Hall, Tulane University, New Orleans.

52. Dufau, "Louisiana College Now Situated in the Parish of St. James."

53. Louis Dufau, "Address of Louis Dufau, President of Louisiana College, Delivered at the Commencement Exercises, Academic year, 1855–1856," 1856, Tulane Manuscripts Department, Special Collections, Howard-Tilton Memorial Library, Jones Hall, Tulane University, New Orleans.

54. George W. Cable, *The Creoles of Louisiana* (1884; reprint, Gretna, LA: Pelican, 2005), 260.

55. Dufau, "Louisiana College Now Situated in the Parish of St. James."

56. Francois J. Gentil, "Francois J. Gentil Class Role Book, 1853," Louisiana State University Special Collections, Hill Memorial Library, Baton Rouge.

57. Dufau, "Louisiana College Now Situated in the Parish of St. James."

58. "Louisiana College," *Daily Picayune*, November 2, 1853, 3.

59. "The College of Louisiana Now Located in the Parish of Saint James," *Le Vigilant, Donaldsonville* (August 20, 1853): 3.

60. "Catalogue of the Officers and Students of Louisiana College, for the Year Ending July, 1853."

61. Dufau, "Louisiana College Now Situated in the Parish of St. James"; "Catalogue of the Officers and Students of Louisiana College."

62. "Catalogue of the Officers and Students of Louisiana College."

63. "Catalogue of the Officers and Students, Regulations and Discipline of Louisiana College for the Academic Year 1854–1855," 17, Educational Institutions, American Antiquarian Society, Worcester, MA.

64. "Catalogue of the Officers and Students of Louisiana College, for the Year Ending July, 1853."

65. "Catalogue of the Officers and Students, Regulations and Discipline of Louisiana College for the Academic Year of 1854–1855."

66. Kappel, "Louis Dufau's Louisiana College," 31; "Catalogue of the Officers and Students of Louisiana College, for the Year Ending July, 1853."

67. "Catalogue of the Officers and Students, Regulations and Discipline of Louisiana College for the Academic Year of 1854–1855. Parish of St. James."

68. "Catalogue of the Officers and Students of Louisiana College. Statutes, Plan of Education, Regulations and Discipline. Act of Incorporation, and Contract between President and the Stockholders. Academic Year 1855–1856."

69. "Catalogue of the Officers and Students of Louisiana College, For the Year Ending July, 1853"; "Catalogue of the Officers and Students, Regulations and Discipline of Louisiana College for the Academic Year of 1854–1855. Parish of St. James, 1854"; "Catalogue of the Officers and Students of Louisiana College. Statutes, Plan of Education, Regulations and Discipline. Act of Incorporation, and Contract between President and the Stockholders. Academic Year 1855–1856."

70. "Catalogue of the Officers and Students of Louisiana College. Statutes, Plan of Education, Regulations and Discipline. Act of Incorporation, and Contract between President and the Stockholders. Academic Year 1855–1856."

71. Ibid.

72. Dufau, "Louisiana College Now Situated in the Parish of St. James."

73. "Catalogue of the Officers and Students, Regulations and Discipline of Louisiana College for the Academic Year of 1854–1855."

74. "The College of Louisiana Now Located in the Parish of Saint James," *Le Vigilant, Donaldsonville* (August 20, 1853): 3.

75. Bourgeois, *Cabanocey: The History, Customs, and Folklore of St. James Parish*, 84; Tinker, *Les Ecrits De Langue Francaise*, 135.

76. "We regret to learn . . . ," *Daily Picayune*, Morning Edition, November 10, 1855, 2.

77. "False Report," *Daily Picayune*, Afternoon Edition, November 10, 1855, 1.

78. Louis Dufau, Louisiana College (St. James, Louisiana) to Archbishop (Anthony Blanc): New Orleans, January 6, 1856, Catholic Church. Archdiocese of New Orleans Collection, University of Notre Dame Archives, Notre Dame, IN.

79. Louis Dufau, Louisiana College (St. James, Louisiana) to Father (Claude Anthony) Tholomier: (St. Michael's Church, Louisiana), January 21, 1856, ibid.; Louis Dufau, Louisiana College (St. James, Louisiana) to Archbishop (Anthony Blanc): New Orleans, January 6, 1856.

80. Roger Baudier, *The Catholic Church in Louisiana* (New Orleans: A. W. Hyatt Stationary Mfg. Co., Ltd., 1939), 374.

81. Louis Dufau, Louisiana College (St. James, Louisiana) to Archbishop (Anthony Blanc): New Orleans, January 6, 1856, Catholic Church. Archdiocese of New Orleans Collection, University of Notre Dame Archives.

82. Louis Dufau, Louisiana College (Louisiana) to Archbishop (Anthony Blanc): New Orleans, Louisiana, February 29, 1856, Catholic Church. Archdiocese of New Orleans Collection, University of Notre Dame Archives.

83. Louis Dufau, Louisiana College (St. James, Louisiana) to Father (Claude Anthony) Tholomier: (St. Michael's Church, Louisiana), January 21, 1856, and Louis Dufau, Louisiana College (Louisiana) to Archbishop (Anthony Blanc): New Orleans, February 29, 1856, Catholic Church. Archdiocese of New Orleans Collection, University of Notre Dame Archives.

84. Louis Dufau, Louisiana College (St. James, Louisiana) to Father (Claude Anthony) Tholomier: (St. Michael's Church, Louisiana), January 7, 1856, and Louis Du-

fau, Louisiana College (St. James, Louisiana) to Father (Claude Anthony) Tholomier: (St. Michael's Church, Louisiana), Archives of the University of Notre Dame, Catholic Church. Archdiocese of New Orleans Collection, University of Notre Dame Archives.

85. Father (Claude Anthony) Tholomier: St. Michael (Louisiana) to Archbishop (Anthony Blanc: New Orleans Louisiana) January, 12, 1856, Archives of the University of Notre Dame, Catholic Church. Archdiocese of New Orleans Collection, University of Notre Dame Archives.

86. Ibid.

87. Louis Dufau, Louisiana College (St. James, Louisiana) to Father (Claude Anthony) Tholomier: (St. Michael's Church, Louisiana), January 21, 1856, Archives of the University of Notre Dame, Catholic Church. Archdiocese of New Orleans Collection, University of Notre Dame Archives.

88. Kappel, "Louis Dufau's Louisiana College," 26.

89. Ibid., 40–41, 44–45.

90. Roger W. Shugg, *Origins of Class Struggle in Louisiana: A Social History of White Farmers and Laborers during Slavery and After, 1840–1875* (1939; reprint, Baton Rouge: Louisiana State University Press, 1966), 74–75; Sarah L. Hyde, *Schooling in the Antebellum South: The Rise of Public and Private Education in Louisiana, Mississippi, and Alabama* (Baton Rouge: Louisiana State University Press, 2016), 128–129.

91. "Louisiana College, Located in the Buildings formerly occupied by Jefferson College, Parish of St. James," *Daily Picayune*, July 27, 1856, 3.

92. "The State Teachers Association," *Times Picayune*, January 12, 1856, 2.

93. "Copes, Joseph S.," Tulane University Special Collections, specialcollections.tulane.edu/archon/?p=creators/creator&id=761, accessed September 15, 2013; Alcée Fortier, *Louisiana: Comprising Sketches of Parishes, Towns, Events, Institutions, and Persons, Aranged in Cyclopedic From* Volume III (Century Historical Association, 1914), 512.

94. W. P. Riddell to Joseph Copes, August 20, 1856, Joseph S. Copes Papers Correspondence, August 1856, Box 11, Folder 13, Louisiana Research Collection, Special Collections, Howard-Tilton Memorial Library, Jones Hall, Tulane University, New Orleans; "The Late Dr. Wm. P. Riddell," *Times Picayune*, May 30, 1872, 1; W. P. Riddell, *A Genealogical Sketch of the Riddell Family, Including a List of the Descendants of the Three Brothers, Hugh, Gawn, and Robert, Who Came to America in 1737* (New Orleans: John F. Trow, 1852), 37.

95. Louis Dufau to Joseph Copes, August 23, 1856, Joseph S. Copes Papers Correspondence, August 1856, Box 11, Folder 19, Louisiana Research Collection, Special Collections, Howard-Tilton Memorial Library, Jones Hall, Tulane University, New Orleans.

96. Dufau, "Address of Louis Dufau, President of Louisiana College, Delivered at the Commencement Exercises, Academic year, 1855–1856."

97. Ibid.

98. Ibid.

99. Ibid.

100. "Collegiate," *Daily Picayune*, September 21, 1856, 4.

101. Kappel, "Louis Dufau's Louisiana College," 24.

102. Earl F. Niehaus, "Jefferson College: The Early Years," *Louisiana Historical Quarterly* 38, no. 4 (October 1955): 76–77 and Appendix H; "1125–1127 Dauphine St.: Chain of Title," Collins C. Diboll Vieux Carré Digital Survey, http://www.hnoc.org/vcs/property_info.php?lot=23065-01, accessed October 29, 2015.

103. "Catalogue of the Officers of the Collegiate and Preparatory Departments of

the University of Louisiana. Statutes, Plan of Education, Regulations and Discipline. Academic Year 1856–1857. New Orleans: Printed by J. Lamare, 96 Exch. Alley," 1856, Louisiana Research Collection, Special Collections, Howard-Tilton Memorial Library, Jones Hall, Tulane University, New Orleans.

104. Minutes, Board of Administrators, University of Louisiana, September 11, 1856, University Archives, Tulane University, New Orleans.

105. Robert C. Reinders, *End of an Era: New Orleans, 1850–1860* (1964; reprint, Gretna, LA: Pelican, 1998), 140.

106. Dyer, *Tulane*, 27–28.

107. Minutes, Board of Administrators, University of Louisiana, September 11, 1856, University Archives, Tulane University, New Orleans.

108. Ibid.; Dyer, *Tulane*, 27–28.

109. "Catalogue of the Officers of the Collegiate and Preparatory Departments of the University of Louisiana."

110. Minutes, Board of Administrators, University of Louisiana, September 11, 1856, University Archives, Tulane University, New Orleans.

111. William Preston Johnston, *A History of Tulane University* (n.p., 1894), in Louisiana Research Collection, Special Collections, Howard-Tilton Memorial Library, Jones Hall, Tulane University, New Orleans; Dyer, *Tulane*, 336.

112. "Catalogue of the Officers of the Collegiate and Preparatory Departments of the University of Louisiana."

113. Johnston, *A History of Tulane University*.

114. "Catalogue of the Officers of the Collegiate and Preparatory Departments of the University of Louisiana. Statutes, Plan of Education, Regulations and Discipline"; "State University of Louisiana, Eighth Annual Session of the Collegiate Department, 1858–59. New Orleans, Published by the University," 1859, Louisiana Research Collection, Special Collections, Howard-Tilton Memorial Library, Jones Hall, Tulane University, New Orleans; "Louisiana University," *Daily Picayune*, January 26, 1859, 2.

115. Letter from Louis Dufau to Joseph S. Copes, March 5, 1857, Letter from Louis Dufau to the chairman and members of the Board of Administration of the University of Louisiana, February 1957, Joseph S. Copes Papers Correspondence, Box 11, Folder 20, Louisiana Research Collection, Special Collections, Howard-Tilton Memorial Library, Jones Hall, Tulane University, New Orleans.

116. *Semi-centennial Register of the Members of the Phi Kappa Sigma Fraternity: Printed by the Order of the Grand Chapter to Commemorate the Fiftieth Anniversary of the Founding of the Fraternity, 1850–1900* (Philadelphia: Avil Printing Co.), iv, 209, 2011.

117. Letter from Louis Dufau to the University of Louisiana Administration Board, July 15, 1857, Joseph S. Copes Papers Correspondence, Box 11, Folder 19, Louisiana Research Collection, Special Collections, Howard-Tilton Memorial Library, Jones Hall, Tulane University, New Orleans.

118. Ibid.

119. "Catalogue of the Officers of the Collegiate and Preparatory Departments of the University of Louisiana. Statutes, Plan of Education, Regulations and Discipline."

120. Letter from Louis Dufau to the chairman and members of the Board of Administration of the University of Louisiana, February 1957, Joseph S. Copes Papers Correspondence, Box 11, Folder 19, Louisiana Research Collection, Special Collections, Howard-Tilton Memorial Library, Jones Hall, Tulane University, New Orleans.

121. Ibid.

122. Ibid.

123. Reinders, *End of an Era*, 140.

124. "The Late Dr. Wm. P. Riddell," *Times Picayune*, May 30, 1872, 1; Johnston, *A History of Tulane University*.

125. Dyer, *Tulane*, 27–28.

126. Bell, *Revolution, Romanticism, and the Afro-Creole Protest Tradition in Louisiana, 1718–1868*, 180–181; Louis Dufau, *Principles de la Franc-Maconnerie Vulgairement Connue Sous le Nom de Rite Ancien et Accepté Ecossais, Indiquant le Sens Philosophique, Politique et Religiux de tous les Degrés jusqu'au 30éme inclusivement* (Nouvelle-Orléans: L. Marchand, Imprimeur, 1859), 3; James Foulhouze, *Historical Inquiry into the Origin of the Ancient and Accepted Scotch Rite; Into the Pretensions of the Dalcho-Mackey and Gourgas-Moore Supreme Councils of Charleston and Boston; Into the Rights of the Supreme Council of the Sovereign and Independent State of Louisiana; and into the Inconsistencies of His Imperial and Royal Highness, Prince Lucien Murat, Pretender to the Crown of Naples, and Grand Master, Pro Tempore, of the Masonic Order in France* (New Orleans: True Delta Job Office, 1859), 10;

127. Bell, *Revolution, Romanticism, and the Afro-Creole Protest Tradition in Louisiana, 1718–1868*, 182.

128. Ibid., 180–181; Michael R. Poll, "Albert Pike's Address Before the Grand Consistory of Louisiana," *Pietre-Stones Review of Freemasonry* 10 (2002): http://www.freemasons-freemasonry.com/pike_address_louisiana.html, accessed February 19, 2015.

129. Michael R. Poll, "Albert Pike's Address before the Grand Consistory of Louisiana," *Pietre-Stones Review of Freemasonry* 10 (2002): http://www.freemasons-freemasonry.com/pike_address_louisiana.html, accessed February 19, 2015.

130. Louis J. B. Brunett, "To All Free-Masons" (February 2, 1860), 180–183, letter printed in *Transactions of the Supreme Council of the 33rd Degree, Ancient and Accepted Rite of Free-masonry for the Southern Jurisdiction of the United States of American, 1857 to 1866* (Washington, DC: Joseph L. Pearson, Printer, 1878), 180; Dufau, *Principles de la Franc-Maconnerie Vulgairement Connue Sous le Nom de Rite Ancien et Accepté Ecossais, Indiquant le Sens Philosophique, Politique et Religiux de tous les Degrés jusqu'au 30éme inclusivement*, 24; James B. Scott, *Outline of the Rise and Progress of Freemasonry in Louisiana from Its Introduction to the Reorganization of the Grand Lodge in 1850* (New Orleans: Order of the Grand Lodge, 1923), 6–7; Frank M. Uter, *Stones beside a River: A History of the Catholic Church on the East Bank of St. James Parish, 1809–2009* (Baton Rouge: Catholic Diocese of Baton Rouge, 2009), 93.

131. Brunett, "To All Free-Masons," 180.

132. Ibid., 180–181.

133. *Proceedings of the M. W. Grand Lodge of F. and A. Masons of the State of Louisiana, at Its Forty-Seventh Annual Communication, Held at New Orleans, February 14, 1859* (New Orleans: Order of the Grand Lodge, 1859), 108–113, 114.

134. Ibid.; Dinstel, "Louis Dufau"; "Camille Dufau," *St. Louis Cemeteries Nos. 1 and 2 Deaths, 1857–1864*, Office of Archives and Records, Archdiocese of New Orleans.

135. *Proceedings of the M. W. Grand Lodge of F. and A. Masons of the State of Louisiana, at its Forty-Seventh Annual Communication, Held at New Orleans, February 14, 1859*, 246–247; "The Union Movement," *New York Times*, June 19, 1862, http://www.nytimes.com/1862/06/19/news/the-union-movement.html, accessed March 11, 2013.

136. "Died: Louis Dufau, the Secretary of the Union Association," *Daily Delta*, July 12, 1862, 2.

137. Ibid.; "Died. Louis Dufau," *New Orleans Bee*, July 12, 1862, 4.

138. Tinker, *Les Ecrits De Langue Francaise*, 148; "Died. Louis Dufau," *New Orleans Bee*, July 12, 1862, 4.

139. Death of Louis Dufau, St. Louis Cemeteries Nos. 1 & 2, New Orleans, Louisiana, Death Record, 1857–1864, page 130. Office of Archives and Records, Archdiocese of New Orleans.

140. "Celebration of the Fourth of July," *New York Times*, July 13, 1862.

141. "John Baptiste Lyman, notary public appearance: Cecile Dufau, widow of Louis Dufau," March 30, 1895, Jefferson Parish Court House, Gretna, Louisiana, Mortgage and Conveyance Records.

142. Hague, "Cecile *Street* Dufau."

143. "1125–1127 Dauphine St.: Chain of Title," The Collins C. Diboll Vieux Carré Digital Survey, http://www.hnoc.org/vcs/property_info.php?lot=23065–01, accessed October 29, 2015; "Sale of Jefferson College to Louis Dufau," as cited in Vernon Elmer Kappel, "Louis Dufau's Louisiana College" (MA thesis, Tulane University, 1949), Appendix 1.

144. "Catalogue of the Officers and Students, Regulations and Discipline of Louisiana College for the Academic Year 1854–1855," p. 6, Educational Institutions, American Antiquarian Society, Worcester, MA; "State University of Louisiana: Eights Annual Session of the Collegiate Department, 1858–'59," p. 7, Educational Institutions, American Antiquarian Society, Worcester, MA; "Catalogue of Straight University for the Years 1869–1870," p. 18, Special Collections and Archives, Dillard University, Will W. Alexander Library, New Orleans; "Jules Dufau," New Orleans City Directory, 1890, p. 318, American Antiquarian Society, MA; "Jules Dufau." Houston City Directory, 1894, p. 198, American Antiquarian Society, Worcester, MA; "Jules Dufau, 10/23/1908," Glenwood Cemetery: Search Our Records, http://www.glenwoodcemetery.org/search/?name=Jules&lastname=Dufau&date_from=&date_to=, accessed November 20, 2016.

145. Kappel, "Louis Dufau's Louisiana College," 19–20.

146. Baudier, *The Catholic Church in Louisiana*, 418.

147. Marie Adrian Persac, *Norman's Chart of the Lower Mississippi River* (New Orleans: B. M. Norman, 1858), Library of Congress Geography and Map Division.

148. "Catalogue of Jefferson College, Convent, La. 1906–07," 1906, Louisiana State University Special Collections, Hill Memorial Library, Baton Rouge.

Chapter 4

1. Michel J. Fortier, Valcour Aime Notes, n.d., Marguerite M. Fortier Papers, 1821–1920, folder: Genealogical Papers, notes, Valcour Aime notes, box 1, Tulane Manuscripts Department, Special Collections, Howard-Tilton Memorial Library, Jones Hall, Tulane University.

2. Dottye Varnado, "Fabulous Valcour Aime: They Called Him 'Louis XIV of Louisiana,'" *Morning Advocate, Baton Rouge* (May 22, 1960): 1E.

3. R. P. Martin, *The Plantation Mansion and Estate of Valcour Aime St. James Parish or A Brief Discussion of the Historical and Architectural Aspects of the Plantation Known as the St. James Refinery* (n.p., 1968), 3, Louisiana Research Collection, Special Collections, Howard-Tilton Memorial Library, Jones Hall, Tulane University, New Orleans; Roulhac B. Toledano, "Louisiana's Golden Age: Valcour Aime in St. James Parish," *Louisiana History: The Journal of the Louisiana Historical Association* 10, no. 3 (Summer 1969): 212; Estelle M. Fortier Cochran, *The Fortier Family and Allied Families* (San Antonio: E. M. F. Cochran, 1963), 55.

4. Dagmar Renshaw LeBreton, *Valcour Aime: A Biography* (n.p., 1948), 15, Louisiana and Special Collections Department, Earl K. Long Library, University of New Orleans.

5. Martin, *The Plantation Mansion and Estate of Valcour Aime St. James*, 3; Cochran, *The Fortier Family and Allied Families*, 3, 55.

6. Eliza Ripley, *Social Life in Old New Orleans: Being Recollections of My Girlhood* (1912; reprint, Gretna, LA: Pelican, 1998), 187; Cochran, *The Fortier Family and Allied Families*, 55; Martin, *The Plantation Mansion and Estate of Valcour Aime St. James Parish*, 5.

7. Toledano, "Louisiana's Golden Age," 219; Cochran, *The Fortier Family and Allied Families*, 55, 60.

8. Varnado, "Fabulous Valcour Aime," 1E; Cochran, *The Fortier Family and Allied Families*, 55; Toledano, "Louisiana's Golden Age," 217; "Sugar Cultivation in Louisiana," in *The Colonial Magazine and East India Review: January to June, 1849, Vol. XVI* (London: John Mortimer, 1849), 227; Alcée Fortier, "The Louisiana Sugar Planters of the Old Regime," *Louisiana Sugar Planter and Sugar Manufacturer* 52 (1914): 76.

9. Ripley, *Social Life in Old New Orleans*, 187.

10. Martin, *The Plantation Mansion and Estate of Valcour Aime St. James Parish or A Brief Discussion of the Historical and Architectural Aspects of the Plantation Known as the St. James Refinery*, 5.

11. Cochran, *The Fortier Family and Allied Families*, 55.

12. Valcour Aime, *Plantation Diary of the Late Mr. Valcour Aime, Formerly Proprietor of the Plantation Known as the St. James Sugar Refinery Situated in the Parish of St. James and Now Owned by Mr. John Burnside* (New Orleans: Lark & Hofeline, 1878), 83 (entry from September 1842), 90 (entry from September 1843).

13. Marc R. Matrana, *Lost Plantations of the South* (Jackson: University Press of Mississippi, 2009), 181–182.

14. Ripley, *Social Life in Old New Orleans*, 188–190.

15. Matrana, *Lost Plantations of the South*, 181–182.

16. Toledano, "Louisiana's Golden Age," 217.

17. Gabriel Aime, *Diary of Gabriel Aime*, vol 1, p. 1–45, 1846, Alcée Fortier Collection, Louisiana State Museum Historical Center, New Orleans; Gabriel Aime, *Diary of Gabriel Aime*, vol. 2, p. 1–66, 1846, Alcée Fortier Collection, Louisiana State Museum Historical Center, New Orleans; Cochran, *The Fortier Family and Allied Families*, 56.

18. Richard Follett, *The Sugar Masters: Planters and Slaves in Louisiana's Cane World, 1820–1860* (Baton Rouge: Louisiana State University Press, 2005), 36.

19. Toledano, "Louisiana's Golden Age," 216.

20. Ripley, *Social Life in Old New Orleans*, foreword, 331.

21. Ibid., 188–190.

22. Matrana, *Lost Plantations of the South*, 183.

23. Toledano, "Louisiana's Golden Age," 219–222; Fred Daspit, *Louisiana Architecture, 1840–1860* (Lafayette: Center for Louisiana Studies, University of Louisiana at Lafayette, 2006), 53; Cochran, *The Fortier Family and Allied Families*, 101; Ripley, *Social Life in Old New Orleans*, 190.

24. Emily Epstein Landau, *Spectacular Wickedness: Sex, Race, and Memory in Storyville, New Orleans* (Baton Rouge: Louisiana State University Press, 2013), 180; Rex Bowman and Carlos Santos, *Rot, Riot, and Rebellion: Mr. Jefferson's Struggle to Save the University That Changed America* (Charlottesville: University of Virginia Press, 2013), 96, 131.

25. Aime, *Diary of Gabriel Aime*, vol 1, p. 10, entries of July 20, 1846, July 28, 1846, Sep-

tember 1, 1846; Aime, *Diary of Gabriel Aime*, vol. 2, entry of April 20, 1847; Glenn Falgoust, interview by R. Eric Platt, Saint Amant, LA, September 11, 2015.

26. John Timbrell, *The Poison Paradox: Chemicals as Friends and Foes* (New York: Oxford University Press, 2005), 138.

27. Aime, *Diary of Gabriel Aime*, vol. 1, p. 1–45, entries of July 20, 1846, July 28, 1846, September 1, 1846; Aime, *Diary of Gabriel Aime*, vol. 2, entry of April 20, 1847; Aime, *Diary of Gabriel Aime*, vol. 2, p. 1–66, entry of April 1, 1847; Alexandre Dumas, *The Count of Monte-Cristo* (reprint, London: Chapman and Hall, 1846), 51–57.

28. Ripley, *Social Life in Old New Orleans*, 186–187.

29. Margaret Humphreys, *Yellow Fever and the South* (New Brunswick, NJ: Rutgers University Press, 1992), 5.

30. "Bladon Springs State Park," *Encyclopedia of Alabama*, http://www.encyclopediaofalabama.org/face/Article.jsp?id=h-2905, accessed September 13, 2014.

31. LeBreton, *Valcour Aime*, 117.

32. Glenn Falgoust, interview by R. Eric Platt, Saint Amant, LA, September 12, 2014.

33. LeBreton, *Valcour Aime*, 135; Toledano, "Louisiana's Golden Age," 222.

34. Obituary of Gabriel Aime, *Daily Picayune*, September 24, 1854, 2.

35. Valcour Aime, *Diary of Valcour Aime*, Louisiana State Archives, New Orleans, quoted in David King Gleason, *Plantation Homes of Louisiana and the Natchez Area* (Baton Rouge: Louisiana State University Press, 1985), 17.

36. Aime, *Plantation Diary of the Late Mr. Valcour Aime, Formerly Proprietor of the Plantation Known as the St. James Sugar Refinery Situated in the Parish of St. James and Now Owned by Mr. John Burnside*, 172 (entries from September 1854 and February 1855).

37. LeBreton, *Valcour Aime*, 133; Cochran, *The Fortier Family and Allied Families*, 101.

38. Ibid., 137.

39. Aime, *Diary of Gabriel Aime*, vol. 1, an example accompanies the entry of June 10, 1846; Aime, *Diary of Gabriel Aime*, vol. 2, quote accompanies the entry of April 15, 1847.

40. Toledano, "Louisiana's Golden Age," 222.

41. LeBreton, *Valcour Aime*, 24.

42. Earl F. Niehaus, "Jefferson College: The Early Years," *Louisiana Historical Quarterly* 38, no. 4 (October 1955): 80.

43. Rodney Cline, *Pioneer Leaders and Early Institutions in Louisiana Education* (Baton Rouge: Claitor Publishing Division, 1969), 259.

44. LeBreton, *Valcour Aime*, 135; Cochran, *The Fortier Family and Allied Families*, 101; "Jesuit Fathers Restoring Ante-Bellum Estate: Old Jefferson College is Now Manresa House," *New Orleans States*, n.d.; Trahan Architects, apac., *Manresa House of Retreats: 2000 Evaluation*, Loyola University, Monroe Library, Archives of the New Orleans Province, Society of Jesus, New Orleans; Toledano, "Louisiana's Golden Age," 218; P. H. Dagneau, *Memories of Seventy-five Years of the Centenary of the Marist Fathers in America* (Atlanta: Keystone Press, 1963), 48.

45. Samuel Wilson Jr., "Manresa—Jefferson College, Convent, Louisiana," *Louisiana Architect* 9 no. 2 (February 1970): 12; "Side Elevation of a One-story Gothic Church Building. Small Tower at Front, Eight Gothic Arched Windows with Diamond Pattern Panes," unsigned, undated, pencil on drawing paper, Collection 143, Sylvester Labrot Collection, #573, Southeastern Architectural Archive, Special Collections Division, Tulane University Libraries, New Orleans, LA.

46. E(leazar) Vignonet, St. Michael, Louisiana to Stephen Rousselon, New Orleans, September 1, 1859, Catholic Church. Archdiocese of New Orleans Collection, University of Notre Dame Archives, Notre Dame, IN.

47. Valcour Aime to the Board of Directors, Jefferson College, St. James Parish, August 10, 1859, Binder, Manresa: Jefferson College, Loyola University, Monroe Library, Archives of the New Orleans Province, Society of Jesus, New Orleans.

48. "Notarized Transfer of Jefferson College to the Marist Fathers," May 6, 1864, Binder, Manresa: Jefferson College, Loyola University, Monroe Library, Archives of the New Orleans Province, Society of Jesus, New Orleans; LeBreton, *Valcour Aime*, 135; Aime to the Board of Directors, Jefferson College, St. James Parish, August 10, 1859; J. W. Dorr, "A Tourist's Description of Louisiana in 1860," *Louisiana Historical Quarterly* 21, no. 4 (1938): 14; Le[opold] LePraire, Baton Rouge, to Archbishop Anthony Blanc, New Orleans, September 5, 1859, Catholic Church. Archdiocese of New Orleans Collection, University of Notre Dame Archives, Notre Dame, IN; Norman R. Shapiro, trans., *Creole Echoes: The Francophone Poetry of Nineteenth-Century Louisiana* (Champaign: University of Illinois Press, 2004), 57; "Jefferson College, Parish of St. James, La., Sixty Miles Above the City of New Orleans," *Daily Picayune*, October 9,1861, 3.

49. Aime to the Board of Directors, Jefferson College, St. James Parish, August 10, 1859.

50. Ibid.

51. Ibid.; Oliver P. Carriere, *A Sketch of the History of Jefferson College and Manresa House of Retreats, Convent, Louisiana: 1830–1973* (Convent, LA: Manresa House of Retreats, 1974), 10.

52. Aime to the Board of Directors, Jefferson College, St. James Parish, August 10, 1859.

53. Jefferson College Stock Receipt for Mr. S. Brown, de la Vergne Papers, 1831–1919, Box 2, Tulane Manuscripts Department, Special Collections, Howard-Tilton Memorial Library, Jones Hall, Tulane University, New Orleans.

54. E[leazer] Vignonet, St. Michael, Louisiana, to Stephen Rousselon, New Orleans, Louisiana, September 1, 1859, Catholic Church. Archdiocese of New Orleans Collection, University of Notre Dame Archives, Notre Dame, IN.

55. Dorr, "A Tourist's Description of Louisiana in 1860," 14.

56. Ted Tunnel, *Crucible of Reconstruction: War, Radicalism, and Race, 1862–1877* (Baton Rouge: Louisiana State University Press, 1984), 13–17.

57. John C. Rodrigue, *Reconstruction in the Cane Fields: From Slavery to Free Labor in Louisiana's Sugar Parishes, 1862–1880* (Baton Rouge: Louisiana State University Press, 2001), 26–27.

58. Ibid., 17.

59. Richard Stringer, *Father Abraham: Lincoln's Relentless Struggle to End Slavery* (New York: Oxford University Press, 2006), 2–4.

60. Tunnel, *Crucible of Reconstruction*, 13–17.

61. Toledano, "Louisiana's Golden Age," 223.

62. Lawrence N. Powell, introduction to *The New Orleans of George Washington Cable: The 1887 Census Office Report* by George Washington Cable (Baton Rouge: Louisiana State University Press, 2008), 12; John M. Sacher, "Louisiana's Secession from the Union," in *KnowLA: Encyclopedia of Louisiana*, ed. David Johnson (Louisiana Endowment for the Humanities, 2011), http://www.knowla.org/entry/559/.

63. "Death of Ex-Gov. A. B. Roman," *Daily Picayune*, January 30, 1866, n.p.; Toledano, "Louisiana's Golden Age," 223.

64. Toledano, "Louisiana's Golden Age," 223; Arthur W. Bergeron Jr., *Guide to Louisiana Confederate Military Units, 1861–1865* (Baton Rouge: Louisiana State University Press, 1989), 169.

65. John D. Winters, *The Civil War in Louisiana* (Baton Rouge: Louisiana State University Press, 1963), 31–32.

66. Ibid., 223–224.

67. Charles P. Roland, *Louisiana Sugar Plantations during the Civil War* (Baton Rouge: Louisiana State University Press, 1997), 56.

68. Ibid., 122, 55, 51; Matthew Pratt Guterl, *American Mediterania: Southern Slaveholders in the Age of Emancipation* (Cambridge, MA: Harvard University Press, 2008), 79–113.

69. Toledano, "Louisiana's Golden Age," 222–224.

70. Winters, *The Civil War in Louisiana*, 30, 31.

71. "Jefferson College: Beginnings, 1831–1842," 1935, Binder, Manresa: Jefferson College, Loyola University, Monroe Library, Archives of the New Orleans Province, Society of Jesus, New Orleans; Oliver P. Carriere, *A Sketch of the History of Jefferson College and Manresa House of Retreats, Convent, Louisiana: 1830–1973* (Convent, LA: Manresa House of Retreats, 1974), 10.

72. LeBreton, *Valcour Aime*, 149.

73. Bourgeois, *Cabanocey: The History, Customs and Folklore of St. James Parish*, 40.

74. Roland, *Louisiana Sugar Plantations during the Civil War*, 38; Robert J. Miller, *Both Prayed to the Same God: Religion and Faith in the American Civil War* (Lanham, MD: Lexington Books, 2007), 143.

75. Eric Foner, *Reconstruction: America's Unfinished Revolution, 1863–1877* (New York: Harper & Row, 1988), 45–48; "The Union Movement," *New York Times*, June 19, 1862, n.p.

76. Rodrigue, *Reconstruction in the Cane Fields*, 35.

77. Foner, *Reconstruction*, 55; Abraham Lincoln, "The Emancipation Proclamation," in *The Gettysburg Address and Other Writings* (New York: Fall River Press, 2013), 49–52.

78. Foner, *Reconstruction*, 55; Edward Royce, *The Origins of Southern Sharecropping* (Philadelphia: Temple University Press, 1993), 39–40.

79. Joe Gray Taylor, *Louisiana Reconstructed: 1863–1877* (Baton Rouge: Louisiana State University Press, 1974), 38–39, 96–97, 329, 330.

80. Bruce Levine, *The Fall of the House of Dixie: The Civil War and the Social Revolution That Transformed the South* (New York: Random House, 2013), 186–189, 298.

81. Heather Cox Richardson, *The Death of Reconstruction: Race, Labor, and Politics in the Post-Civil War North, 1865–1901* (Cambridge, MA: Harvard University Press, 2001), 138; Foner, *Reconstruction*, 207.

82. Kathryn Page, "A First Born Child of Liberty: The Constitution of 1864," in *In Search of Fundamental Law: Louisiana's Constitutions, 1812–1974*, ed. Warren M. Billings and Edward F. Haas (Lafayette: Center for Louisiana Studies, University of Southwestern Louisiana, 1993), 52, 58–59.

83. Winters, *The Civil War in Louisiana*, 398.

84. John David Smith, introduction to *Louisiana Sugar Plantations during the Civil War* by Charles P. Roland (Baton Rouge: Louisiana State University Press, 1997), xiii.

85. William Kauffman Scarborough, *Masters of the Big House: Elite Slaveholders of the Mid-Nineteenth-Century South* (Baton Rouge: Louisiana State University Press, 2003), 405.

86. Ibid.

87. Ferry, *Journals*, vol. 2, 35–37, as cited in Roulhac B. Toledano, "Louisiana's Golden Age: Valcour Aime in St. James Parish," *Louisiana History: The Journal of the Louisiana Historical Association* 10, no. 3 (Summer 1969): 221.

88. Robert F. Pace, *Halls of Honor: College Men in the Old South* (Baton Rouge: Louisiana State University Press, 2004), 116.

89. Follett, *The Sugar Masters: Planters and Slaves in Louisiana's Cane World, 1820–1860*, 235; Michael Wayne, "Transformation of the Plantation," in *The Louisiana Bicentennial Series in Louisiana History: Volume VI, Reconstructing Louisiana*, ed. Lawrence N. Powell (Lafayette: Center for Louisiana Studies, University of Louisiana at Lafayette, 2001), 519.

90. J. Carlyle Sitterson, "A New Labor System, 1865–1880," in *The Louisiana Bicentennial Series in Louisiana History: Volume VI, Reconstructing Louisiana*, ed. Lawrence N. Powell, 503–518 (Lafayette: Center for Louisiana Studies, University of Louisiana at Lafayette, 2001), 475.

91. ROGER W. Shugg, "Survival of the Plantation System in Louisiana," in *The Louisiana Bicentennial Series in Louisiana History: Volume VI, Reconstructing Louisiana*, ed. Lawrence N. Powell (Lafayette: Center for Louisiana Studies, University of Louisiana at Lafayette, 2001), 469, 474.

92. J. Carlyle Sitterson, *Sugar Country: The Cane Sugar Industry in the South, 1753–1950* (Westport, CT: Greenwood, 1973), 231–233.

93. Michael David Cohen, *Reconstructing the Campus: Higher Education and the American Civil War* (Charlottesville: University of Virginia Press, 2012), 129–130.

94. Richard Follett and Rick Halpern, "From Slavery to Freedom in Louisiana's Sugar Country: Changing Labor Systems and Workers' Power, 1861–1913," in *Sugar, Slavery, and Society: Perspectives on the Caribbean, India, the Mascarenes, and the United States*, ed. Bernard Moitt (Gainesville: University Press of Florida, 2004), 149.

95. Lillian C. Bourgeois, *Cabanocey: The History, Customs and Folklore of St. James Parish* (Gretna, LA: Pelican, 1957), 53–54.

96. Carl A. Brasseaux, "Declining Economic Fortunes in Postbellum Louisiana," in *The Louisiana Bicentennial Series in Louisiana History: Volume VI, Reconstructing Louisiana*, ed. Lawrence N. Powell (Lafayette: Center for Louisiana Studies, University of Louisiana at Lafayette, 2001), 588.

97. Rien Fertel, *Imagining the Creole City: The Rise of Literary Culture in Nineteenth-Century New Orleans* (Baton Rouge: Louisiana State University Press, 2014).

98. Winters, *The Civil War in Louisiana*, 207; James Marten, *The Children's War* (Chapel Hill: University of North Carolina Press, 1998), 152.

99. Winters, *The Civil War in Louisiana*, 131–132.

100. Ibid., 131.

101. Powell, introduction to *The New Orleans of George Washington Cable*, 12.

102. Sitterson, *Sugar Country*, 311.

103. "Historic Marist College," *Morning Star* (1907): 10.

104. LeBreton, *Valcour Aime*, 148.

105. Roger Baudier, *The Catholic Church in Louisiana* (New Orleans: A. W. Hyatt Stationary Mfg. Co. Ltd., 1939), 433.

106. Ronald E. Butchart, *Schooling the Freed People: Teaching, Learning, and the Struggle for Black Freedom, 1861–1876* (Chapel Hill: University of North Carolina Press, 2010), 5, 43, 47, 89, 158.

107. Joe Gray Taylor, *Louisiana Reconstructed: 1863–1877* (Baton Rouge: Louisiana State University Press, 1974), 456; Barry A. Crouch, "Black Education in Civil War and Reconstruction Louisiana: George T. Ruby, the Army, and the Freedmen's Bureau," in *The Louisiana Purchase Bicentennial Series in Louisiana History: Volume XI, the African*

American Experience in Louisiana, Part B: From the Civil War to Jim Crow, ed. Charles Vincent (Lafayette: Center for Louisiana Studies, University of Louisiana at Lafayette, 2000), 263; Howard A. White, *The Freedmen's Bureau in Louisiana* (Baton Rouge: Louisiana State University Press, 1970), 168–169.

108. Howard Ashley White, "Redeeming of the People from Ignorance," in *The Louisiana Purchase Bicentennial Series in Louisiana History: Volume XI, the African American Experience in Louisiana, Part B: From the Civil War to Jim Crow*, ed. Charles Vincent (Lafayette: Center for Louisiana Studies, University of Louisiana at Lafayette, 2000), 300; Barry A. Crouch, "Black Education in Civil War and Reconstruction Louisiana: George T. Ruby, the Army, and the Freedmen's Bureau," in *The Louisiana Purchase Bicentennial Series in Louisiana History: Volume XI, the African American Experience in Louisiana, Part B: From the Civil War to Jim Crow*, ed. Charles Vincent (Lafayette: Center for Louisiana Studies, University of Louisiana at Lafayette, 2000), 261, 264, 269, 271.

109. George R. Bentley, *A History of the Freedmen's Bureau* (London: Oxford University Press, 1955), 14, 21, 89; David H. Goldfield, *The American Journey* (New York: Pearson, 2011), 424–426.

110. Howard A. White, *The Freedmen's Bureau in Louisiana* (Baton Rouge: Louisiana State University Press, 1970), 175.

111. Rodrigue, *Reconstruction in the Cane Fields: From Slavery to Free Labor in Louisiana's Sugar Parishes, 1862–1880*, 61; Howard A. White, *The Freedmen's Bureau in Louisiana* (Baton Rouge: Louisiana State University Press, 1970), 175.

112. René de Sennegy, *Une Paroisse Louisianaise: St. Michel du Comté d'Acadie* (New Orleans: Mr. Capo, 1877), 121.

113. Ibid.

114. Henry L. Abbot, *New Orleans to Vicksburg: Prepared by Order of Maj. Gen. N. P. Banks* (Washington, DC: Department of the Gulf, 1863), Library of Congress, Geography and Map Division.

115. Charles O. Perrine, *Perrine's New Topographical War Map of the Southern States: Taken from the Latest Government Surveys and Official Reports* (Buffalo: E. R. Jewett & Co., Engravers, 1863), Library of Congress, Geography and Map Division.

116. Arthur and de Kernion, *Old Families of Louisiana*, 48.

117. "Historic Marist College," 10; "Jefferson College: Beginnings, 1831–1842."

118. De Sennegy, *Une Paroisse Louisianaise*, 121.

119. "Celebrations around the Country: Marists Served New Orleans Area for 150 Years," Society of Mary: Marists in the U.S., http://www.societyofmaryusa.org/about/celebrations-clarionherald.html, accessed September 13, 2014; Bourgeois, *Cabanocey: The History, Customs and Folklore of St. James Parish*, 43.

120. *François Drouilly, 15 Days of Prayer with Jean-Claude Colin* (Hyde Park: New City Press, 2012), 17–25.

121. Donal Kerr, *Jean-Claude Colin, Marist: A Founder in an Era of Revolution and Restoration: The Early Years 1790–1836* (Dublin, Ireland: Columba Press, 2000), 15.

122. J. Coste, *Lectures on Society of Mary History (Marist Fathers), 1786–1854* (Rome: Via Alessandro Poerio, 1965), 81–83, 144–145, 223–235.

123. Ibid., 176.

124. Ibid., 81–83, 144–145, 223–235.

125. Kerr, *Jean-Claude Colin, Marist*, 137, 261.

126. J. Francis Georgelin, *The Catholic Church in Algiers, Louisiana: The Marist Fathers and the Holy Name of Mary Church*, ed. Rev. Charles F. Brogley, SM, MA (New Orleans: Privately printed by the Society of Mary, 1949), 16.

127. Edwin Whitfield Fay, *The History of Education in Louisiana* (Washington, DC: Government Printing Office, 1898), 142; C. M. Dubray, "The Society of Mary and Education," *Catholic Educational Review* 3 (April 1911): 316.

128. John L. Peytavin, "Jefferson College and Education: Address Delivered on June 20th, 1906," *Jefferson College Record* (July 1906), 139, Box: American Province Jefferson College 3, Marist School, Atlanta, Georgia, Provincial Archives of the Society of Mary.

129. Fay, *The History of Education in Louisiana*, 142.

130. De Sennegy, *Une Paroisse Louisianaise*, 123.

131. "Jefferson College: Beginnings, 1831–1842"; "Historic Marist College," 10.

132. "Jefferson College: Beginnings, 1831–1842."

133. De Sennegy, *Une Paroisse Louisianaise*, 123.

134. Valcour Aime to President, Board of Directors Jefferson College, Parish, St. James, March 6, 1864, Binder, Manresa: Jefferson College, Loyola University, Monroe Library, Archives of the New Orleans Province, Society of Jesus, New Orleans.

135. "Notarized Transfer of Jefferson College to the Marist Fathers."

136. Bourgeois, *Cabanocey*, 50.

137. Fay, *The History of Education in Louisiana*, 142; Dubray, "The Society of Mary and Education," 316.

138. Ibid.

139. "Jefferson College: Beginnings, 1831–1842."

140. "Notarized Transfer of Jefferson College to the Marist Fathers."

141. "Jefferson College: Beginnings, 1831–1842"; "Notarized Transfer of Jefferson College to the Marist Fathers."

142. Leroy E. Willie, *A Look at Louisiana's First Century, 1804–1903* (Baton Rouge: SAR Books, 2002), 57; White, *The Freedmen's Bureau in Louisiana*, 175–176.

143. Claude F. Jacobs and Andrew J. Kaslow, *The Spiritual Churches of New Orleans: Origins, Beliefs, and Rituals of an African-American Religion* (Knoxville: University of Tennessee Press, 1991), 22–23.

144. Heather Andrea Williams, *Self-taught: African American Education in Slavery and Freedom* (Chapel Hill: University of North Carolina Press, 2005), 14.

145. Frank M. Uter, *Stones beside a River: A History of the Catholic Church on the East Bank of St. James Parish, 1809–2009* (Baton Rouge: Catholic Diocese of Baton Rouge, 2009), 51.

146. Ibid.

147. Baudier, *The Catholic Church in Louisiana*, 433.

148. Randall M. Miller, "Slaves and Southern Catholicism," in *Masters and Slaves in the House of the Lord: Race and Religion in the American South, 1740–1870*, ed. John B. Boles (Lexington: University Press of Kentucky, 1988), 130.

149. Claude F. Jacobs and Andrew J. Kaslow, *The Spiritual Churches of New Orleans: Origins, Beliefs, and Rituals of an African-American Religion* (Knoxville: University of Tennessee Press, 1991), 23–24.

150. Miller, "Slaves and Southern Catholicism," 151.

151. Roger Baudier, *The Catholic Church in Louisiana* (New Orleans: A. W. Hyatt Stationary Mfg. Co. Ltd., 1939), 433; R. Eric Platt, *Sacrifice and Survival: Identity, Mission, and Jesuit Higher Education in the American South* (Tuscaloosa: University of Alabama Press, 2014), 5, 69.

152. Sally K. Reeves, "The Society of the Sacred Heart in New Orleans," in *Cross, Crozier, and Crucible: A Volume Celebrating the Bicentennial of a Catholic Diocese in Loui-*

siana, ed. Glenn R. Conrad, 219–232 (New Orleans: Archdiocese of New Orleans in cooperation with the Center for Louisiana Studies, 1993), 224.

153. Uter, *Stones beside a River: A History of the Catholic Church on the East Bank of St. James Parish, 1809–2009*, 51–54; Roger Baudier, *The Catholic Church in Louisiana*, 433; White, *The Freedmen's Bureau in Louisiana*, 193.

154. Annemarie Kasteel, *Francis Janssens, 1843–1897: A Dutch-American Prelate* (Lafayette: Center for Louisiana Studies, University of Southwestern Louisiana, 1992), 287; Baudier, *The Catholic Church in Louisiana*, 433.

155. White, *The Freedmen's Bureau in Louisiana*, 193.

156. De Sennegy, *Une Paroisse Louisianaise*, 125.

157. Ibid.

158. Ibid., 124; "Catalogue of Jefferson College, Convent, La. 1906–07"; "History of Jefferson College Dates Back to the Year 1831," *State Times*, October 19, 1926, 13.

159. Niehaus, "Jefferson College in St. James Parish, Louisiana: 1830–1875," 99; De Sennegy, *Une Paroisse Louisianaise*, 124.

160. *Catalogue of the Officers and Students of St. Mary Jefferson College, Distribution of Premiums for the Academic Year of 1872–73. St. James, Louisiana* (New Orleans: Morning Star Office, 1873), 10, Louisiana State University Special Collections, Hill Memorial Library, Baton Rouge.

161. "St. Mary-Jefferson College, Under the Direction of the MARIST FATHERS, 60 Miles above New Orleans and near the Convent of the Sacred Heart, St. James, Louisiana,"1870, Folder: Jefferson College, Convent, LA, Department of the Archives at the Catholic Diocese of Baton Rouge.

162. De Sennegy, *Une Paroisse Louisianaise*, 124.

163. Ibid., 123–124.

164. "Historic Marist College," 10; "Catalogue of Jefferson College, Convent, La. 1906–07."

165. Varnado, "Fabulous Valcour Aime," 1E.

166. "Historic Marist College," 11; Cline, *Pioneer Leaders and Early Institutions in Louisiana Education*, 262.

167. Michael O'Brien, *Intellectual Life and the American South, 1810–1860* (Chapel Hill: University of North Carolina Press, 2010), 91.

168. Father Thouvenin, Jefferson College, St. James Parish to Madame Fortier, December 28, 1897, Folder: Marist Fathers Jefferson College, November 12, 1888–May 31, 1922, Box: Central Files Prior to 1935, Religious Men, Office of Archives and Records Archdiocese of New Orleans, New Orleans.

169. LeBreton, *Valcour Aime*, 137.

170. Ibid., 138; Pamela Folse, *St. James Catholic Church: Remembering Our Story* (St. James, LA: St. James Catholic Church, n.d.), 7.

171. "Aged Valcour Aime Portrait Photograph," Binder, Manresa: Jefferson College, Loyola University, Monroe Library, Archives of the New Orleans Province, Society of Jesus, New Orleans.

172. Cochran, *The Fortier Family and Allied Families*, 76.

173. "Death of Ex-Gov. A. B. Roman," *Daily Picayune*, January 30, 1866, n.p.

174. Ibid.

175. Toledano, "Louisiana's Golden Age," 224.

176. "Died: On Tuesday, January 1, on his estate in the parish of St. James, Valcour Aime, aged 69 years, a native of Louisiana," *New Orleans Daily Crescent*, January 2, 1867, 6.

177. "Schooner Gabriel Aime," *New Orleans Times*, October 6, 1867, n.p.

178. "The Valcour Aime Plantation," *Daily Picayune*, December 21, 1871, 2; "Valcour Aim's Place," *Daily Picayune*, January 6, 1872.

179. Scarborough, *Masters of the Big House*, 126–127, 137, 399; Daspit, *Louisiana Architecture, 1840–1860*, 169.

180. Alcée Fortier, *Louisiana Studies: Literature, Customs and Dialects, History and Education* (New Orleans: F. F. Hansell & Bro., 1894), 304.

181. Clarence L. Mohr and Joseph E. Gordon, *Tulane: The Emergence of a Modern University, 1945–1980* (Baton Rouge: Louisiana State University Press, 2001), xxii.

182. "Jefferson College at Convent is Celebrating Its 95th Birthday," *State Times*, July 22, 1926, 15.

183. "St. Mary-Jefferson College, Under the Direction of the Marist Fathers, 60 Miles above New Orleans and near the Convent of the Sacred Heart, St. James, Louisiana," 1870, Folder: Jefferson College, Convent, Louisiana, Department of the Archives at the Catholic Diocese of Baton Rouge; St. Mary's Jefferson College Postcard, Campus Scene with College Seal, 1918, Author's Collection, Hattiesburg, Mississippi; Peter F. Quinn to Archbishop John William Shaw, December 2, 1921, Folder: Marist Fathers Jefferson College, November 12, 1888–May 31, 1922, Box: Central Files Prior to 1935, Religious Men, Office of Archives and Records Archdiocese of New Orleans, New Orleans.

184. Toledano, "Louisiana's Golden Age," 224.

185. Cline, *Pioneer Leaders and Early Institutions in Louisiana Education*, 263.

186. "J.M.J. St. Mary Jefferson College, Catalogue of Students and Distribution of Premiums for Academic Year 1869–70. Non Coronatur Nisi Legitime Ceravert"; *Catalogue of the Officers and Students of St. Mary Jefferson College, Distribution of Premiums for the Academic Year of 1872–73. St. James, Louisiana.*

187. Samuel H. Lock, *Louisiana As It Is: A Geographical and Topographical Description of Louisiana*, ed. Lauren C. Post (Baton Rouge: Louisiana State University Press, 1969), 41.

188. "St. Mary-Jefferson College, Under the Direction of the Marist Fathers."

189. *Catalogue of the Officers and Students of St. Mary Jefferson College, Distribution of Premiums for the Academic Year of 1872–73. St. James, Louisiana.*

190. P. H. Dagneau, *Memories of Seventy-Five Years of the Centenary of the Marist Fathers in America* (Atlanta: Keystone Press, 1963), 61.

191. "J.M.J. St. Mary Jefferson College, Catalogue of Students and Distribution of Premiums for Academic Year 1869–70. Non Coronatur Nisi Legitime Ceravert."

192. Dagneau, *Memoirs of Seventy-five Years for the Centenary of the Marist Fathers in America*, 58; "Jefferson College: Beginnings, 1831–1842."

193. R. P. Lowry, "A History of the Public Schools of St. James Parish from 1865 to the Present Day" (MA thesis, Louisiana State University, 1928), appendix.

194. Niehaus, "Jefferson College in St. James Parish, Louisiana: 1830–1875," 102.

195. *Catalogue of the Officers and Students of St. Mary Jefferson College, Distribution of Premiums for the Academic Year of 1872–73. St. James, Louisiana.*

196. *Annual Catalogue of Jefferson College (St. Mary's.) St. James, Louisiana, 1890–91* (New Orleans: T. Fitzwilliam & Co., 1891), 20, Louisiana State University Special Collections, Hill Memorial Library, Baton Rouge; "St. Mary-Jefferson College, Under the Direction of the Marist Fathers."

197. "Catalogue of Jefferson College, Convent, La. 1906–07," 1906, 10, Louisiana State University Special Collections, Hill Memorial Library, Baton Rouge.

198. "St. Mary-Jefferson College, Under the Direction of the Marist Fathers."

199. "Catalogue of Jefferson College, Convent, La. 1906–07."

200. "J.M.J. St. Mary Jefferson College, Catalogue of Students and Distribution of Premiums for Academic Year 1869–70. Non Coronatur Nisi Legitime Ceravert"; Elisabeth Joan Doyle, 1993, "Mightier than the Sword: The Catholic Press of Louisiana," in *Cross, Crozier, and Crucible: A Volume Celebrating the Bicentennial of a Catholic Diocese in Louisiana*, ed. Glenn R. Conrad (New Orleans: Archdiocese of New Orleans with the Center for Louisiana Studies), 252–253.

201. "Historic Marist College," 11.

202. *Catalogue of the Officers and Students of St. Mary Jefferson College, Distribution of Premiums for the Academic Year of 1872–73. St. James, Louisiana.*

203. "Catalogue of Jefferson College, Convent, La. 1916–17," 1916, Louisiana State University Special Collections, Hill Memorial Library, Baton Rouge.

204. Dagneau, *Memoirs of Seventy-five Years for the Centenary of the Marist Fathers in America*, 61.

205. Ibid., 11–18.

206. Ibid., 6; "Catalogue of Jefferson College, Convent, La. 1906–07"; *Catalogue of Jefferson College (St. Mary's.) 1901–1902* (New Orleans: T. Fitzwilliam & Co., 1891), 70–71, Louisiana State University Special Collections, Hill Memorial Library, Baton Rouge.

207. "St. Mary-Jefferson College, Under the Direction of the Marist Fathers."

208. *Annual Catalogue of Jefferson College (St. Mary's.) St. James, Louisiana. 1890–91*, 21.

209. "Catalogue of Jefferson College, Convent, La. 1906–07."

210. Ibid.

211. *Biographical and Historical Memoirs of Louisiana: Embracing an Authentic and Comprehensive Account of the Chief Events in the History of the State, a Special Sketch of Every Parish and a Record of the Lives of Many of the Most Worthy and Illustrious Families and Individuals*, vol. 2, 347.

212. Miriam G. Reeves, *The Governors of Louisiana* (Gretna, LA: Pelican, 1998), 53.

213. "St. Mary-Jefferson College, Under the Direction of the Marist Fathers."

214. Niehaus, "Jefferson College in St. James Parish, Louisiana: 1830–1875," 104.

215. De Sennegy, *Une Paroisse Louisianaise*, 125.

216. Niehaus, "Jefferson College in St. James Parish, Louisiana: 1830–1875," 104.

217. Randall M. Miller, "A Church in Cultural Captivity: Some Speculations on Catholic Identity in the Old South," in *Catholics in the Old South*, ed. Randall Miller and John Wakelyn (Macon, GA: Mercer University Press, 1999), 50.

218. Niehaus, "Jefferson College in St. James Parish, Louisiana: 1830–1875," 104.

219. Ibid.; Transfer L'Auction, Du College Jefferson, Parish St. Jacques, Louisiana, 1867, Copy of 1867, Box: American Province Jefferson College 2, Folder: Transfer Document, Marist School, Atlanta, Georgia, Provincial Archives of the Society of Mary; Sale by Rev. L Chauvin to Rev. H Bellanger, L. E. Poché Deputy Recorder, June 2, 1876, Box: American Province Jefferson College 2, Folder: Jefferson College Property Deeds, 1872–1881, Marist School, Atlanta, Georgia, Provincial Archives of the Society of Mary; Copy of 1867 Sale of land—by Ludger to Rev. O. Renaudier, May 30, 1884, Box: American Province Jefferson College 2, Folder: Jefferson College Property Deeds, 1872–1881, Marist School, Atlanta, Georgia, Provincial Archives of the Society of Mary.

220. De Sennegy, *Une Paroisse Louisianaise*, 126–127.

221. Transfer L'Auction, Du College Jefferson, Parish St. Jacques, Louisiana; Sale by

Rev. L Chauvin to Rev. H Bellanger, L. E. Poché Deputy Recorder, June 2, 1876; Copy of 1867 Sale of land—by Ludger to Rev. O. Renaudier, May 30, 1884.

222. State of Louisiana, 4th Judicial District Court, Parish of St. James, Box: American Province Jefferson College 2, Folder: Jefferson College Property Deeds, 1864–1867, Marist School, Atlanta, Georgia, Provincial Archives of the Society of Mary; "Jefferson College: Beginnings, 1831–1842"; Niehaus, "Jefferson College in St. James Parish, Louisiana: 1830–1875," 105–106.

223. "An Act to Incorporate the Society of the 'Fathers of the Society of Mary,'" Box: American Province Jefferson College 1, Folder: Marist Society of La—Act of Incorporation, 1875, Marist School, Atlanta, Georgia, Provincial Archives of the Society of Mary.

224. Ibid.

225. Ibid.

226. "A List of the Professors and Boys at Jefferson College, 1865 to 1927," Box: American Province Jefferson College 1, Folder: Faculty and Students, No. of, 1865–1927; "An Act to Incorporate the Society of the 'Fathers of the Society of Mary,'" Box: American Province Jefferson College 1, Folder: Marist Society of La—Act of Incorporation, 1875, Marist School, Atlanta, Georgia, Provincial Archives of the Society of Mary; Georgelin, *The Catholic Church in Algiers, Louisiana (1848–1948)*, 16; John L. White and Mark Edwards, *The Society of Mary: Its Foundation and First American Mission*, ed. Nicholas A. Weber (privately printed, 1949), 59–61; Bernard Bourtot, archivist for the European Provincial Archives of the Society of Mary, e-mail to Susan Illis, archivist for the U.S. Provincial Archives of the Society of Mary, November 4, 2016; Bernard Bourtot, archivist for the European Provincial Archives of the Society of Mary, e-mail to Susan Illis, archivist for the U.S. Provincial Archives of the Society of Mary, September 26, 2016.

227. "An Act to Incorporate the Society of the 'Fathers of the Society of Mary,'" Box: American Province Jefferson College 1, Folder: Marist Society of La—Act of Incorporation, 1875, Marist School, Atlanta, Georgia, Provincial Archives of the Society of Mary.

228. De Sennegy, *Une Paroisse Louisianaise*, 128.

Chapter 5

1. Martha R. Field, *Louisiana Voyages: The Travel Writings of Catharine Cole*, ed. Joan B. McLaughlin and Jack McLaughlin (Jackson: University Press of Mississippi, 2006), ix, 41.

2. Ibid., 42.

3. Ted Tunnel, *Crucible of Reconstruction: War, Radicalism, and Race, 1862–1877* (Baton Rouge: Louisiana State University Press, 1984), 162.

4. Carl A. Brasseaux, *Acadian to Cajun: Transformation of a People, 1803–1877* (Jackson: University Press of Mississippi, 1992), 180–181.

5. Michael David Cohen, *Reconstructing the Campus: Higher Education and the American Civil War* (Charlottesville: University of Virginia Press, 2012), 127, 188.

6. Roger W. Shugg, *Origins of Class Struggle in Louisiana: A Social History of White Farmers and Laborers during Slavery and After, 1840–1875* (1939; Baton Rouge: Louisiana State University Press, 1966), 296–297.

7. Michael Perman, *The Road to Redemption: Southern Politics, 1869–1879* (Chapel Hill: University of North Carolina Press, 1984), 245.

8. Shugg, *Origins of Class Struggle in Louisiana*, 296–297.

9. Margaret Humphreys, *Yellow Fever and the South* (New Brunswick, NJ: Rutgers University Press, 1992), 82.

10. John C. Rodrigue, *Reconstruction in the Cane Fields: From Slavery to Free Labor in Louisiana's Sugar Parishes, 1862–1880* (Baton Rouge: Louisiana State University Press, 2001), 115; Richard Follett, "The Rise and Fall of American Sugar," in *Plantation Kingdom: The American South and Its Global Commodities*, ed. Richard Follett, Sven Beckert, Peter Coclanis, and Barbara Hahn (Baltimore: Johns Hopkins University Press, 2016), 87–88.

11. Agreement between Pére Favre and Archbishop Odin, New Orleans, September 13, 1869, Folder: Marist Fathers Jefferson College, November 12, 1888–May 31, 1922, Box: Central Files Prior to 1935, Religious Men, Office of Archives and Records Archdiocese of New Orleans.

12. "A List of the Professors and Boys at Jefferson College, 1865 to 1927," Box: American Province Jefferson College 1, Folder: Faculty and Students, No. of, 1865–1927.

13. P. H. Dagneau, *Memories of Seventy-five Years of the Centenary of the Marist Fathers in America* (Atlanta: Keystone Press, 1963), 48.

14. Ibid., 58.

15. Field, *Louisiana Voyages*, 42, 45.

16. Ibid., 45.

17. "Historic Marist College," *Morning Star* (1907): 11.

18. L. A. Chassé to A. de la Peichardiére, November 6, 1875, Peichardiére Papers, Louisiana State University Special Collections, Hill Memorial Library, Baton Rouge.

19. "St. Mary-Jefferson College, Under the Direction of the Marist Fathers, 60 Miles above New Orleans and near the Convent of the Sacred Heart, St. James, Louisiana,"1870, Folder: Jefferson College, Convent, Louisiana, Department of the Archives at the Catholic Diocese of Baton Rouge.

20. Nathan A. Buman, "Two Histories, One Future: Louisiana Sugar Planters, Their Slaves, and the Anglo-Creole Schism, 1815–1865" (PhD diss., Louisiana State University, 2013), 136.

21. Marcia Gaudet, "Christmas Bonfires in South Louisiana: Tradition and Innovation," in *Mardi Gras, Gumbo, and Zydeco: Readings in Louisiana Culture*, ed. Marcia Gaudet and James C. McDonald (Jackson: University Press of Mississippi, 2003), 60; "A List of the Professors and Boys at Jefferson College, 1865 to 1927"; Justin D. Poché, "Bonfires on the Levee: Place, Memory, and the Sacred in River Road Catholicism," in *Gods of the Mississippi (Religion in North America)*, ed. Michael Pasquire (Bloomington: Indiana University Press, 2013), 161–164.

22. "Bonfires and Blazing Hearth Date Far Back in Time," *Daily Advertiser*, December 25, 2000.

23. Letter from Sidney Montegin, student at Jefferson College to an unknown recipient, December, 27, 1881, Folder: Montegin (Sidney) Letter 222, Louisiana Research Collection, Special Collections, Howard-Tilton Memorial Library, Jones Hall, Tulane University, New Orleans.

24. "A List of the Professors and Boys at Jefferson College, 1865 to 1927."

25. Edward L. Ayers, *The Promise of the New South: Life after Reconstruction, 15th Anniversary Edition* (Oxford: Oxford University Press, 2007), 191–192; "A List of the Pro-

fessors and Boys at Jefferson College, 1865 to 1927"; Follett, "The Rise and Fall of American Sugar," 82–83.

26. Charles P. Roland, *Louisiana Sugar Plantations during the Civil War* (Baton Rouge: Louisiana State University Press, 1997), 89–90, 139.

27. "Jefferson College, Statement of Properties, March 3rd, 1898," Box: American Province Jefferson College 1, Folder: Inventory, Marist School, Atlanta, Georgia, Provincial Archives of the Society of Mary.

28. Dagneau, *Memoirs of Seventy-five Years for the Centenary of the Marist Fathers in America*, 59.

29. "Lay Professors at Jefferson College from 1864 to 1927—May 31," Box: American Province Jefferson College 1, Folder: Lay Professors—Number of Years Served, 1864–1927, Marist School, Atlanta, Georgia, Provincial Archives of the Society of Mary.

30. Dagneau, *Memoirs of Seventy-five Years for the Centenary of the Marist Fathers in America*, 56–58.

31. Lillian C. Bourgeois, *Cabanocey: The History, Customs and Folklore of St. James Parish* (Gretna, LA: Pelican, 1957), 56.

32. Joseph H. Fichter, 1987, "First Black Students at Loyola University: A Strategy to Obtain Teacher Certification," in *The Louisiana Purchase Bicentennial Series in Louisiana: Volume XVII*, ed. Michael G. Wade (Lafayette: Center for Louisiana Studies, University of Louisiana at Lafayette, 1999), 381; Mary Lee Muller, "New Orleans Public School Desegregation," in *The Louisiana Purchase Bicentennial Series in Louisiana History: Volume XI, The African American Experience in Louisiana, Part C, From Jim Crow to Civil Rights*, ed. Charles Vincent (Lafayette: Center for Louisiana Studies, University of Southwestern Louisiana, 2002), 339; James G. Dauphine, *A Question of Inheritance: Religion, Education, and Louisiana's Cultural Boundary* (Lafayette: Center for Louisiana Studies, University of Southwestern Louisiana, 1993), 52–65; Melissa Kean, *Desegregating Private Higher Education in the South: Duke, Emory, Rice, Tulane, and Vanderbilt* (Baton Rouge: Louisiana State University Press, 2008), 81n86.

33. Walter L. Fleming, *Louisiana State University: 1860–1896* (Baton Rouge: Louisiana State University Press, 1936), 167–168.

34. Ibid., 232.

35. *St. Mary's Brass Band, Jefferson College, La.*, record book, "History of St. Mary's B[rass] B[and]," n.d., Box: American Province Jefferson College 3, Marist School, Atlanta, Georgia, Provincial Archives of the Society of Mary.

36. Ibid.

37. Ibid.

38. Ibid.; *St. Mary's Brass Band, Jefferson College, La.*, record book, "Visitant Francis T. Nichols," November 3, 1887, "The Archbishop's Visit," May 2, 1898, Box: American Province Jefferson College 3, Marist School, Atlanta, Georgia, Provincial Archives of the Society of Mary.

39. Ibid.

40. Dagneau, *Memoirs of Seventy-five Years for the Centenary of the Marist Fathers in America*, 59.

41. "Catalogue of Jefferson College, Convent, La. 1906–07," 1906, Louisiana State University Special Collections, Hill Memorial Library, Baton Rouge.

42. "Commencement," *St. James Courier*, July 4, 1889.

43. *Catalogue of the Officers and Students of St. Mary Jefferson College, Distribution of Premiums for the Academic Year of 1878–79. St. James, Louisiana* (New Orleans: Morning

Star Office, 1879), 24, Louisiana State University Special Collections, Hill Memorial Library, Baton Rouge.

44. "President's Feast," in *Jefferson College Record* (January 1905), 237, Louisiana State University Special Collections, Hill Memorial Library, Baton Rouge.

45. "Amusement," in *Jefferson College Record* (December 1905), 47, Louisiana State University Special Collections, Hill Memorial Library, Baton Rouge.

46. "Historic Marist College," 10; Rodney Cline, *Pioneer Leaders and Early Institutions in Louisiana Education* (Baton Rouge: Claitor Publishing Division, 1969), 263.

47. Mary Bernardine Hill, "The Influence of James Hubert Blenk on Catholic Education in the Archdiocese of New Orleans, 1885–1917" (EdD diss., Louisiana State University, 1964), 49–51, 53–54, 57–58.

48. Ibid., 59–61, 63, 65, 67–68, and 73–74.

49. Ibid., 59–61, 63, 69–71.

50. Ibid., 74, 75–82, 187.

51. "Catalogue of Jefferson College (St. Mary's.) 1901–1902," 1901, Louisiana State University Special Collections, Hill Memorial Library, Baton Rouge, 69.

52. "Catalogue of Jefferson College, Convent, La. 1906–07," 1906, Louisiana State University Special Collections, Hill Memorial Library, Baton Rouge.

53. Ibid.

54. Dagneau, *Memoirs of Seventy-five Years for the Centenary of the Marist Fathers in America*, 50–51; R. P. Lowry to Reverend P. H. Dagneau, August 30, 1932, Binder, Manresa: Jefferson College, Loyola University, Monroe Library, Archives of the New Orleans Province, Society of Jesus, New Orleans.

55. Dagneau, *Memoirs of Seventy-five Years for the Centenary of the Marist Fathers in America*, 61.

56. John W. Madden, *Handbook of the Carnival: A Wonderful Account of the History, Krewes, and Events Surrounding Mardi Gras in New Orleans* (New Orleans: Cornerstone, 2011), 94–119; Stephen W. Hales, *Rex: An Illustrated History of the School of Design* (Mandeville: Arthur Hardy Enterprises, 2010), 20, 34–35.

57. "Mardi Gras," in *Jefferson College Record* (March 1905), 163–164, Box: American Province Jefferson College 3, Marist School, Atlanta, Georgia, Provincial Archives of the Society of Mary.

58. "Catalogue of Jefferson College (St. Mary's.) 1901–1902," 1901, Louisiana State University Special Collections, Hill Memorial Library, Baton Rouge, 26.

59. "Historic Marist College," 11.

60. "Catalogue of Jefferson College (St. Mary's.) 1901–1902," 26.

61. Ibid., 8.

62. Ibid.

63. "Capt. Kantz," *Jefferson College Record, Convent, Louisiana* (March 1905), 97, Louisiana State University Special Collections, Hill Memorial Library, Baton Rouge.

64. Edwin Whitfield Fay, *The History of Education in Louisiana* (Washington, DC: Government Printing Office, 1898), 142.

65. "The Competitive Drill," *Jefferson College Record, Convent, Louisiana* (January 1905), 240–241, Louisiana State University Special Collections, Hill Memorial Library, Baton Rouge.

66. "Saint Mary's Jefferson College Catalogue for the Academic Year of 1912–13," 1912, Binder, Manresa: Jefferson College, Loyola University, Monroe Library, Archives of the New Orleans Province, Society of Jesus, New Orleans; See R. Eric Platt, Melandie

McGee, and Amanda King, "Marching in Step: Patriotism and the Southern Catholic Cadet Movement," *Catholic Historical Review* 102, no. 3 (Summer 2016): 517–544.

67. Dagneau, *Memoirs of Seventy-five Years for the Centenary of the Marist Fathers in America*, 65–67.

68. "Catalogue of Jefferson College (St. Mary's), 1913–1914," 1913, Louisiana State University Special Collections, Hill Memorial Library, Baton Rouge, 55; W. M. Haulsee, F. G. Howe, and A. C. Doyle, *Soldiers of the Great War*, vol. 3 (Washington, DC: Soldiers Record Publishing Association, 1920), 420; C. Plaisance, "Henry Noel [J.] Plaisance," *Find a Grave*, http://www.findagrave.com/cgi-bin/fg.cgi?page=gr&GSln=plaisance&GSfn=henry&GSbyrel=all&GSdy=1918&GSdyrel=in&GSob=n&GRid=116459793&df=all&, accessed August 14, 2016.

69. Certificate of Service, the World War, 1921, Box: American Province Jefferson College 2, Folder: Certificate of Service W.W.I., 1921, Marist School, Atlanta, Georgia, Provincial Archives of the Society of Mary.

70. "Jefferson 5, Pelicans 7" and "Jefferson 1, L. S. U. 11," *Jefferson College Record* (July 1906), 176, Box: American Province Jefferson College 3, Marist School, Atlanta, Georgia, Provincial Archives of the Society of Mary.

71. "Introduction," *Jefferson College Record* (1922), 9, Box: American Province Jefferson College 3, Marist School, Atlanta, Georgia, Provincial Archives of the Society of Mary.

72. "Local Notes," *Jefferson College Record* (July 1905), 45, Louisiana State University Special Collections, Hill Memorial Library, Baton Rouge.

73. "Football Season," *Jefferson College Record* (1922), 45, Box: American Province Jefferson College 3, Marist School, Atlanta, Georgia, Provincial Archives of the Society of Mary; "Football! Football! 'The Real Home Coming Day,'" Box: American Province Jefferson College 1, Folder: Miscellaneous, Marist School, Atlanta, Georgia, Provincial Archives of the Society of Mary.

74. Dagneau, *Memoirs of Seventy-five Years for the Centenary of the Marist Fathers in America*, 50.

75. "Baseball," *Jefferson College Record* (July 1905), 225, Louisiana State University Special Collections, Hill Memorial Library, Baton Rouge.

76. Dagneau, *Memoirs of Seventy-five Years for the Centenary of the Marist Fathers in America*, 50.

77. "Basketball," in *Sugar Bowl* (1923), 43, Box: American Province Jefferson College 3, Marist School, Atlanta, Georgia, Provincial Archives of the Society of Mary; "Basketball," in *Sugar Bowl* (1924), 64, Box: American Province Jefferson College 3, Marist School, Atlanta, Georgia, Provincial Archives of the Society of Mary.; "Athletics," "Athletics," *Jefferson College Record* (1906), n.p., Box: American Province Jefferson College 3, Marist School, Atlanta, Georgia, Provincial Archives of the Society of Mary;

78. "Just for Fun," *Jefferson College Record* (July 1905), 35, 63, Box: American Province Jefferson College 3, Marist School, Atlanta, Georgia, Provincial Archives of the Society of Mary.

79. Caedmus, "Ode to Jefferson College," *Jefferson College Record* (January 1905), 1, Louisiana State University Special Collections, Hill Memorial Library, Baton Rouge.

80. *Oxford English Dictionary*, "pile, n.5, I.2.a.," http://www.oed.com.lynx.lib.usm.edu/view/Entry/143824?rskey=HTATb9&result=5&isAdvanced=false#eid, accessed November 19, 2016.

81. Caedmus, "Ode to Jefferson College."

82. "Advertisements," *Jefferson College Record* (1922), 59–79, Box: American Province Jefferson College 3, Marist School, Atlanta, Georgia, Provincial Archives of the Society of Mary.

83. Ibid., 66, 74, 75, 77–79, and inside back cover.

84. Dagneau, *Memoirs of Seventy-five Years for the Centenary of the Marist Fathers in America*, 47, 49, 55–56.

85. Joseph G. Tregle Jr., "Creoles and Americans," in *Creole New Orleans: Race and Americanization*, ed. Arnold R. Hirsch and Joseph Logsdon (Baton Rouge: Louisiana State University Press, 1992), 132.

86. "Catalogue of Jefferson College, Convent, La. 1906–07," 1906, Louisiana State University Special Collections, Hill Memorial Library, Baton Rouge.

87. Ibid.

88. "Historic Marist College," 11.

89. "Catalogue of Jefferson College (St. Mary's.) 1901–1902," 9, 1901, Louisiana State University Special Collections, Hill Memorial Library, Baton Rouge.

90. "Historic Marist College," 11.

91. *Annual Catalogue of Jefferson College (St. Mary's.) St. James, Louisiana, 1892–93* (New Orleans: T. Fitzwilliam & Co., Printers, 62 Camp Street, 1891), 7, Louisiana State University Special Collections, Hill Memorial Library, Baton Rouge.

92. Matthew Pratt Guterl, *American Mediterranean: Southern Slaveholders in the Age of Emancipation* (Cambridge, MA: Harvard University Press, 2008), 108–109.

93. Paul W. Schott, "The Jesuit House That Never Should Have Been," *Southern Jesuits* 1, no. 1, September 1981, 10; Dagneau, *Memoirs of Seventy-five Years for the Centenary of the Marist Fathers in America*, 59; John Wilds, Charles L. Dufour, and Walter G. Cowan, *Louisiana Yesterday and Today: A Historical Guide to the State* (Baton Rouge: Louisiana State University Press, 1996), 107.

94. Rien Fertel, *Imagining the Creole City: The Rise of Literary Culture in Nineteenth-Century New Orleans* (Baton Rouge: Louisiana State University Press, 2014), 4–5.

95. Dagneau, *Memoirs of Seventy-five Years for the Centenary of the Marist Fathers in America*, 59–60.

96. Ibid., 60.

97. "Advertisements," *Jefferson College Record* (1922), 66, 74, 75, 77, 78, 79, and inside back cover, Box: American Province Jefferson College 3, Marist School, Atlanta, Georgia, Provincial Archives of the Society of Mary.

98. "Catalogue of Jefferson College, Convent, La. 1906–07," 52.

99. Dagneau, *Memoirs of Seventy-five Years for the Centenary of the Marist Fathers in America*, 52.

100. Ibid.

101. Ibid.

102. "Building News," *American Architect* 98, no. 1809 (1910): 8.

103. "Historic Marist College," 11.

104. Oliver P. Carriere, *A Sketch of the History of Jefferson College and Manresa House of Retreats, Convent, Louisiana: 1830–1973* (Convent, LA: Manresa House of Retreats, 1974), 11.

105. R. H. Smith, SM, Jefferson College, St. James Parish, Convent, La., to Most Rev. Archbishop Blenk, April 24, 1912, Folder: Marist Fathers Jefferson College, November 12, 1888–May 31, 1922, Box: Central Files Prior to 1935, Religious Men, Office of Archives and Records Archdiocese of New Orleans.

106. Ibid.

107. R. Eric Platt, *Sacrifice and Survival: Identity, Mission, and Jesuit Higher Education in the American South* (Tuscaloosa: University of Alabama Press, 2014), 7.

108. James Blenk to Albert Biever, March 14, 1912, AR Files: 1431, Loyola University, Monroe Library, Archives of the New Orleans Province, Society of Jesus, New Orleans.

109. "Saint Mary's Jefferson College Catalogue for the Academic Year of 1912–13," Louisiana State University Special Collections, Hill Memorial Library, Baton Rouge.

110. Ibid.

111. "Catalogue of Jefferson College (St. Mary's), 1918–19," 1918, Louisiana State University Special Collections, Hill Memorial Library, Baton Rouge.

112. Ibid.

113. "High School Diploma, Jefferson College High School," Marist School, Atlanta, Georgia, Provincial Archives of the Society of Mary.

114. "Jefferson College, Convent, Louisiana," *Morning Star*, May 17, 1917, 50, no. 14.

115. Dagneau, *Memoirs of Seventy-five Years for the Centenary of the Marist Fathers in America*, 68.

116. Cline, *Pioneer Leaders and Early Institutions in Louisiana Education*, 267.

117. Teddy (Student at St. Mary's Jefferson College), Convent, Louisiana, to "Mother," Macacona, Costa Rica, March 27, 1918, author's collection, Hattiesburg, Mississippi.

118. E. G., "Dedication," in *Sugar Bowl* (1924), 5, Box: American Province Jefferson College 3, Marist School, Atlanta, Georgia, Provincial Archives of the Society of Mary.

119. P. F. Quinn, Jefferson College, St. James Parish, Convent, Louisiana, to Most Rev. J. M. Shaw DD, Archbishop of New Orleans, August 16, 1920, Folder: Marist Fathers Correspondence, May 26, 1912–August 15, 1934, Box: Central Files Prior to 1935, Religious Men, Office of Archives and Records Archdiocese of New Orleans, New Orleans.

120. Archbishop Shaw, New Orleans, to Father Quinn, Jefferson College, Convent, Louisiana, Folder: Marist Fathers Appointments, March 31, 1919–September 1925, Box: Central files prior to 1935, Religious men, NODA, Office of Archives and Records Archdiocese of New Orleans, New Orleans.

121. Dagneau, *Memoirs of Seventy-five Years for the Centenary of the Marist Fathers in America*, 68; H. de La Chapelle, Provincial, Society of Mary to Archbishop John W. Shaw, New Orleans, August 13, 1920, Folder: Marist Father Appointments, Box: Central Files Prior to 1935, Religious Men, Office of Archives and Records Archdiocese of New Orleans, New Orleans.

122. Letter from Marist Fathers, Jefferson College to Archbishop Shaw, New Orleans, December 2, 1921, Folder: Marist Fathers Jefferson College, November 12, 1888–May 31, 1922, Box: Central Files Prior to 1935, Religious Men, Office of Archives and Records Archdiocese of New Orleans; "Laymen's Retreat," in *Sugar Bowl* (1923), 68–69, Box: American Province Jefferson College 3, Marist School, Atlanta, Georgia, Provincial Archives of the Society of Mary.

123. Cline, *Pioneer Leaders and Early Institutions in Louisiana Education*, 267; Dagneau, *Memoirs of Seventy-five Years for the Centenary of the Marist Fathers in America*, 72; H. Harris, Louisiana State Superintendent of Public Education and Chas. F. Trudeau, State High School Supervisor to Rev. P. H. Dagneau, September 20, 1932, Box: American Province Jefferson College 1, Folder: Early History, Marist School, Atlanta, Georgia, Provincial Archives of the Society of Mary; "Alumni," in *Sugar Bowl* (1923), 53–58, Box: American Province Jefferson College 3, Marist School, Atlanta, Georgia, Provincial Archives of the Society of Mary.

124. "Catalogue of Jefferson College, Convent, La. 1906–07."

125. Carriere, *A Sketch of the History of Jefferson College and Manresa House of Retreats, Convent, Louisiana*, 11; "Historic Marist College," 11.

126. "History of Jefferson College Dates Back to the Year 1831," *State Item*, October 19, 1926, 13; "Catalogue of Jefferson College, Convent, La. 1906–07."

127. Dagneau, *Memoirs of Seventy-five Years for the Centenary of the Marist Fathers in America*, 49.

128. "Canonical Visitation of St. Michael's Church, Convent, Louisiana," 1923, Department of the Archives at the Catholic Diocese of Baton Rouge.

129. Adam Fairclough, *Race and Democracy: The Civil Rights Struggle in Louisiana, 1915–1972*, 2nd ed. (Athens: University of Georgia Press, 2008), 5–6; Nikki Brown, "Jim Crow," in *KnowLA: Encyclopedia of Louisiana*, ed. David Johnson, http://www.knowla.org/entry/735/, accessed December 27, 2015.

130. Cline, *Pioneer Leaders and Early Institutions in Louisiana Education*, 268.

131. Ibid.; Dagneau, *Memoirs of Seventy-five Years for the Centenary of the Marist Fathers in America*, 51.

132. Ibid., 69.

133. "The College County Fair," *Jefferson College Record* (1922), 51–52, Box: American Province Jefferson College 3, Marist School, Atlanta, Georgia, Provincial Archives of the Society of Mary.

134. "Articles of Agreement for Deed between Louisiana Real Estate & Development Company and R. H. Smith, Pres. Jefferson College Convent, La., Dated June 9, 1909," Box: American Province Jefferson College 1, Folder: Caddo Parish Property, Marist School, Atlanta, Georgia, Provincial Archives of the Society of Mary; "La. R.E. & D.Co. to R.H. Smith: Sale of Land, Filed for Record, 1910," Box: American Province Jefferson College 1, Folder: Caddo Parish Property, Marist School, Atlanta, Georgia, Provincial Archives of the Society of Mary.

135. Tonja Koob Marking and Jennifer Snape, *Images of America: Louisiana's Oil Heritage* (Mount Pleasant, SC: Arcadia Publishing, 2012), 7; Dan R. Frost and Kou K. Nelson, *The LSU College of Engineering, Volume 1: Origins and Establishment, 1860–1908* (Baton Rouge: Louisiana State University College of Engineering, 1995), 149.

136. "W. Shropshire, Louisiana Real Estate & Development Co. to R. H. Smith, Convent, La., Sept. 3, 1910," Box: American Province Jefferson College 1, Folder: Caddo Parish Property, Marist School, Atlanta, Georgia, Provincial Archives of the Society of Mary.

137. "J. F. Black to R. H Smith, Convent, La., September 16, 1918," Box: American Province Jefferson College 1, Folder: Caddo Parish Property, Marist School, Atlanta, Georgia, Provincial Archives of the Society of Mary; "J. F. Black, Shreveport, La. P.O. Box 681," Box: American Province Jefferson College 1, Folder: Caddo Parish Property, Marist School, Atlanta, Georgia, Provincial Archives of the Society of Mary; "J. F. Black Room 206 Merchants Bldg., P. O. Box 681, Shreveport, La., property superscription letter," Box: American Province Jefferson College 1, Folder: Caddo Parish Property, Marist School, Atlanta, Georgia, Provincial Archives of the Society of Mary.

138. Ibid., 70; Dagneau, *Memoirs of Seventy-five Years for the Centenary of the Marist Fathers in America*, 69.

139. "Father Rapier Heads This School," *Times Picayune*, February 6, 1927, 23.

140. Dagneau, *Memoirs of Seventy-five Years for the Centenary of the Marist Fathers in America*, 70.

141. Ibid.

142. John M. Barry, *Rising Tide: The Great Mississippi Flood of 1927 and How It Changed America* (New York: Simon and Schuster, 1997), 240.

143. Jim Bradshaw, "Great Flood of 1927," in *KnowLA: Encyclopedia of Louisiana*, ed. David Johnson (Louisiana Endowment for the Humanities, 2010–, 2011), http://www.knowla.org/entry/763/; Barry, *Rising Tide*, 222.

144. Bradshaw, "Great Flood of 1927."

145. Lyle Saxon, *Father Mississippi: The Story of the Great Flood of 1927* (New York: Century Co., 1927), 331, 341.

146. Ibid., 392.

147. Bradshaw, "Great Flood of 1927."

148. Barry, *Rising Tide*, 409.

149. Dagneau, *Memoirs of Seventy-five Years for the Centenary of the Marist Fathers in America*, 70; "St. Stanislaus at Bay St. Louis Opens Registration," *Times Picayune*, August 27, 1927, 15.

150. *Catalogue of Jefferson College (St. Mary's) 1926–1927* (New Orleans: T. Fitzwilliam & Co., 1891), 5–7, Louisiana State University Special Collections, Hill Memorial Library, Baton Rouge.

151. "Away from City Distractions: Jefferson College, Boarding School for Boys," *Times Picayune*, August 8, 1927, 20.

152. Rodney Cline, "Early Colleges and Schools in Louisiana," in *The Louisiana Bicentennial Series in Louisiana History: Volume XVII, Education in Louisiana*, ed. Michael G. Wade (Lafayette: Center for Louisiana Studies, University of Louisiana at Lafayette, 1999), 235.

153. Letter from Father Rapier, president of Jefferson College, Convent, Louisiana, to Archbishop Shaw, New Orleans, August 15, 1927, Folder: Marist Fathers Correspondence, May 26, 1912–August 15, 1934, Box: Central files prior to 1935, Religious men, NODA, Office of Archives and Records Archdiocese of New Orleans.

154. Ibid.

155. "Jefferson College Will Not Reopen: Institution to Close Doors after Sixty-three Years," *Times Picayune*, August 18, 1927, 16.

156. Cline, "Early Colleges and Schools in Louisiana," 235.

157. "R. H. Smith to Father George, November 21, 1927," Box: American Province Jefferson College 1, Folder: Closing & Sale, Marist School, Atlanta, Georgia, Provincial Archives of the Society of Mary.

158. R. H. Smith to Father George, July 26, 1928," Box: American Province Jefferson College 1, Folder: Closing & Sale, Marist School, Atlanta, Georgia, Provincial Archives of the Society of Mary.

159. Ibid.

160. Washington Province Records, 1924–1999, Provincial Chapter Promulgation, 1928. Box 1, Folder 5, Marist School, Atlanta, Georgia, Provincial Archives of the Society of Mary.

161. E. Pfleger to Rev. N. A. Weber, Prov. S. M., June 12, 1945, Box: American Province Jefferson College 1, Folder: Closing & Sale, Marist School, Atlanta, Georgia, Provincial Archives of the Society of Mary.

162. "Graduates of 'Jefferson College' Convent, Louisiana 1871 to 1927," Box: American Province Jefferson College 1, Folder: List of Graduates 1871–1927, Marist School, Atlanta, Georgia, Provincial Archives of the Society of Mary; Jefferson College Student

Information, Box: American Province Jefferson College Student Records A-K, 9, Marist School, Atlanta, Georgia, Provincial Archives of the Society of Mary.

163. R. P. Lowry, Superintendent of Education, St. James Parish, Louisiana, to Rev. P. H. Dagneau, S. M., August 30, 1932, Box: American Province Jefferson College 1, Folder: Early History, Marist School, Atlanta, Georgia, Provincial Archives of the Society of Mary.

164. Roger Baudier, *The Catholic Church in Louisiana* (New Orleans: A. W. Hyatt Stationary Mfg. Co. Ltd., 1939), 418–419.

165. Dagneau, *Memoirs of Seventy-five Years for the Centenary of the Marist Fathers in America*, 60.

166. R. P. Lowry to Reverend P. H. Dagneau, August 30, 1932.

167. Cline, "Early Colleges and Schools in Louisiana," 235; William V. Bangert, *A History of the Jesuits* (St. Louis: Institute of Jesuit Resources, 1972), 7–9.

Conclusion

1. Charles Gayarré, *The Creoles of History and the Creoles of Romance: A Lecture Delivered in the Hall of the Tulane University by Charles Gayarré, on the 25th of April, 1885* (New Orleans: C. E. Hopkins, 1885), 14; Carl A. Brasseaux, *Acadian to Cajun: Transformation of a People, 1803–1877* (Jackson: University Press of Mississippi, 1992), 109, 151–152.

2. Glenn R. Conrad, "The Faces of French Louisiana," in *Cross, Crozier, and Crucible: A Volume Celebrating the Bicentennial of a Catholic Diocese in Louisiana*, ed. Glenn R. Conrad, 5–15 (New Orleans: Archdiocese of New Orleans in cooperation with the Center for Louisiana Studies, 1993), 13.

3. Lawrence N. Powell, introduction to *The New Orleans of George Washington Cable: The 1887 Census Office Report* by George Washington Cable (Baton Rouge: Louisiana State University Press, 2008), 9.

4. Arlin Turner, *George Washington Cable* (Baton Rouge: Louisiana State University Press, 1966), 1; Emily Clark, *The Strange History of the American Quadroon: Free Women of Color in the Revolutionary Atlantic World* (Chapel Hill: University of North Carolina Press, 2013), 189.

5. Richard Campanella, *Bourbon Street: A History* (Baton Rouge: Louisiana State University Press, 2014), 83.

6. Rien Fertel, *Imagining the Creole City: The Rise of Literary Culture in Nineteenth-Century New Orleans* (Baton Rouge: Louisiana State University Press, 2014), 84, 101; Will H. Coleman, *Historical Sketch Book and Guide to New Orleans and Environs* (New York: Will H. Coleman, 1885), 319–325.

7. Thomas Ruys Smith, *Southern Queen: New Orleans in the Nineteenth Century* (New York: Continuum International, 2011), 148.

8. Joseph G. Tregle Jr., "Creoles and Americans," in *Creole New Orleans: Race and Americanization*, ed. Arnold R. Hirsch and Joseph Logsdon (Baton Rouge: Louisiana State University Press, 1992), 175–177; Lawrence N. Powell, *The Accidental City: Improvising New Orleans* (Cambridge, MA: Harvard University Press, 2012), 116, 304.

9. William Stubbs, *The Cultivation of Sugar Cane* (Savannah, GA: D. G. Purse, 1900), 7; Gayarré, *The Creoles of History and the Creoles of Romance*, 1; Smith, *Southern Queen*, 141.

10. Charles Gayarré, *Fernando de Lemos: Truth and Fiction, a Novel* (New York: G. W. Carleton & Co., Publishers, 1872), 18.

11. Gayarré, *The Creoles of History and the Creoles of Romance*, 3, 6, 17–18; V. Faye

Phillips, "Charles Gayarré," in *KnowLA: Encyclopedia of Louisiana*, ed. David Johnson (Louisiana Endowment for the Humanities, 2010), http://www.knowla.org/entry/744/&view=summary; Marcel Giraud, *A History of French Louisiana, Volume 5: The Company of the Indies, 1723–1731*, trans. Brian Pearce (Baton Rouge: Louisiana State University Press, 1991), 325.

12. Gayarré, *The Creoles of History and the Creoles of Romance*, 7.

13. Ibid., 9.

14. Ibid., 22, 30.

15. Turner, *George Washington Cable*, 270 n2.

16. Powell, introduction to *The New Orleans of George Washington Cable*, 20–21.

17. Turner, *George Washington Cable*, 127.

18. Fertel, *Imagining the Creole City: The Rise of Literary Culture in Nineteenth-Century New Orleans*, 93.

19. Clark, *The Strange History of the American Quadroon*, 189.

20. Leonard V. Huber, "New Orleans Cemeteries: A Brief History," in *New Orleans Architecture, Volume III: The Cemeteries*, ed. Mary Louise Christovich, 3–62 (Gretna, LA: Pelican, 1974), 7.

21. Tregle, "Creoles and Americans," 175–177.

22. Mary Gehman, *The Free People of Color of New Orleans: An Introduction*, 5th ed. (Chelsea, MI: Sheridan Books, 2009), 79.

23. Fertel, *Imagining the Creole City: The Rise of Literary Culture in Nineteenth-Century New Orleans*, 74.

24. Gehman, *The Free People of Color of New Orleans*, 79.

25. Conrad, "The Faces of French Louisiana," 13.

26. Gwendolyn Midlo Hall, *Africans in Colonial Louisiana: The Development of Afro-Creole Culture in the Eighteenth Century* (Baton Rouge: Louisiana State University Press, 1995), 157.

27. "Fathers and Brothers of the Society of Mary Who Served People of God at Jefferson College and St. Michael Catholic Church, Convent, Louisiana," n.d., Memorial Plaque, St. Michael's Church Cemetery, River Road, Convent, LA.

28. Mary Bernardine Hill, "The Influence of James Hubert Blenk on Catholic Education in the Archdiocese of New Orleans, 1885–1917" (EdD diss., Louisiana State University, 1964), 187.

29. Douglas Keister, *Stories in Stone: A Field Guide to Cemetery Symbolism and Iconography* (Layton, UT: Gibbs Smith, 2004), 137.

30. La-Cemeteries, "Louisiana Governors," http://www.la-cemeteries.com/Governor/Roman,%20Andre%20Bienvenu/Roman,%20Andre%20Bienvenu.shtml, accessed December 23, 2013.

31. Mary Ann Sternberg, *River Road Rambler: A Curious Traveler along Louisiana's Historic Byway* (Baton Rouge: Louisiana State University Press, 2013), 27–34.

32. Cochran, *The Fortier Family and Allied Families*, 57–58.

33. Death of Louis Dufau, St. Louis Cemeteries Nos. 1 & 2, New Orleans, Louisiana, Death Record, 1857–1864, p. 130. Office of Archives and Records, Archdiocese of New Orleans.

34. John R. Thelin, *A History of American Higher Education*, 2nd ed. (Baltimore: Johns Hopkins University Press), x.

35. Ibid.

36. Dagmar Renshaw LeBreton, *Valcour Aime: A Biography* (n.p., 1948), 16, Louisiana and Special Collections Department, Earl K. Long Library, University of New Orleans.

37. Michael O'Brien, *Intellectual Life and the American South, 1810–1860* (Chapel Hill: University of North Carolina Press, 2010), 33–34, 91.

38. Conrad, "The Faces of French Louisiana," 13.

39. Tregle, "Creoles and Americans," 173.

40. Robert C. Reinders, *End of an Era: New Orleans, 1850–1860* (1964; reprint, Gretna, LA: Pelican, 1998), 131, 138; Stuart Grayson Noble, "Early School Superintendents in New Orleans," *Journal of Educational Research* 24, no. 4 (November 1931): 227.

41. Joseph Logsdon and Caryn Cossé Bell, "The Americanization of Black New Orleans 1850–1900," in Hirsch and Logsdon, *Creole New Orleans: Race and Americanization*, 242.

42. Reinders, *End of an Era: New Orleans, 1850–1860*, 138.

43. John C. Rodrigue, *Reconstruction in the Cane Fields: From Slavery to Free Labor in Louisiana's Sugar Parishes, 1862–1880* (Baton Rouge: Louisiana State University Press, 2001), 59; Roger W. Shugg, *Origins of Class Struggle in Louisiana: A Social History of White Farmers and Laborers during Slavery and After, 1840–1875* (1939; reprint, Baton Rouge: Louisiana State University Press, 1966), 474.

44. LeBreton, *Valcour Aime: A Biography*, 150.

45. John L. Peytavin, "Jefferson College and Education: Addressed Delivered on June 20th, 1906," *Jefferson College Record* (July 1906), 140, 141, Box: American Province Jefferson College 3, Marist School, Atlanta, Georgia, Provincial Archives of the Society of Mary.

46. Elizabeth Barrouquere, "Last of the Marists Catholic Order Leaving Convent after Arriving in 1862," *The Advocate (Baton Rouge)* July 13, 2002, 1F.

47. T. H. Harris and Chas F. Trudeau to Reverend P. H. Dagneau, September 20, 1932, Binder, Manresa: Jefferson College, Loyola University, Monroe Library, Archives of the New Orleans Province, Society of Jesus, New Orleans.

48. Harnett T. Kane, *Plantation Parade: The Grand Manner in Louisiana* (New York: William Morrow, 1945), 308.

49. Charles P. Roland, *Louisiana Sugar Plantations during the Civil War* (Baton Rouge: Louisiana State University Press, 1997), 139.

Epilogue

1. "T. H. Harris State Superintendent to E. C. Carrere, Ernest A. Carrere's Sons, November 20, 1928," Box: American Province Jefferson College 1, Folder: Closing & Sale, Marist School, Atlanta, Georgia, Provincial Archives of the Society of Mary; Edward Allen Fay, ed., *American Annals of the Deaf, Volume XL* (Washington, DC: Conference of Superintendents and Principles of American Schools for the Deaf, 1895), 60.

2. Elinor J. Gast to Very Rev. Butin, SM, Provincial, Marist Fathers, Brookland, Washington, DC, July 9, 1930, Box: American Province Jefferson College 1, Folder: Closing & Sale, Marist School, Atlanta, Georgia, Provincial Archives of the Society of Mary; Butin, SM, Provincial, Marist Fathers, Brookland, Washington, DC, to Elinor J. Gast, July 28, 1930, Box: American Province Jefferson College 1, Folder: Closing & Sale, Marist School, Atlanta, Georgia, Provincial Archives of the Society of Mary; Linda Robin, "Emmett Louis Hardy," *Find a Grave*, http://www.findagrave.com/cgi-bin/fg.cgi?page=gr&GSln=hardy&GSfn=emmett&GSbyrel=all&GSdyrel=all&GSst=20&GScntry=4&GSob=n&GRid=107267143&df=all&, accessed November 30, 2015.

3. "Ed Carrere, Ernest A. Carrere's Sons Real Estate and Insurance Mortgages,

to Rev. R. H. Smith, SM, May 21, 1930," Box: American Province Jefferson College 1, Folder: Closing & Sale, Marist School, Atlanta, Georgia, Provincial Archives of the Society of Mary.

4. "R. Butin, SM, to Ernest A. Carrer's Sons, New Orleans, July 28, 1930," Box: American Province Jefferson College 1, Folder: Closing & Sale, Marist School, Atlanta, Georgia, Provincial Archives of the Society of Mary.

5. William D. Reeves, *Manresa on the Mississippi: For the Greater Glory of God* (Convent, LA: Manresa House of Retreats, 1996), 26.

6. "Jefferson College Will Change Hands: Property in St. James Parish to Pass to Control of Jesuit Fathers," *Times-Picayune*, September 9, 1930, 24; Charles I. Denechaud, Attorney at Law, to Rev. H. George, St. Joseph's Church, Paulina, La., December 2, 1930. Box: American Province Jefferson College 1, Folder: Closing & Sale, Marist School, Atlanta, Georgia, Provincial Archives of the Society of Mary.

7. "Seventy-Third Meeting of the Corporation of: The Father of the Society of Mary. January 5th. 1931. Rectification of Title. Sale of Jefferson College. Election of New Members," Box: American Province Jefferson College 1, Folder: Closing & Sale, Marist School, Atlanta, Georgia, Provincial Archives of the Society of Mary.

8. Ibid.; Harry S. Crane, SJ, to A. William Crandell, SJ, Provincial of the New Orleans Province, Society of Jesus, April 19, 1955, Box: Manresa, Convent, La. Ownership Documents 1954–1955, Folder: I, Loyola University, Monroe Library, Archives of the New Orleans Province, Society of Jesus, New Orleans.

9. "Manresa—Institutional Considerations: Preliminary Outline of the Financial Relationship of Jesuit High School as Founded by Provisions of the Institution," Box: Manresa, Convent, La. Ownership Documents 1954–1955, Folder: I, Loyola University, Monroe Library, Archives of the New Orleans Province, Society of Jesus, New Orleans; "Jesuit High—Manresa Settlement," Box: Manresa, Convent, La. Ownership Documents 1954–1955, Folder: I, Loyola University, Monroe Library, Archives of the New Orleans Province, Society of Jesus, New Orleans; Harry L. Crane, SJ, to A. William Crandell, SJ, Provincial of the New Orleans Province, Society of Jesus, Box: Manresa, Convent, La. Ownership Documents 1954–1955, Folder: I, Loyola University, Monroe Library, Archives of the New Orleans Province, Society of Jesus, New Orleans; Harry L. Crane, SJ, to James D. Carroll, SJ, Provincial Archivist, November 10, 1954, Box: Manresa, Convent, La. Ownership Documents 1954–1955, Folder: I, Loyola University, Monroe Library, Archives of the New Orleans Province, Society of Jesus, New Orleans; Cash Sale of Property By Catholic Society of Religious and Literary Education to Manresa House of Retreats, March 14, 1950, Box: Manresa, Convent, La. Ownership Documents 1954–1955, Folder: II, Loyola University, Monroe Library, Archives of the New Orleans Province, Society of Jesus, New Orleans; "Keith Will is Filed," *Black Diamond* 35, no. 18 (November 1905): 27; Marty Denzer, "Orphanages Cared for Homeless Children," *Catholic Key Online*, January 19, 2012, catholickey.org/2012/01/19/orphanages-cared-for-homeless-children/, accessed December 1, 2016.

10. Harry L. Crane, SJ, "Excerpts on Manresa," n.d., 42, Box: Manresa, Convent, La. Ownership Documents, 1954–1955 Folder: III, Loyola University, Monroe Library, Archives of the New Orleans Province, Society of Jesus, New Orleans; Jerome Neyrey, and Thomas Clancy, *Southern Jesuit Biographies: Pastors and Preachers, Builders and Teachers of the New Orleans Province* (Lafayette, LA: Acadian House Publishing, 2015), 138.

11. Michael Kenny, *Jesuits in our Southland, 1566–1946: Origin and Growth of New Orleans Province*, Kenny Papers, Archives of the New Orleans Province, Society of Jesus, Loyola University Monroe Library, New Orleans, 189.

12. Paul W. Schott, "The Jesuit House That Never Should Have Been," *Southern Jesuits* 1, no. 1 (September 1981): 11.

13. Reeves, *Manresa on the Mississippi*, 21.

14. Schott, "The Jesuit House That Never Should Have Been," 11.

15. Ibid.

16. Trahan Architects, apac., *Manresa House of Retreats: 2000 Evaluation*, Loyola University, Monroe Library, Archives of the New Orleans Province, Society of Jesus, New Orleans; "A.M.D.G. Register of Retreatants, Manresa House, Convent, LA, February 26, 1931 to December 9, 1944," Manresa House of Retreats, Convent, LA, St. Peter Canisus Hall, Library.

17. Frank M. Uter, *Stones beside a River: A History of the Catholic Church on the East Bank of St. James Parish, 1809–2009* (Baton Rouge: Catholic Diocese of Baton Rouge, 2009), 54.

18. Arthur Lippa, Humble Oil & Refining Company, to Father N. A. Weber, Marist Provincial, Washington DC, July 12, 1944, Box: American Province Jefferson College 1, Folder: Mineral Rights, Marist School, Atlanta, Georgia, Provincial Archives of the Society of Mary; Sigur Martin, Martin and Himel Law Offices, to Father Thomas Bolduc, August 15, 1944, Box: American Province Jefferson College 1, Folder: Mineral Rights, Marist School, Atlanta, Georgia, Provincial Archives of the Society of Mary; Thomas Buldoc, SM, to Nicholas Weber, SM, August 17, 1944, Box: American Province Jefferson College 1, Folder: Mineral Rights, Marist School, Atlanta, Georgia, Provincial Archives of the Society of Mary.

19. Oliver P. Carriere, *A Sketch of the History of Jefferson College and Manresa House of Retreats, Convent, Louisiana: 1830–1973* (Convent, LA: Manresa House of Retreats, 1974), 23.

20. Harry L. Crane, SJ, "Excerpts on Manresa," n.d., 44, Box: Manresa, Convent, La. Ownership Documents, 1954–1955 Folder: III, Loyola University, Monroe Library, Archives of the New Orleans Province, Society of Jesus, New Orleans.

21. "Manresa—Institutional Considerations"; "Jesuit High—Manresa Settlement"; Harry L. Crane, SJ, to A. William Crandell, SJ, Provincial of the New Orleans Province, Society of Jesus, Box: Manresa, Convent, La. Ownership Documents 1954–1955, Folder: I, Loyola University, Monroe Library, Archives of the New Orleans Province, Society of Jesus, New Orleans; Harry L. Crane, SJ, to James D. Carroll, SJ, Provincial Archivist, November 10, 1954, Box: Manresa, Convent, La. Ownership Documents 1954–1955, Folder: I, Loyola University, Monroe Library, Archives of the New Orleans Province, Society of Jesus, New Orleans; Cash Sale of Property By Catholic Society of Religious and Literary Education to Manresa House of Retreats, March 14, 1950, Box: Manresa, Convent, La. Ownership Documents 1954–1955, Folder: II, Loyola University, Monroe Library, Archives of the New Orleans Province, Society of Jesus, New Orleans.

22. Harry L. Crane, SJ, "Excerpts on Manresa," n.d., 45, Box: Manresa, Convent, La. Ownership Documents, 1954–1955 Folder: III, Loyola University, Monroe Library, Archives of the New Orleans Province, Society of Jesus, New Orleans.

23. "Constitution of Manresa House, Convent, Louisiana," n.d., Box 3: Manresa Retreat House, Convent, La., Folder: Legal Documents/Opinions, Loyola University, Monroe Library, Archives of the New Orleans Province, Society of Jesus, New Orleans; "Restatement of Articles of Incorporation as Needed of the Administrators of the Rev. John W. Hynes, SJ, Manresa Memorial Endowment Fund, Inc. Box 3: Manresa Retreat House, Convent, La., Folder: St. Mary's Hall Renovations, Loyola University, Monroe

Library, Archives of the New Orleans Province, Society of Jesus, New Orleans; Cash Sale of Property by Catholic Society of Religious and Literary Education to Manresa House of Retreats, March 14, 1950, Box: Manresa, Convent, La. Ownership Documents 1954–1955, Folder: II, Loyola University, Monroe Library, Archives of the New Orleans Province, Society of Jesus, New Orleans.

24. "Constitution of Manresa House, Convent, Louisiana," n.d.

25. Carriere, *A Sketch of the History of Jefferson College and Manresa House of Retreats, Convent, Louisiana*, 4–6.

26. Schott, "The Jesuit House That Never Should Have Been," 11.

27. "Excerpts on Manresa, 1949," Box: Manresa, Convent, La. Ownership Documents 1954–1955, Folder: III, Loyola University, Monroe Library, Archives of the New Orleans Province, Society of Jesus, New Orleans.

28. "Remodeled Manresa Will Open Sunday," *Catholic Action of the South*, March 16, 1950, 7.

29. Schott, "The Jesuit House That Never Should Have Been," 11.

30. "National Register of Historic Places Database [Louisiana]," *Office of Cultural Development: Division of Historic Preservation*, http://www.crt.state.la.us/dataprojects/hp/nhl/index.asp, accessed December 1, 2016.

31. Tim Murphy, General Manager, Manresa House of Retreats, to Hynes Board Members Concerning Renovation Work, St. Mary's Hall, November 25, 2002, Box 3: Manresa Retreat House, Convent, La., Folder: St. Mary's Hall Renovations, Loyola University, Monroe Library, Archives of the New Orleans Province, Society of Jesus, New Orleans; "St. Mary's Hall Renovation Program: Manresa Retreat House, November 18, 2002," Box 3: Manresa Retreat House, Convent, La., Folder: St. Mary's Hall Renovations, Loyola University, Monroe Library, Archives of the New Orleans Province, Society of Jesus, New Orleans.

32. Tim Murphy, interview by R. Eric Platt, Convent, LA, August 29, 2014.

33. *Manresa "Good News" Letter* 39, no. 4 (Fall 2013): 1, 3, http://manresala.org/Manresa-Retreat-House-News/Newsletters.aspx.

34. Tim Murphy, interview by R. Eric Platt.

35. *Manresa on the Mississippi: Retreat Manual* (Convent, LA: Manresa House of Retreats, n.d.), 10.

36. Ibid.

37. Ibid., 6.

38. *Manresa Retreat House, Convent, Louisiana, Conducted by the Jesuit Fathers: Schedule of Retreats for Laymen* (Convent, LA: Manresa House of Retreats, 2008), cover-2; *Manresa on the Mississippi*, 2–3.

39. Louis H. Yarrut, *The Reflections of a Jew at a Catholic Retreat* (n.p., 1964), 14.

Appendix A

Earl Francis Niehaus, "Jefferson College in St. James Parish, Louisiana: 1830–1875" (MA thesis, Tulane University, 1954), 33–36, 40, 45–46, 64; Louis Dufau, "Louisiana College Now Situated in the Parish of St. James, in the Extensive and Commodious Buildings, Formerly Known as the Jefferson College," 1853, Tulane Manuscripts Department, Special Collections, Howard-Tilton Memorial Library, Jones Hall, Tulane University, New Orleans; Vernon Elmer Kappel, "Louis Dufau's Louisiana College" (MA thesis, Tulane University, 1949), 24; Valcour Aime to the Board of Directors, Jefferson College,

St. James Parish, August 10, 1859, Binder, Manresa: Jefferson College, Loyola University, Monroe Library, Archives of the New Orleans Province, Society of Jesus, New Orleans; "A List of the Professors and Boys at Jefferson College, 1865 to 1927," Box: American Province Jefferson College 1, Folder: Faculty and Students, 1865–1927; "An Act to Incorporate the Society of the 'Fathers of the Society of Mary,'" Box: American Province Jefferson College 1, Folder: Marist Society of La—Act of Incorporation, 1875, Marist School, Atlanta, Georgia, Provincial Archives of the Society of Mary; P. H. Dagneau, *Memories of Seventy-Five Years of the Centenary of the Marist Fathers in America* (Atlanta: Keystone Press, 1963), 48, 65–67, 69; Mary Bernardine Hill, "The Influence of James Hubert Blenk on Catholic Education in the Archdiocese of New Orleans, 1885–1917" (EdD diss., Louisiana state University, 1964), 49–51; "Historic Marist College," *Morning Star* (1907): 11; P. F. Quinn, Jefferson College, St. James Parish, Convent, La. to Most Rev. J. M. Shaw DD, Archbishop of N. Orleans, August 16, 1920, Folder: Marist Fathers Correspondence 5/26/12–8/15/34, Box: Central Files Prior to 1935, Religious Men, Office of Archives and Records Archdiocese of New Orleans, New Orleans, Louisiana; "R. H. Smith to Father George, November 21, 1927," Box: American Province Jefferson College 1, Folder: Closing & Sale, Marist School, Atlanta, Georgia, Provincial Archives of the Society of Mary.

Appendix B

Henry A. Bullard and Thomas Curry, *A New Digest of the Statutes and Laws of the State of Louisiana, from the Change of Government to the year 1841, Inclusive, Vol. I* (New Orleans: E. Johns & Co. Stationers' Hall, 1842), 300–302.

Appendix C

"Sale of Jefferson College to Louis Dufau," as cited in Vernon Elmer Kappel, "Louis Dufau's Louisiana College" (MA thesis, Tulane University, 1949), Appendix I.

Appendix D

Valcour Aime to the Board of Directors, Jefferson College, St. James Parish, August 10, 1859, Binder, Manresa: Jefferson College, Loyola University, Monroe Library, Archives of the New Orleans Province, Society of Jesus, New Orleans.

Appendix E

"Notarized Transfer of Jefferson College to the Marist Fathers," May 6, 1864, Binder, Manresa: Jefferson College, Loyola University, Monroe Library, Archives of the New Orleans Province, Society of Jesus, New Orleans.

Appendix F

"An Act to Incorporate the Society of the 'Fathers of the Society of Mary,'" Box: American Province Jefferson College 1, Folder: Marist Society of La—Act of Incorporation, 1875, Marist School, Atlanta, Georgia, Provincial Archives of the Society of Mary.

Appendix G

"Seventy-Third Meeting of the Corporation of: The Father of the Society of Mary. January 5th. 1931. Rectification of Title. Sale of Jefferson College. Election of New Members," Box: American Province Jefferson College 1, Folder: Closing & Sale, Marist School, Atlanta, Georgia, Provincial Archives of the Society of Mary.

Bibliography

NEWSPAPERS

Advocate, Baton Rouge, LA
Courier, New Orleans, LA
Daily Advertiser, Lafayette, LA
Daily Delta, New Orleans, LA
Daily Item, New Orleans, LA
Daily Picayune, New Orleans, LA
Morning Advocate, Baton Rouge, LA
Morning Star, New Orleans, LA
New Orleans Daily Crescent, New Orleans, LA
New Orleans States, New Orleans, LA
New Orleans Times, New Orleans, LA
New York Times, New York, NY
State Item, New Orleans, LA
St. James Courier, Convent, LA
Sunday Advocate, Baton Rouge, LA
Times Picayune, New Orleans, LA
Vigilant, Donaldsonville, LA

ARCHIVAL SOURCES

Atlanta, GA

Archives of the Society of Mary, Marist School

Lincoln, Rixford J. "Jefferson College [Poem]." Box: American Province Jefferson College 3

St. Mary's Jefferson College Collection

Baton Rouge, LA

Department of the Archives at the Catholic Diocese of Baton Rouge

Canonical Visitation Files. St. Michael's Church, Convent, Louisiana
J.M.J. St. Mary Jefferson College, Catalogue of Students and Distribution of Premiums for Academic Year 1869–70. Non Coronatur Nisi Legitime Ceravert. 1870
St. Mary-Jefferson College. Under the Direction of the Marist Fathers, 60 Miles above New Orleans and Near the Convent of the Sacred Heart, St. James, Louisiana. 1870. Folder: Jefferson College, Convent, LA

Louisiana State University Special Collections, Hill Memorial Library

Valcour Aime, Ledgers
Annual Catalogue of Jefferson College (St. Mary's.) St. James, Louisiana. 1890–91
Everett, Alexander Hill. *Address Delivered at Jefferson College, St. James Parish, La. June 30, 1841, on Assuming the Functions of President of That Institution*
Alcée Fortier Collection
Francois J. Gentil Class Role Book, 1853
Jefferson College Catalog Series (1831–1847, 1864–1927)
Jefferson College Record
Robert M. Lusher Papers. Folder: History of Early System of Education in Louisiana
Peichardiére Papers
Statutes, Plan of Education, and Rules, Adopted in Jefferson College, Louisiana

Convent, LA

St. Peter Canisus Hall, Library, Manresa House of Retreats

A.M.D.G. Register of Retreatants, Manresa House, Convent, Louisiana, February 26, 1931, to December 9, 1944

Gretna, LA

Jefferson Parish Court House, Mortgage and Conveyance Records

John Baptiste Lyman, Notary Public Appearance: Cecile Dufau, Widow of Louis Dufau, March 30, 1895

Lafayette, LA

Edith Garland Dupré Library, Special Collection and Archives, University of Louisiana at Lafayette
René de Sennegy Papers

New Orleans, LA

Archives of the New Orleans Province of the Society of Jesus, Monroe Library, Loyola University

Aime, Valcour. Letter to President and Board of Directors, Jefferson College, Parish, St. James. August 10, 1859, March 6, 1864. Binder, Manresa: Jefferson College
Blenk, James. Letter to Albert Biever. March 14, 1912. AR Files: 1431
Jordan, J. Letter to Father Abbadie concerning the opening of colleges. October 8, 1847. AR Files: 1049E

Kenny, Michael. *Jesuits in Our Southland 1566–1946: Origin and Growth of New Orleans Province*
Maisounabe, Jean Baptiste. Diary. 1847. Maisounabe Papers
Manresa: Jefferson College. Binder and Files
Saint Mary's Jefferson College Catalogue for the Academic Year of 1912–13. 1912. Binder, Manresa: Jefferson College
Trahan Architects. *Manresa House of Retreats: 2000 Evaluation*

Louisiana State Museum Collections, Historical Center

Valcour Aime and Gabrielle Aime Diaries

Louisiana and Special Collections Department, Earl K. Long Library, University of New Orleans

Dagmar Renshaw LeBreton Files
LeBreton, Dagmar Renshaw. *Valcour Aime: A Biography*, 1948

Office of Archives and Records Archdiocese of New Orleans

"An Act to Incorporate the Society of the Fathers of the Society of Mary." April 20, 1875. Folder: Holy Name of Mary 1870 Charter Jefferson College. Box: Holy Name of Mary Historical Documents
Archbishop James Hubert Blenk Papers
Chapelle, H. de La, Provincial, Society of Mary. Letter to Archbishop John W. Shaw, New Orleans, August 13, 1920. Folder: Marist Father Appointments. Box: Central Files Prior to 1935, Religious Men
St. Louis Cemeteries Nos. 1 and 2 Deaths, 1857–1864
Father Thouvenin (Jefferson College, St. James Parish). Letter to Madame Fortier, December 28, 1897. Folder: Marist Fathers Jefferson College 11/12/1888–5/31/1922, Box: Central Files Prior to 1935, Religious Men

Southeastern Architectural Archive, Special Collections Division, Tulane University

"Side Elevation of a One-story Gothic Church Building. Small Tower at Front, Eight Gothic Arched Windows with Diamond Pattern Panes." Unsigned. Undated. Pencil on Drawing Paper. Collection 143. Sylvester Labrot Collection, #573

Tulane Manuscripts Department, Howard-Tilton Memorial Library, Tulane University

Address of Louis Dufau, President of Louisiana College, Delivered at the Commencement Exercises, Academic Year, 1855–1856
Board of Administrators, University of Louisiana. Minutes, 1849, 1856
The Citizens' League: A History of the Great Reform Movement in New Orleans, April 21st 1896, with Biographical Sketches of Those Who Took Prominent Part in the Movement
S. Copes Papers
Catalogue of the Officers and Students of Louisiana College, for the Year Ending July 1853
Catalogue of the Officers and Students of Louisiana College. Statutes, Plan of Education, Regulations and Discipline. Act of Incorporation, and Contract between President and the Stockholders. Academic Year 1855–1856. Parish of St. James (La.), 1855

Catalogue of the Officers of the Collegiate and Preparatory Departments of the University of Louisiana. Statutes, Plan of Education, Regulations and Discipline. Academic Year 1856–1857

De la Vergne Papers

Dufau, Louis. *Louisiana College Now Situated in the Parish of St. James, in the Extensive and Commodious Buildings, Formerly Known as the Jefferson College*

Marguerite M. Fortier Papers

Johnston, William Preston. "A History of Tulane University"

State University of Louisiana, Eighth Annual Session of the Collegiate Department, 1858–59

Notre Dame, IN

University of Notre Dame Archives

Archdiocese of New Orleans Collection

Washington, DC

Library of Congress, Geography and Map Division

Abbot, Henry L. *New Orleans to Vicksburg: Prepared by Order of Maj. Gen. N. P. Banks*, 1863

Hergesheimer, Edwin. *Map Showing the Distribution of the Slave Population of the Southern States of the United States. Compiled from the Census of 1860*, 1861

La Tourrette, John. *Tourrette's Reference Map of the State of Louisiana: From the Original Surveys of the United States, which Show the Townships, Sections, or Mile Squares, Spanish Grants, Settlement Rights & C., Also the Plantations with the Owners Names Engraved Thereon*, 1848

Perrine, Charles O. *Perrine's New Topographical War Map of the Southern States: Taken from the Latest Government Surveys and Official Reports*, 1863

Persac, Marie Adrian. *Norman's Chart of the Lower Mississippi River*, 1858

National Archives

Seventh Census of the United States. 1850. National Archives Microfilm Publication M432, 1,009 rolls. Records of the Bureau of the Census. Record Group 29

SELECT PRIMARY SOURCES

"1125–1127 Dauphine St.: Chain of Title." Collins C. Diboll Vieux Carré Digital Survey. http://www.hnoc.org/vcs/property_info.php?lot=23065-01. Accessed October 29, 2015.

1852 Orleans Parish Death Index. Vol. 13, p. 248. Louisiana Archives, Baton Rouge. http://files.usgwarchives.net/la/orleans/vitals/deaths/index/1852di.txt. Accessed January 3, 2015.

Acts Passed by the First Legislature of the State of Louisiana, Held and Begun in the Town of Baton Rouge, on the 17th January, 1853. New Orleans: Emile La Sere, 1853.

Aime, Valcour. *Plantation Diary of the Late Mr. Valcour Aime, Formerly Proprietor of the Plantation Known as the St. James Sugar Refinery Situated in the Parish of St. James and Now Owned by Mr. John Burnside*. New Orleans: Lark & Hofeline, 1878.

Brunett, Louis J. B. "To All Free-Masons (2 February 1860)." *Transactions of the Supreme Council of the 33rd Degree, Ancient and Accepted Rite of Free-masonry for the Southern*

Jurisdiction of the United States of American, 1857 to 1866, 180–183. Washington, DC: Joseph L. Pearson, 1878.

Chevet, F. "To the President of the Directors of Jefferson College." *Journal of the Senate. Second Session of the Sixteenth Legislature of Louisiana* (1844): viii–xiiii [*sic*].

De Bow, J. D. B., ed. "A. B. Roman, of Louisiana, Agriculturalist." In *De Bow's Review of the Southern and Western States, Volume 11*, 436–443. New Orleans: Office, 1851.

De Sennegy, René. *Une Paroisse Louisianaise: St. Michel du Comté d'Acadie*. New Orleans: Mr. Capo, Printer-Bookseller, 1877.

Dufau, Louis. *Manuel des Verbes Français, Réguliers et Irréguliers*. New Orleans: Chez Leliévre, 1851.

———. *Principles de la Franc-Maconnerie Vulgairement Connue Sous le Nom de Rite Ancien et Accepté Ecossais, Indiquant le Sens Philosophique, Politique et Religieux de tous les Degrés jusqu'au 30éme inclusivement*. Nouvelle-Orléans: L. Marchand, Imprimeur, 1859.

"Fathers and Brothers of the Society of Mary Who Served People of God at Jefferson College and St. Michael Catholic Church, Convent, Louisiana." N.d. Memorial Plaque, St. Michael's Church Cemetery, River Road, Convent, LA.

Fay, Edward Allen, ed., *American Annals of the Deaf, Volume XL*. Washington, DC: Conference of Superintendents and Principles of American Schools for the Deaf, 1895.

Fay, Edwin Whitfield. *The History of Education in Louisiana*. Washington, DC: US Government Printing Office, 1898.

Foulhouze, James. *Historical Inquiry into the Origin of the Ancient and Accepted Scotch Rite; Into the Pretensions of the Dalcho-Mackey and Gourgas-Moore Supreme Councils of Charleston and Boston; Into the Rights of the Supreme Council of the Sovereign and Independent State of Louisiana; and into the Inconsistencies of His Imperial and Royal Highness, Prince Lucien Murat, Pretender to the Crown of Naples, and Grand Master, Pro Tempore, of the Masonic Order in France*. New Orleans: True Delta Job Office, 1859.

Gayarré, Charles. *Fernando de Lemos: Truth and Fiction, a Novel*. New York: G. W. Carleton & Co., 1872.

———. *The Creoles of History and the Creoles of Romance: A Lecture Delivered in the Hall of the Tulane University by Charles Gayerré, on the 25th of April, 1888*. New Orleans: C. E. Hopkins, 1885.

Lislet, Par L. Moreau. *Digeste Général Des Actes de la Législature de la Louisiane, Passés Depuis L'Anée 1804 Jusqu'en 1827, Volume 1*. New Orleans: Imprimé Par Benjamin Levy, 1828.

"Manresa House of Retreats." N.d. *Louisiana Historical Marker Program*. Historical Marker, River Road, Convent, LA.

Order of the Board of Trustees. *The Charts and Statues of Jefferson College, Washington, Mississippi, as Revised and Amended: Together with a Historical Sketch of the Institution from Its Establishment to the Present Time: To Which Is Prefixed a List of the Trustees, Officers and Faculty, The Acts of Congress and of the Legislature Relating to the Institution, and a Catalogue of Its Library, Apparatus, &c.* Natchez, MS: Book and Job Office, 1840.

Riddell, W. P. *A Genealogical Sketch of the Riddell Family, Including a List of the Descendants of the Three Brothers, Hugh, Gawn, and Robert, Who Came to America in 1737*. New Orleans: John F. Trow, Printer, 1852.

Semi-centennial Register of the Members of the Phi Kappa Sigma Fraternity: Printed by the Order of the Grand Chapter to Commemorate the Fiftieth Anniversary of the Founding of the Fraternity, 1850–1900. Philadelphia: Avil Printing Co., 1900.

"Sugar Cultivation in Louisiana." In *The Colonial Magazine and East India Review: January to June, 1849, Vol. XVI*, 222–230. London: John Mortimer, 1849.

Teddy (student at St. Mary's Jefferson College, Convent, LA). Letter to "Mother," Macacona, Costa Rica. Author's collection, Hattiesburg, MS.

Testut, Charles. *Portraits Litteraires: Nouvelle-Orleans.* Nouvelle-Orleans: Imprimerie des Veillees Louisianaises, 1850.

Yearbook of Education for 1879, The. Being the Second Annual Supplement to the Cyclopedia of Education: Dictionary of Information for the Use of Teachers, School Officers, Parents, and Others. New York: E. Stieger, 1879.

SECONDARY SOURCES

"About Us." Manresa House of Retreats. http://manresala.org/About-Us.aspx. Accessed July 19, 2012.

Angers, Trent. *Grand Coteau: The Holy Land of the South.* Lafayette, LA: Acadian House, 2005.

Arthur, Stanley Clisby, and George Campbell Huchet dé Kernion. *Old Families of Louisiana.* Gretna, LA: Pelican, 1998.

Asbury, Herbert. *The French Quarter: An Informal History of the New Orleans Underworld.* New York: Thunder's Mouth Press, 1939.

Ayers, Edward L. *The Promise of the New South: Life after Reconstruction, 15th Anniversary Edition.* Oxford: Oxford University Press, 2007.

Bangert, William V. *A History of the Jesuits.* St. Louis: Institute of Jesuit Resources, 1972.

Barry, John M. *Rising Tide: The Great Mississippi Flood of 1927 and How It Changed America.* New York: Simon & Schuster, 1997.

Baudier, Roger. *The Catholic Church in Louisiana.* New Orleans: A. W. Hyatt Stationary Mfg. Co. Ltd., 1939.

Bauer, Craig A. *Creole Genesis: The Bringier Family and Antebellum Plantation Life in Louisiana.* Lafayette: University of Louisiana at Lafayette Press, 2011.

———. *A Leader among Peers: The Life and Times of Duncan Farrar Kenner.* Lafayette: University of Louisiana at Lafayette Press, 1989.

Bell, Caryn Cossé. *Revolution, Romanticism, and the Afro-Creole Protest Tradition in Louisiana, 1718–1868.* Baton Rouge: Louisiana State University Press, 1997.

Bentley, George R. *A History of the Freedmen's Bureau.* London: Oxford University Press, 1955.

Bergeron, Arthur W. Jr. *Guide to Louisiana Confederate Military Units, 1861–1865.* Baton Rouge: Louisiana State University Press, 1989.

Biever, Albert H. *The Jesuits in New Orleans and the Mississippi Valley: Jubilee Memorial.* New Orleans: Hauser Printing Company, 1924.

Biographical and Historical Memoirs of Louisiana: Embracing an Authentic and Comprehensive Account of the Chief Events in the History of the State, a Special Sketch of Every Parish and a Record of the Lives of Many of the Most Worthy and Illustrious Families and Individuals, Vol. 2. Chicago: Goodspeed Publishing Company, 1892.

"Bladon Springs State Park." *Encyclopedia of Alabama.* http://www.encyclopediaofalabama.org/face/Article.jsp?id=h-2905. Accessed September 13, 2014.

Blain, William T. *Education in the Old Southwest: A History of Jefferson College Washington, Mississippi.* Natchez, MS: Judson Printing Company, 1976.

Bourgeois, Lillian C. *Cabanocey: The History, Customs, and Folklore of St. James Parish.* Gretna: Pelican, 1957.

Bourne, Edward Gaylord. *History of the Surplus Revenue of 1837 Being an Account of Its Origin, Its Distribution among the State, and the Uses to Which It Was Applied.* New York: G. P. Putnam's Sons, 1885.

Bowman, Rex, and Carlos Santos. *Rot, Riot, and Rebellion: Mr. Jefferson's Struggle to Save the University That Changed America.* Charlottesville: University of Virginia Press, 2013.

Brackenridge, Henry Marie. *Views of Louisiana, Together with a Journal of a Voyage up the Mississippi River in 1811.* 1814. Reprint, Chicago: Quadrangle Books, 1962.

Bradshaw, Jim. "Great Flood of 1927." In *KnowLA Encyclopedia of Louisiana*, ed. David Johnson. Louisiana Endowment for the Humanities, May 13, 2011. http://www.knowla.org/entry/763/.

Brasseaux, Carl A. *Acadian to Cajun: Transformation of a People, 1803–1877.* Jackson: University Press of Mississippi, 1992.

———. "Acadian Education: From Cultural Isolation to Mainstream America." In *The Louisiana Bicentennial Series in Louisiana History: Volume XVII, Education in Louisiana*, 236–242. Lafayette: Center for Louisiana Studies, University of Louisiana at Lafayette, 1999.

———. "Declining Economic Fortunes in Postbellum Louisiana." In *The Louisiana Bicentennial Series in Louisiana History: Volume VI, Reconstructing Louisiana*, ed. Lawrence N. Powell, 582–594. Lafayette: Center for Louisiana Studies, University of Louisiana at Lafayette, 2001.

———. *In Search of Evangeline: Birth and Evolution of the Evangeline Myth.* Thibodaux, LA: Blue Heron Press, 1988.

Breaux, Joseph A. "Some Early Colleges and Schools of Louisiana." In *Publications of the Louisiana Historical Society: New Orleans, Louisiana, Volume VII—1913–14*, 136–142. New Orleans: Louisiana Historical Society, 1915.

Brown, John Howard, ed. *The Cyclopedia of American Biography: Comprising the Men and Women of the United States Who Have Been Identified with the Growth of the Nation, Volume 3.* Boston: James H. Lamb Company, 1900.

Brown, Nikki. "Jim Crow." In *KnowLA Encyclopedia of Louisiana*, ed. David Johnson. Louisiana Endowment for the Humanities, January 21, 2015. http://www.knowla.org/entry/735/.

Brunett, Louis J. B. "To All Free-Masons, 2 February 1860." In *Transactions of the Supreme Council of the 33rd Degree, Ancient and Accepted Rite of Free-masonry for the Southern Jurisdiction of the United States of American, 1857 to 1866*, 180–183. Washington, DC: Joseph L. Pearson, 1878.

"Building News." *American Architect* 48, no. 1809 (1910): 8.

Bullard, Henry A., and Thomas Curry. *A New Digest of the Statutes and Laws of the State of Louisiana, from the Change of Government to the Year 1841, Inclusive, Vol. I.* New Orleans: E. Johns & Co. Stationers' Hall, 1842.

Buman, Nathan A. "Two Histories, One Future: Louisiana Sugar Planters, Their Slaves, and the Anglo-Creole Schism, 1815–1865." PhD diss., Louisiana State University, 2013.

Butchart, Ronald E. *Schooling the Freed People: Teaching, Learning, and the Struggle for Black Freedom, 1861–1876.* Chapel Hill: University of North Carolina Press, 2010.

Cable, George W. *The Creoles of Louisiana*. New York: Charles Scribner's Sons, 1884. Reprint, Gretna, LA: Pelican, 2005.

Campanella, Richard. *Bourbon Street: A History*. Baton Rouge: Louisiana State University Press, 2014.

Carriere, Oliver P. *A Sketch of the History of Jefferson College and Manresa House of Retreats, Convent, Louisiana: 1830–1973*. Convent, LA: Manresa House of Retreats, 1974.

Carter, Clarence Edwin, ed. *The Territorial Papers of the United States: Volume V, The Territory of Mississippi, 1798–1817*. Washington, DC: US Government Printing Office, 1937.

———. *The Territorial Papers of the United States: Volume VI, The Territory of Mississippi, 1809–1817 Continued*. Washington, DC: US Government Printing Office, 1937.

———. *The Territorial Papers of the United States: Volume IX, The Territory of Orleans, 1803–1812*. Washington, DC: US Government Printing Office, 1940.

Castellanos, Henry C. *New Orleans As It Was: Episodes of Louisiana Life*. New Orleans: L. Graham & Son, Ltd., 1895.

Caulfield, Ruby Van Allen. *French Literature of Louisiana*. Gretna, LA: Pelican, 1998.

"Celebrations around the Country: Marists Served New Orleans Area for 150 Years." Society of Mary: Marists in the US. http://www.societyofmaryusa.org/about/celebrations-clarionherald.html. Accessed September 13, 2014.

"Centenary State Historic Site." Louisiana State Parks. http://www.crt.state.la.us/louisiana-state-parks/historic-sites/centenary-state-historic-site/index. Accessed November 10, 2016.

Chandler, J. A. C. *The South in the Building of the Nation: A History of Southern States Designed to Record the South's Part in the Making of the American Nation; to Portray the Character and Genius, to Chronicle the Achievements and Progress and to Illustrate the Life and Traditions, Vol. 12*. Richmond, VA: Southern Historical Publication Society, 1909.

Chevet, F. "Experiments on Bichlorure [*sic*] of Sulphur [*sic*] and Certain Carbures [*sic*] of Hydrogen, made in the Laboratory of Jefferson College (Louisiana)." *American Journal of Science and Arts* 42 (April 1842): 71–74.

———. "On Bichlorure [*sic*] Sulphur [*sic*]." *American Journal of Science and Arts* and *The Annals of Electricity, Magnetism, and Chemistry; and Guardian of Experimental Science* 8 (January 1842): 493–495.

Cline, Rodney. "Early Colleges and Schools in Louisiana." In *The Louisiana Bicentennial Series in Louisiana History: Volume XVII, Education in Louisiana*, ed. Michael G. Wade, 230–235. Lafayette: Center for Louisiana Studies, University of Louisiana at Lafayette, 1999.

———. *Pioneer Leaders and Early Institutions in Louisiana Education*. Baton Rouge, LA: Claitor's Publishing Division, 1969.

Clark, Emily. *The Strange History of the American Quadroon: Free Women of Color in the Revolutionary Atlantic World*. Chapel Hill: University of North Carolina Press, 2013.

———. *Voices from an Early American Convent: Marie Madeleine and the Development of a New Society, 1724–1834*. Baton Rouge: Louisiana State University Press, 2007.

Cochran, Estelle M. Fortier. *The Fortier Family and Allied Families*. San Antonia: E. M. F. Cochran, 1963.

Cohen, Michael David. *Reconstructing the Campus: Higher Education and the American Civil War*. Charlottesville: University of Virginia Press, 2012.

Coleman, Will H. *Historical Sketch Book and Guide to New Orleans and Environs*. New York: Will H. Coleman, 1885.

Conrad, Glenn R. "The Faces of French Louisiana." In *Cross, Crozier, and Crucible: A Volume Celebrating the Bicentennial of a Catholic Diocese in Louisiana*, ed. Glenn R. Conrad, 5–15. New Orleans: Archdiocese of New Orleans in cooperation with the Center for Louisiana Studies, 1993.

"Copes, Joseph S." Tulane University Special Collections. specialcollections.tulane.edu/archon/?p=creators/creator&id=761. Accessed September 15, 2013.

Coste, J. *Lectures on Society of Mary History (Marsit Fathers), 1786–1854*. Rome: Via Alessandro Poerio, 1965.

Cowan, Walter Greaves, and Jack B. McGuire. *Louisiana Governors: Rules, Rascals, and Reformers*. Jackson: University Press of Mississippi, 2008.

Crété, Liliane. *Daily Life in Louisiana, 1815–1830*. Trans. Patrick Gregory. Baton Rouge: Louisiana State University Press, 1981.

Crouch, Barry A. "Black Education in Civil War and Reconstruction Louisiana: George T. Ruby, the Army, and the Freedmen's Bureau." In *The Louisiana Purchase Bicentennial Series in Louisiana History: Volume XI, the African American Experience in Louisiana, Part B: From the Civil War to Jim Crow*, ed. Charles Vincent, 260–275. Lafayette: Center for Louisiana Studies, University of Louisiana at Lafayette, 2000.

Dagneau, P. H. *Memoirs of Seventy-five Years for the Centenary of the Marist Fathers in America*. Atlanta: Keystone Press, 1963.

Dargo, George. *Jefferson's Louisiana: Politics and the Clash of Legal Traditions*. Rev. ed. Clark: Lawbook Exchange, 2009.

Daspit, Fred. *Louisiana Architecture, 1714–1820*. Lafayette: Center for Louisiana Studies, University of Louisiana at Lafayette, 2004.

———. *Louisiana Architecture, 1840–1860*. Lafayette: Center for Louisiana Studies, University of Louisiana at Lafayette, 2006.

Dauphine, James G. "Catholic Education in North Louisiana: The Pre-World War One Foundations of Catholic Life in a Southern Diocese." In *Cross, Crozier, and Crucible: A Volume Celebrating the Bicentennial of a Catholic Diocese in* Louisiana, ed. Glenn R. Conrad, 289–304. New Orleans: Archdiocese of New Orleans in cooperation with the Center for Louisiana Studies, 1993.

———. *A Question of Inheritance: Religion, Education, and Louisiana's Cultural Boundary*. Lafayette: Center for Louisiana Studies, University of Southwestern Louisiana, 1993.

Dawdy, Shannon Lee. *Building the Devil's Empire: French Colonial New Orleans*. Chicago: University of Chicago Press, 2008.

Dawson, John Charles. *Lakanal, The Regicide: A Biographical and Historical Study of the Career of Joseph Lakanal*. Tuscaloosa: University of Alabama Press, 1946.

Deléry, Simone de la Souchére. *Napoleon's Soldiers in America*. 1950. Reprint, Gretna, LA: Pelican, 1998.

Denzer, Marty. "Orphanages Cared for Homeless Children." *Catholic Key Online*. January 19, 2012. catholickey.org/2012/01/19/orphanages-cared-for-homeless-children/. Accessed December 1, 2016.

De Pedro, José. *The Spanish in New Orleans and Louisiana*. Trans. Richard E. Chandler. Gretna, LA: Pelican, 2000.

Dessens, Nathalie. *Creole City: A Chronicle of Early American New Orleans*. Gainesville: University Press of Florida, 2015.

———. *From Saint-Domingue to New Orleans, Migration and Influences*. Gainesville: University Press of Florida, 2007.

DeVore, Donald E., and Joseph Logsdon. *Crescent City Schools: Public Education in New*

Orleans, 1841–1991. Lafayette: Center for Louisiana Studies, University of Southwestern Louisiana, 1991.

Dickason, Julie. "James Harrison Dakin." *Find a Grave*. Accessed November 9, 2016. http://www.findagrave.com/cgi-bin/fg.cgi/page/gr/fg.cgi?page=gr&Grid=142282254.

Dinstel, Donna. "Louis Dufau." *Find a Grave*. http://www.findagrave.com/cgi-bin/fg.cgi?page=gr&GRid=148604074. Accessed November 29, 2015.

Dominguez, Virginia R. *White by Definition: Social Classification in Creole Louisiana*. New Brunswick, NJ: Rutgers University Press, 1986.

Donlon, Jocelyn Hazelwood. "Louisiana's Three Fold Regions." http://www.louisianafolklife.org/LT/Articles_Essays/la_3_folk_reg.html. Accessed February 26, 2013.

Dorr, J. W. "A Tourist's Description of Louisiana in 1860." *Louisiana Historical Quarterly* 21, no. 4 (October 1938): 1110–1214.

Doyle, Elisabeth Joan. "Mightier than the Sword: The Catholic Press of Louisiana." In *Cross, Crozier, and Crucible: A Volume Celebrating the Bicentennial of a Catholic Diocese in Louisiana*, ed. Glenn R. Conrad, 249–261. New Orleans: Archdiocese of New Orleans with the Center for Louisiana Studies, 1993.

Drouilly, François. *15 Days of Prayer with Jean-Claude Colin*. Hyde Park, NY: New City Press, 2012.

Dubray, C. M. "The Society of Mary and Education." *Catholic Educational Review* 3 (April 1911): 309–320.

Duffy, John. *The Tulane Medical Center: One Hundred and Fifty Years of Medical Education*. Baton Rouge: Louisiana State University Press, 1984.

Dumas, Alexandre. *The Count of Monte Cristo*. Reprint, London: Chapman and Hall, 1846.

Duncan, Patricia L. "From Creole College to Religious Retreat: The Story of a Louisiana Landmark." *Preservation in Print* (August 9, 1994): 1–4.

Dyer, John P. "Education in New Orleans." In *The Past as Prelude: New Orleans, 1718–1968*, ed. Hodding Carter, 116–145. Gretna, LA: Firebird Press, 1968.

———. *Tulane: The Biography of a University, 1834–1965*. New York: Harper & Row, 1966.

Evans, Elizabeth. "Alexander Hill Everett: Man of Letters." PhD diss., University of North Carolina, 1970.

Evans, Freddi Williams. *Congo Square: African Roots in New Orleans*. Lafayette: University of Louisiana at Lafayette Press, 2011.

Fairclough, Adam. *Race and Democracy: The Civil Rights Struggle in Louisiana, 1915–1972*. 2nd ed. Athens: University of Georgia Press, 2008.

Ferguson, John C. "Charles and James Dakin." In *KnowLA Encyclopedia of Louisiana*, ed. David Johnson. Louisiana Endowment for the Humanities, 2010–. Article published January 6, 2011. http://www.knowla.org/entry/572/.

Fertel, Rien. *Imagining the Creole City: The Rise of Literary Culture in Nineteenth-Century New Orleans*. Baton Rouge: Louisiana State University Press, 2014.

Fichter, Joseph H. "First Black Students at Loyola University: A Strategy to Obtain Teacher Certification." In *The Louisiana Purchase Bicentennial Series in Louisiana: Volume XVII*, ed. Michael G. Wade, 371–382. Lafayette: Center for Louisiana Studies, 1999.

Ficklen, John R. "History of Education in New Orleans." In *Standard History of New Orleans: Giving a Description of the Natural Advantages, Natural History in Regard to the Flora and Birds, Settlement, Indians, Creoles, Municipal and Military History, Mercantile and Commercial Interests, Banking, Transportation, Struggles Against High Water, the Press, Educational, Literature and Art, the Churches, Old Burying Grounds, Bench and Bar, Medical, Public and Charitable Institutions, the Carnival, Amusements, Clubs, Societies, Associations, Etc.*, ed. Henry Rightor. Chicago: Lewis Publishing Company, 1900.

Field, Martha R. *Louisiana Voyages: The Travel Writings of Catherine Cole*. Ed. Joan B. McLaughlin and Jack McLaughlin. Jackson: University Press of Mississippi, 2006.

Fleming, Walter L. *Louisiana State University*. Baton Rouge: Louisiana State University Press, 1936.

Follett, Richard. "The Rise and Fall of American Sugar." In *Plantation Kingdom: The American South and Its Global Commodities*, ed. Richard Follett, Sven Beckert, Peter Coclanis, and Barbara Hahn, 61–90. Baltimore: Johns Hopkins University Press, 2016.

———. *The Sugar Masters: Planters and Slaves in Louisiana's Cane World, 1820–1860*. Baton Rouge: Louisiana State University Press, 2005.

Follett, Richard, and Rick Halpern. "From Slavery to Freedom in Louisiana's Sugar Country: Changing Labor Systems and Workers' Power, 1861–1913." In *Sugar, Slavery, and Society: Perspectives on the Caribbean, India, the Mascarenes, and the United States*, ed. Bernard Moitt, 135–156. Gainesville: University Press of Florida, 2004.

Folse, Pamela. *St. James Catholic Church: Remembering Our Story*. St. James, LA: St. James Catholic Church, n.d.

Foner, Eric. *Reconstruction: America's Unfinished Revolution, 1863–1877*. New York: Harper & Row, 1988.

Fortier, Alcée. *A History of Louisiana In Four Volumes: Volume III, The American Domination*. Paris: Goupil & Co., 1904.

———. *Louisiana: Comprising Sketches of Parishes, Towns, Events, Institutions, and Persons, Arranged in Cyclopedic Form, Volume I*. Madison, WI: Century Historical Association, 1914.

———. *Louisiana: Comprising Sketches of Parishes, Towns, Events, Institutions, and Persons, Arranged in Cyclopedic Form, Volume III*. Madison, WI: Century Historical Association, 1914.

———. *Louisiana Studies: Literature, Customs and Dialects, History and Education*. New Orleans: F. F. Hansell and Bro., 1894.

———. "The Louisiana Sugar Planters of the Old Regime." *Louisiana Sugar Planter and Sugar Manufacturer* 52 (1914): 76–77.

Fossier, Albert E. *New Orleans the Glamour Period, 1800–1840: A History of the Conflicts of Nationalities, Languages, Religion, Morals, Cultures, Laws, Politics and Economics during the Formative Period of New Orleans*. Gretna, LA: Pelican, 1998.

Frost, Dan R. *Thinking Confederates: Academia and the Idea of Progress in the New South*. Knoxville: University of Tennessee Press, 2000.

Frost, Dan R., and Kou K. Nelson. *The LSU College of Engineering, Volume I: Origins and Establishment, 1860–1908*. Baton Rouge: Louisiana State University College of Engineering, 1995.

Gale, Robert L. *A Henry Wadsworth Longfellow Companion*. Westport, CT: Greenwood, 2003.

Gallier, James. *Autobiography of James Gallier, Architect*. New York: Da Capo Press, 1973.

Gaudet, Marcia. "Christmas Bonfires in South Louisiana: Tradition and Innovation." In *Mardi Gras, Gumbo, and Zydeco: Readings in Louisiana Culture*, ed. Marcia Gaudet and James C. McDonald, 59–70. Jackson: University Press of Mississippi, 2003.

Gehman, Mary. *The Free People of Color of New Orleans: An Introduction*. 5th ed. Chelsea, MI: Sheridan Books, 2009.

Geiger, Roger L. "Introduction: New Themes in the History of Nineteenth-Century College." In *The American College in the Nineteenth Century*, ed. Roger L. Geiger, 1–36. Nashville: Vanderbilt University Press, 2000.

Genovese, Eugene D. *Roll, Jordan, Roll: The World the Slaves Made*. New York: Random House, 1976.

Georgelin, Francis J. *The Catholic Church in Algiers, Louisiana: The Marist Fathers and the Holy Name of Mary Church*. Ed. Rev. Charles F. Brogley, SM, MA New Orleans: Privately Printed by the Society of Mary, 1949.

Giraud, Marcel. *A History of French Louisiana, Volume 5: The Company of the Indies, 1723–1731*. Trans. Brian Pearce. Baton Rouge: Louisiana State University Press, 1991.

Gleason, David King. *Plantation Homes of Louisiana and the Natchez Area*. Baton Rouge: Louisiana State University Press, 1985.

Glover, Lorri. *Southern Sons: Becoming Men in the New Nation*. Baltimore: Johns Hopkins University Press, 2007.

Glymph, Thavolia, Ira Berlin, Steven F. Miller, Joseph P. Reidy, Leslie S. Rowland, and Julie Saville. "The Wartime Genesis of Free Labor in Southern Louisiana, 1861–1865." In *The Louisiana Bicentennial Series in Louisiana History: Volume VI, Reconstructing Louisiana*. Ed. Lawrence N. Powell, 107–134. Lafayette: Center for Louisiana Studies, University of Louisiana at Lafayette, 2001.

Goldfield, David H. *The American Journey*. New York: Pearson, 2011.

Goldschmidt, Debra. "Colleges Come to Terms with Slave-owning Pasts." *CNN*. May 23, 2011. http://www.cnn.com/2011/US/05/23/university.slavery/. Accessed August 12, 2016.

Green, Jennifer R. *Military Education and the Emerging Middle Class in the Old South*. Cambridge: Cambridge University Press, 2008.

Guenin-Lelle, Dianne. *The Story of French New Orleans: History of a Creole City*. Jackson: University Press of Mississippi, 2016.

Guterl, Matthew Pratt. *American Mediterranean: Southern Slaveholders in the Age of Emancipation*. Cambridge, MA: Harvard University Press, 2008.

Hague, Robert. "Cecile *Street* Dufau." *Find a Grave*. http://www.findagrave.com/cgi-bin/fg.cgi?page=gr&GRid=79127533&ref=acom. Accessed August 19, 2014.

Hales, Stephen W. *Rex: An Illustrated History of the School of Design*. Mandeville, LA: Arthur Hardy Enterprises, 2010.

Hall, Gwendolyn Midlo. *Africans in Colonial Louisiana: The Development of Afro-Creole Culture in the Eighteenth Century*. Baton Rouge: Louisiana State University Press, 1995.

———. "The Formation of Afro-Creole Culture." In *Creole New Orleans: Race and Americanization*, ed. Arnold R. Hirsch and Joseph Logsdon, 58–87. Baton Rouge: Louisiana State University Press, 1992

Hanger, Kimberly S. *Bounded Lives, Bounded Places: Free Black Society in Colonial New Orleans*. Durham, NC: Duke University Press, 1997.

Hauck, Philomena. *Bienville: Father of Louisiana, Louisiana Life Series, No. 10*. Lafayette: Center for Louisiana Studies, University of Southwestern Louisiana, 1998.

Haulsee, W. M., F. G. Howe, and A. C. Doyle. *Soldiers of the Great* War, vol. 3. Washington, DC: Soldiers Record Publishing Association, 1920.

Hebert, Adam Otis Jr. "History of Education in Colonial Louisiana." In *The Louisiana*

Purchase Bicentennial Series in Louisiana History, Volume XVIII: Education in Louisiana, 9–50. Lafayette: Center for Louisiana Studies, University of Louisiana at Lafayette, 1999.

Hill, Mary Bernardine. "The Influence of James Hubert Blenk on Catholic Education in the Archdiocese of New Orleans, 1885–1917." EdD diss., Louisiana State University, 1964.

Huber, Leonard V. *The Battle of New Orleans: New Orleans As It Was in 1814–1815*. Gretna, LA: Pelican, 2011.

———. "New Orleans Cemeteries: A Brief History." In *New Orleans Architecture, Volume III: The Cemeteries*, ed. Mary Louise Christovich, 3–62. Gretna, LA: Pelican, 1974.

Humphreys, Margaret. *Yellow Fever and the South*. New Brunswick, NJ: Rutgers University Press, 1992.

Hunt, Alfred N. *Haiti's Influence on Antebellum America: Slumbering Volcano in the Caribbean*. Baton Rouge: Louisiana State University Press, 1988.

Hunter, Robert F., and Edwin L. Dooley Jr. *Claudius Crozet: French Engineer in America, 1790–1864*. Charlottesville: University Press of Virginia, 1989.

Hyde, Sarah L. "Punctuated Progress: Education Developments in the Antebellum Gulf South." In *The Enigmatic South: Toward Civil War and Its Legacies*, ed. Samuel C. Hyde Jr., 24–47. Baton Rouge: Louisiana State University Press, 2014.

———. *Schooling in the Antebellum South: The Rise of Public and Private Education in Louisiana, Mississippi, and Alabama*. Baton Rouge: Louisiana State University Press, 2016.

———. "'Teach Us Incessantly': Lessons and Learning in the Antebellum South." PhD diss., Louisiana State University, 2010.

Jacobs, Claude F., and Andrew J. Kaslow. *The Spiritual Churches of New Orleans: Origins, Beliefs, and Rituals of an African-American Religion*. Knoxville: University of Tennessee Press, 1991.

Jeansonne, Glen, and David Luhrssen. "William C. C. Claiborne." In *KnowLA Encyclopedia of Louisiana*, ed. David Johnson. Louisiana Endowment for the Humanities, May 7, 2013. http://www.knowla.org/entry/921/.

"Jefferson College Today." MDAH: Historic Jefferson College. http://mdah.state.ms.us/hprop/hjccurrent.html. Accessed August 3, 2014.

"Jefferson Military College Foundation." MDAH: Historic Jefferson College, http://mdah.state.ms.us/hprop/jmcf.html. Accessed August 3, 2014.

"Jules Dufau, 10/23/1908." Glenwood Cemetery: Search Our Records. http://www.glenwoodcemetery.org/search/?name=Jules&lastname=Dufau&date_from=&date_to=. Accessed November 20, 2016.

Kane, Harnett T. *Plantation Parade: The Grand Manner in Louisiana*. New York: William Morrow, 1945.

Kappel, Vernon Elmer. "Louis Dufau's Louisiana College." MA thesis, Tulane University, 1949.

Kasteel, Annemarie. *Francis Janssens, 1843–1897: A Dutch-American Prelate*. Lafayette: Center for Louisiana Studies, University of Southwestern Louisiana, 1992.

Kean, Melissa. *Desegregating Private Higher Education in the South: Duke, Emory, Rice, Tulane, and Vanderbilt*. Baton Rouge: Louisiana State University Press, 2008.

Keister, Douglas. *Stories in Stone: A Field Guide to Cemetery Symbolism and Iconography*. Layton, UT: Gibbs Smith, 2004.

"Keith Will Is Filed." *Black Diamond* 35, no. 18 (November 1905): 27.

Kenny, Michael. *The Torch on the Hill: Centenary Story of Spring Hill College, 1830–1930*. New York: American Press, 1931.

Kerr, Donal. *Jean-Claude Colin, Marist: A Founder in an Era of Revolution and Restoration: The Early Years 1790–1836*. Dublin, Ireland: Columba Press, 2000.

Kilpatrick, Thaddeus Rogers. "A Conservation Study of the Decorative Paintings at Whitney Plantation, St. John the Baptist Parish, Louisiana." MS thesis, University of Pennsylvania, 1992.

King, Grace Elizabeth. *Creole Families of New Orleans*. New Orleans: Cornerstone Book, 2010.

Kingsley, Karen. *Buildings of Louisiana*. New York: Oxford University Press, 2003.

Knight, Edgar W., ed. *A Documentary of Education in the South before 1860 in Five Volumes: Volume V, Educational Theories and Practices*. Chapel Hill: University of North Carolina Press, 1953.

La-Cemeteries. "Louisiana Governors." http://www.la-cemeteries.com/Governor/Roman,%20Andre%20Bienvenu/Roman,%20Andre%20Bienvenu.shtml. Accessed December 23, 2013.

Lachance, Paul F. "The Foreign French." In *Creole New Orleans: Race and Americanization*, ed. Arnold R. Hirsch and Joseph Logsdon, 101–130. Baton Rouge: Louisiana State University Press, 1992.

Landau, Emily Epstein. *Spectacular Wickedness: Sex. Race, and Memory in Storyville, New Orleans*. Baton Rouge: Louisiana State University Press, 2013.

Leeper, Clare D'artois. *Louisiana Place Names: Popular, Unusual, and Forgotten Stories of Towns, Cities, Plantations, Bayous, and Even Some Cemeteries*. Baton Rouge: Louisiana State University Press, 2012.

Leliévre, Jacques-Felix. *Jacques-Felix Leliévre's New Louisiana Gardner*. Trans. with an introduction by Sally Kittredge Reeves. Baton Rouge: Louisiana State University Press, 2001.

Levine, Bruce. *The Fall of the House of Dixie: The Civil War and the Social Revolution That Transformed the South*. New York: Random House, 2013.

Lewis, Richard Anthony. *Robert W. Tebbs, Photographer to Architects: Louisiana Plantations in 1926*. Baton Rouge: Louisiana State University Press, 2011.

Lincoln, Abraham. "The Emancipation Proclamation." In *The Gettysburg Address and Other Writings*, 49–52. New York: Fall River Press, 2013.

Lock, Samuel H. *Louisiana as It Is: A Geographical and Topographical Description of Louisiana*. Ed. Lauren C. Post. Baton Rouge: Louisiana State University Press, 1969.

Logsdon, Joseph, and Caryn Cossé Bell. "The Americanization of Black New Orleans 1850–1900." In *Creole New Orleans: Race and Americanization*, ed. Arnold R. Hirsch and Joseph Logsdon. Baton Rouge: Louisiana State University Press, 1992.

Long, Alicia P. *The Great Southern Babylon: Sex, Race, and Respectability in New Orleans, 1865–1920*. Baton Rouge: Louisiana State University Press, 2005.

Longfellow, Henry Wadsworth. *Evangeline: A Tale of Acadie*. Reprint, Boston: Houghton, Mifflin, 1866.

Lowry, R. P. "A History of the Public Schools of St. James Parish from 1865 to the Present Day." MA thesis, Louisiana State University, 1928.

Ludlow, John Malcolm Forbes. *A Sketch of the United States from Independence to Secession*. Cambridge: Macmillan and Co., 1862.

Macchione, Mikko. *Napoleon House*. New Orleans: Vissi d'Arte Books, 2006.

Madden, John W. *Handbook of the Carnival: A Wonderful Account of the History, Krewes, and Events Surrounding Mardi Gras in New Orleans*. New Orleans: Cornerstone Book Publishers, 2011.

Manning, Curtis A. *The History of Higher Education in Louisiana*. Bloomington, IN: Xlibris Corporation, 2006.

Manresa "Good News" Letter 39, no. 4 (Fall 2013): 1–4. http://manresala.org/Manresa-Retreat-House-News/Newsletters.aspx.

"Manresa Iconography." Manresa House of Retreats, Convent, LA: Society of Jesus USA. http://manresala.org/Prayers/Iconography.aspx. Accessed September 7, 2014.

Manresa on the Mississippi: Retreat Manual. Convent, LA: Manresa House of Retreats, n.d.

Manresa Retreat House, Convent, Louisiana, Conducted by the Jesuit Fathers: Schedule of Retreats for Laymen. Convent, LA: Manresa House of Retreats, 2008.

"The Marine Pavilion at Rockaway, Long Island." *NOAA Photo Library*. www.photolib.noaa.gov/brs/clind50.htm. Accessed August 3, 2014.

Marking, Tonja Koob, and Jennifer Snape. *Images of America: Louisiana's Oil Heritage*. Mount Pleasant, SC: Arcadia Publishing, 2012.

Marten, James. *The Children's War*. Chapel Hill: University of North Carolina Press, 1998.

Martin, R. P. *The Plantation Mansion and Estate of Valcour Aime St. James Parish or A Brief Discussion of the Historical and Architectural Aspects of the Plantation Known as the St. James Refinery*. 1968. Louisiana Research Collection, Special Collections, Howard-Tilton Memorial Library, Jones Hall, Tulane University.

Matrana, Marc R. *Lost Plantations of the South*. Jackson: University Press of Mississippi, 2009.

McCulloch, Gary, and William Richardson. *Historical Research in Educational Settings*. Buckingham: Open University Press, 2000.

McGill, Anna Blanche. *The Sisters of Charity of Nazareth Kentucky*. New York: Encyclopedia Press, 1917.

Miller, Randall M. "A Church in Cultural Captivity: Some Speculations on Catholic Identity in the Old South." In *Catholics in the Old South*, ed. Randall Miller and Jon Wakelyn, 11–52. Macon, GA: Mercer University Press, 1999.

———. "Slaves and Southern Catholicism." In *Masters and Slaves in the House of the Lord: Race and Religion in the American South, 1740–1870*, ed. John B. Boles, 127–152. Lexington: University Press of Kentucky, 1988.

Miller, Robert J. *Both Prayed to the Same God: Religion and Faith in the American Civil War*. Lanham, MD: Lexington Books, 2007.

Mohr, Clarence L., and Joseph E. Gordon. *Tulane: The Emergence of a Modern University, 1945–1980*. Baton Rouge: Louisiana State University Press, 2001.

Morgan, Lee. *Centenary College of Louisiana, 1825–2000: The Biography of an American Academy*. Shreveport: Centenary College of Louisiana Press, 2008.

Muller, Mary Lee. "New Orleans Public School Desegregation." *The Louisiana Purchase Bicentennial Series in Louisiana History, Vol XI, The African American Experience in Louisiana, Part C, From Jim Crow to Civil Rights*, ed. Charles Vincent, 336–351. Lafayette: Center for Louisiana Studies, 2002.

"National Register of Historic Places Database [Louisiana]." *Office of Cultural Development: Division of Historic Preservation*. http://www.crt.state.la.us/dataprojects/hp/nhl/index.asp. Accessed December 1, 2016.

Nelson, William Hamilton. *A Burning Touch and a Flaming Fire: The Story of Centenary College of Louisiana*. Nashville: Methodist Publishing House, 1931.

Neyrey, Jerome, and Thomas Clancy. *Southern Jesuit Biographies: Pastors and Preachers, Builders and Teachers of the New Orleans Province*. Lafayette, LA: Acadian House, 2015.

Niehaus, Earl F. "Jefferson College: The Early Years." *Louisiana Historical Quarterly* 4 (October 1955): 63–89.

———. "Jefferson College in St. James Parish, Louisiana: 1830–1875." MA thesis, Tulane University, 1954.

Noble, Stuart Grayson. "Early School Superintendents in New Orleans." *Journal of Educational Research* 24, no. 4 (November, 1931): 274–279.

———. "Governor Claiborne and the Public School System of the Territorial Government of Louisiana." In *The Louisiana Bicentennial Series in Louisiana History: Volume XVII, Education in Louisiana*, ed. Michael G. Wade, 51–64. Lafayette: Center for Louisiana Studies, University of Louisiana at Lafayette, 1999.

Northup, Solomon. *Twelve Years a Slave*. Ed. Sue Eakin and Joseph Logsdon. 1853. Reprint, Baton Rouge: Louisiana State University Press, 1996.

O'Brien, Michael. *Intellectual Life and the American South, 1810–1860*. Chapel Hill: University of North Carolina Press, 2010.

Oneband, Donna McGee. *The House That Sugarcane Built: The Louisiana Burguières*. Jackson: University Press of Mississippi, 2014.

Pace, Robert F. *Halls of Honor: College Men in the Old South*. Baton Rouge: Louisiana State University Press, 2004.

Page, Kathryn. "A First Born Child of Liberty: The Constitution of 1864." In *Search of Fundamental Law: Louisiana's Constitutions, 1812–1974*, ed. Warren M. Billings and Edward F. Haas, 52–68. Lafayette: Center for Louisiana Studies, University of Southwestern Louisiana, 1993.

Perman, Michael. *The Road to Redemption: Southern Politics, 1869–1879*. Chapel Hill: University of North Carolina Press, 1984.

Peterkin, Darryl L. "'Within These Walls': Reading and Writing Institutional Histories." In *The History of U.S. Higher Education: Methods for Understanding the Past*, ed. Marybeth Gasman, 9–18. New York: Routledge, 2010.

Phillips, V. Faye. "Charles Gayarré." In *KnowLA Encyclopedia of Louisiana*, ed. David Johnson. Louisiana Endowment for the Humanities, December 22, 2010. http://www.knowla.org/entry/744/&view=summary.

Plaisance, C. "Henry Noel [J.] Plaisance." *Find a Grave*. http://www.findagrave.com/cgi-bin/fg.cgi?page=gr&GSln=plaisance&GSfn=henry&GSbyrel=all&GSdy=1918&GSdyrel=in&GSob=n&GRid=116459793&df=all&. Accessed August 14, 2016.

Platt, R. Eric. *Sacrifice and Survival: Identity, Mission, and Jesuit Higher Education in the American South*. Tuscaloosa: University of Alabama Press, 2014.

Platt, R. Eric, Melandie McGee, and Amanda King. "Marching in Step: Patriotism and the Southern Catholic Cadet Movement." *Catholic Historical Review* 102, no. 3 (Summer 2016): 517–544.

Poché, Justin D. "Bonfires on the Levee: Place, Memory, and the Sacred in River Road Catholicism." In *Gods of the Mississippi (Religion in North America)*, ed. Michael Pasquire, 161–181. Bloomington: Indiana University Press, 2013.

Poesch, Jessie, and Barbara Sorelle Bacot. *Louisiana Buildings 1720–1940: The Historic American Building Survey*. Baton Rouge: Louisiana State University Press, 1997.

Poll, Michael R. "Albert Pike's Address before the Grand Consistory of Louisiana."

Pietre-Stones Review of Freemasonry 10 (2002). http://www.freemasons-freemasonry.com/pike_address_louisiana.html. Accessed February 19, 2015.

Powell, Lawrence N. *The Accidental City: Improvising New Orleans*. Cambridge, MA: Harvard University Press, 2012.

———. Introduction to *The New Orleans of George Washington Cable: The 1887 Census Office Report* by George Washington Cable. Baton Rouge: Louisiana State University Press, 2008.

Power, Edward John. *A History of Catholic Higher Education in the United States*. Milwaukee: Bruce Publishing Company, 1958.

Purvis, Thomas L. "The Seven Years War and Its Political Legacy." In *A Companion to the American Revolution*, ed. Jack P. Green and J. R. Pole, 112–117. Malden, MA: Blackwell Publishing, 2004.

Rasmussen, Daniel. *American Uprising: The Untold Story of American's Largest Slave Revolt*. New York: HarperCollins, 2011.

Reed, Germaine A. "David Boyd, LSU, and Louisiana Reconstruction." *Louisiana Studies* 14 (Fall 1975): 259–276.

Reeves, Miriam G. *The Felicianas of Louisiana*. Baton Rouge: Claitor's Book Store, 1967.

———. *The Governors of Louisiana*. Gretna, LA: Pelican, 1998.

Reeves, Sally K. "The Society of the Sacred Heart in New Orleans." In *Cross, Crozier, and Crucible: A Volume Celebrating the Bicentennial of a Catholic Diocese in Louisiana*, ed. Glenn R. Conrad, 219–232. New Orleans: Archdiocese of New Orleans in cooperation with the Center for Louisiana Studies, 1993.

Reeves, William D. *Historic Louisiana: An Illustrated History*. San Antonio: Historical Publishing Network, 2003.

———. *Manresa on the Mississippi: For the Greater Glory of God*. Convent, LA: Manresa House of Retreats, 1996.

Rehder, John B. *Delta Sugar: Louisiana's Vanishing Plantation Landscape*. Baltimore: Johns Hopkins University Press, 1999.

Reinders, Robert C. *End of an Era: New Orleans, 1850–1860*. 1964. Reprint, Gretna, LA: Pelican, 1998.

Richardson, Heather Cox. *The Death of Reconstruction: Race, Labor, and Politics in the Post-Civil War North, 1865–1901*. Cambridge, MA: Harvard University Press, 2001.

Ripley, Eliza. *Social Life in Old New Orleans: Being Recollections of my Girlhood*. New York: D. Appleton and Company, 1912. Reprint, Gretna, LA: Pelican, 1998.

Robertson, Henry O. *"In the Habit of Acting Together": The Emergence of the Whig Party in Louisiana, 1838–1840*. Lafayette: Center for Louisiana Studies, University of Louisiana at Lafayette, 2007.

Robin, Linda. "Emmett Louis Hardy." *Find a Grave*. http://www.findagrave.com/cgi-bin/fg.cgi?page=gr&GSln=hardy&GSfn=emmett&GSbyrel=all&GSdyrel=all&GSst=20&GScntry=4&GSob=n&GRid=107267143&df=all&. Accessed November 30, 2015.

Rodrigue, John C. *Reconstruction in the Cane Fields: From Slavery to Free Labor in Louisiana's Sugar Parishes, 1862–1880*. Baton Rouge: Louisiana State University Press, 2001.

Roland, Charles P. *Louisiana Sugar Plantations during the Civil War*. Baton Rouge: Louisiana State University Press, 1997.

Royce, Edward. *The Origins of Southern Sharecropping*. Philadelphia: Temple University Press, 1993.

Rudd, Robert Mark. "Apocryphal Grandeur: Belle Grove Plantation in Iberville Parish, Louisiana." MA thesis, University of Delaware, 2002.

Rudolph, Fredrick. *The American College and University: A History*. Athens: University of Georgia Press, 1990.

Sacher, John M. "Louisiana's Secession from the Union." In *KnowLA Encyclopedia of Louisiana*, ed. David Johnson. Louisiana Endowment for the Humanities, July 27, 2011. http://www.knowla.org/entry/559/.

———. *A Perfect War of Politics: Parties, Politicians, and Democracy in Louisiana, 1824–1861*. Baton Rouge: Louisiana State University Press, 2003.

Savage, Neil J. *Extraordinary Tenure: Massachusetts and the Making of a Nation, from President Adams to Speaker O'Neill*. Worcester, MA: Ambassador Books, 2004.

Scarborough, William Kauffman. *Masters of the Big House: Elite Slaveholders of the Mid-Nineteenth-century South*. Baton Rouge: Louisiana State University Press, 2003.

Schmandt, Raymond H. "An Overview of Institutional Establishments in the Antebellum Southern Church." In *Catholics in the Old South*, ed. Randall Miller and John Wakelyn. 53–76. Macon, GA: Mercer University Press, 1999.

Schott, Paul W. "The Jesuit House That Never Should Have Been." *Southern Jesuits* 1, no. 1 (September 1981): 10–11.

Schulten, Susan. *Mapping the Nation: History and Cartography in Nineteenth-Century American*. Chicago: University of Chicago Press, 2012.

Scott, James B. *Outline of the Rise and Progress of Freemasonry in Louisiana from Its Introduction to the Reorganization of the Grand Lodge in 1850*. New Orleans: Order of the Grand Lodge, 1923.

Scully, Arthur Jr. *James Dakin, Architect: His Career in New York and the South*. Baton Rouge: Louisiana State University Press, 1973.

Shapiro, Norman R., trans. *Creole Echoes: The Francophone Poetry of Nineteenth-Century Louisiana*. Urbana: University of Illinois Press, 2004.

Short, Simine. *Locomotive to Aeromotive: Octave Chanute and the Transportation Revolution*. Urbana: University of Illinois Press, 2011.

Shugg, Roger W. *Origins of Class Struggle in Louisiana: A Social History of White Farmers and Laborers during Slavery and After, 1840–1875*. 1939. Reprint, Baton Rouge: Louisiana State University Press, 1966.

———. "Survival of the Plantation System in Louisiana." In *The Louisiana Bicentennial Series in Louisiana History: Volume VI, Reconstructing Louisiana*, ed. Lawrence N. Powell, 469–480. Lafayette: Center for Louisiana Studies, University of Louisiana at Lafayette, 2001.

Sitterson, J. Carlyle. "A New Labor System, 1865–1880." In *The Louisiana Bicentennial Series in Louisiana History: Volume VI, Reconstructing Louisiana*, ed. Lawrence N. Powell, 503–518. Lafayette: Center for Louisiana Studies, University of Louisiana at Lafayette, 2001.

———. *Sugar Country: The Cane Sugar Industry in the South, 1753–1950*. Westport, CT: Greenwood, 1973.

Smith, J. Frazer. *White Pillars: The Architecture of the South*. New York: Bramhall Publishing, 1970.

Smith, John David. Introduction to *Louisiana Sugar Plantations during the Civil War* by Charles P. Roland. Baton Rouge: Louisiana State University Press, 1997.

Smith, Thomas Ruys. *Southern Queen: New Orleans in the Nineteenth Century*. New York: Continuum, 2011.

Snyder, Thomas D., ed. *120 Years of American Education: A Statistical Report*. Washington, DC: National Center for Education Statistics, US Department of Education, Office of Educational Research and Improvement, 1993.

Spear, Jennifer M. *Race, Sex, and Social Order in Early New Orleans*. Baltimore: Johns Hopkins University Press, 2009.

Sternberg, Mary Ann. *River Road Rambler: A Curious Traveler along Louisiana's Historic Byway*. Baton Rouge: Louisiana State University Press, 2013.

Stringer, Richard. *Father Abraham: Lincoln's Relentless Struggle to End Slavery*. New York: Oxford University Press, 2006.

Stubbs, William. *The Cultivation of Sugarcane*. Savannah, GA: D. G. Purse, 1900.

Sublette, Ned. *The World That Made New Orleans: From Spanish Silver to Congo Square*. Chicago: Lawrence Hill Books, 2008.

Sublette, Ned, and Constance Sublette. *The American Slave Coast: A History of the Slave-Breeding Industry*. Chicago: Lawrence Hill Books, 2016.

Taylor, Joe Gray. *Louisiana Reconstructed: 1863–1877*. Baton Rouge: Louisiana State University Press, 1974.

———. *Negro Slavery in Louisiana*. Baton Rouge: Louisiana State University Press, 1963.

Tewksbury, Donald G. *The Founding of American Colleges and Universities before the Civil War*. New York: Bureau of Publications, Teachers College, Columbia University, 1932. Reprint, Eastford, CT: Martino, 2011.

Thelin, John R. *A History of American Higher Education*. 2nd ed. Baltimore: Johns Hopkins University Press, 2011.

Timbrell, John. *The Poison Paradox: Chemicals as Friends and Foes*. New York: Oxford University Press, 2005.

Tinker, Edward Larocque. "Charles Gayarré 1805–96." *The Papers of the Bibliographical Society of American* 27, no. 1 (1933): 24–64.

———. *Les Ecrits De Langue Francaise: En Louisiane Au XIXe Siecle*. 1923. Reprint, Nendeln, Liechtenstein: Kraus Reprint, 1970.

Toledano, Roulhac B. "Louisiana's Golden Age: Valcour Aime in St. James Parish." *Louisiana History: The Journal of the Louisiana Historical Association* 10, no. 3 (Summer 1969): 211–224.

Toledano, Roulhac B., and Mary Louise Christovich. *New Orleans Architecture, Volume VI: Faubourg Tremé and the Bayou Road, North Rampart Street to North Broad Street, Canal Street to St. Bernard Avenue*. Gretna, LA: Pelican, 2003.

Trask, Benjamin H. *Fearful Ravages: Yellow Fever in New Orleans, 1796–1905*. Lafayette: Center for Louisiana Studies, University of Louisiana at Lafayette, 2005.

Tregle, Joseph G. "Creoles and Americans." In *Creole New Orleans: Race and Americanization*, ed. Arnold R. Hirsch and Joseph Logsdon, 131–185. Baton Rouge: Louisiana State University Press, 1992.

———. "Democracy Triumphant." In *The Louisiana Purchase Bicentennial Series in Louisiana History: Volume IV, Antebellum Louisiana, 1830–1860. Part B: Politics*, 105–121. Lafayette: Center for Louisiana Studies, 2004.

Tunnel, Ted. *Crucible of Reconstruction: War, Radicalism, and Race, 1862–1877*. Baton Rouge: Louisiana State University Press, 1984.

Turner, Arlin. *George Washington Cable*. Baton Rouge: Louisiana State University Press, 1966.

Uter, Frank M. *Stones beside a River: A History of the Catholic Church on the East Bank of*

St. James Parish, 1809–2009. Baton Rouge, LA: Catholic Diocese of Baton Rouge, 2009.

Varnado, Otto S. "A History of the Early Institutions of Higher Learning in Louisiana." MA thesis, Louisiana State University, 1927.

Vella, Christina. *Intimate Enemies: The Two Worlds of the Baroness De Pontalba*. Baton Rouge: Louisiana State University Press, 1997.

Vogel, Claude L. *The Capuchins in French Louisiana (1722–1766)*. New York: Joseph F. Wagner, 1928.

Wakelyn, Jon L. "Catholic Elites in the Slaveholding South." In *Catholics in the Old South*, ed. Randall Miller and Jon Wakelyn, 211–239. Macon, GA: Mercer University Press, 1999.

Washington, Henry Augustine. "The Social System of Virginia." *SLM* 14 (February 1848): 65–81. Quoted in *All Clever Men, Who Make Their Way: Critical Discourse in the Old South*, ed. Michael O'Brien. Athens: University of Georgia Press, 1992.

Wayne, Michael. "Transformation of the Plantation." In *The Louisiana Bicentennial Series in Louisiana History: Volume VI, Reconstructing Louisiana*, ed. Lawrence N. Powell, 519–546. Lafayette: Center for Louisiana Studies, University of Louisiana at Lafayette, 2001.

White, Howard A. *The Freedmen's Bureau in Louisiana*. Baton Rouge: Louisiana State University Press, 1970.

———. "Redeeming of the People from Ignorance." In *The Louisiana Purchase Bicentennial Series in Louisiana History: Volume XI, The African American Experience in Louisiana, Part B: From the Civil War to Jim Crow*, ed. Charles Vincent, 293–314. Lafayette: Center for Louisiana Studies, University of Louisiana at Lafayette, 2000.

White, John L., and Mark Edwards. *The Society of Mary: Its Foundation and First American Mission*. Ed. Nicholas A. Weber. Privately printed, 1949.

Wilder, Craig Steven. *Ebony and Ivey: Race, Slavery, and the Troubling History of America's Universities*. New York: Bloomsbury Press, 2013.

Wilds, John, Charles L. Dufour, and Walter G. Cowan. *Louisiana Yesterday and Today: A Historical Guide to the State*. Baton Rouge: Louisiana State University Press, 1996.

Williams, Heather Andrea. *Self-taught: African American Education in Slavery and Freedom*. Chapel Hill: University of North Carolina Press, 2005.

Willie, Leroy E. *A Look at Louisiana's First Century, 1804–1903*. Baton Rouge, LA: SAR Books, 2002.

Wilson, Samuel Jr. "Introduction." In James Gallier, *Autobiography of James Gallier, Architect*. New York: Da Capo Press, 1973.

———. "Manresa—Jefferson College, Convent, Louisiana." *Louisiana Architect* 9, no. 2 (February 1970): 10–12. Louisiana Research Collection, Tulane University.

Winters, John D. *The Civil War in Louisiana*. Baton Rouge: Louisiana State University Press, 1963.

Yarrut, Louis H. *The Reflections of a Jew at a Catholic Retreat*. N.p., 1964.

Index

Page numbers in italics refer to figures

Academy of Sciences, 84
Academy of the Sacred Heart (Convent, Louisiana), 42, 102, 115–16, 135, 150
Acadia, County of, 2, 109
Acadians, 2, 19, 26, 107–8, 136; within Creole class system, 20, 21–22, 107–8, 170; higher education, 46–47, 136, 172; the 1927 flood, 162–63
act of incorporation, Jefferson College, 187, 198
act of sale, Jefferson College, 92
act of support, Jefferson College, 187
act of transfer, Jefferson College, 112
Ad Majorem Dei Glorium, 12
admission practices: at Jefferson College, 44, 46; at St. Mary's Jefferson College, 116, 163
Aedes aegypti (mosquito), 74. *See also* yellow fever
African Americans, 16, 140; education of, 108–10, 114; racial discrimination, 21, 107, 160; first New Orleans Catholic church devoted to, 36. *See also* Creoles of Color; *placage*; servants; slaves
agricultural economic collapse of 1837, 4, 56
Aime, Edwige (daughter of Valcour Aime), 95, 98, 104
Aime, Félicie (daughter of Valcour Aime), 95–96, 98–99
Aime, Félicité Emma (daughter of Valcour Aime), 95, 98–99
Aime, Francois II (father of Valcour Aime), 94
Aime, Gabriel (son of Valcour Aime), 62, 95, 98; contracted yellow fever, 99–100
Aime, Josephine (daughter of Valcour Aime), 95, 98
Aime, Josephine Roman (wife of Valcour Aime), 95, 100
Aime, Michel (brother of Valcour Aime), 95
Aime, Valcour (Francois Gabriel), 3, 4, 12, 14, 42, 62, 106; act of transfer, 197–99; Civil War, role during, 103–4, 108; death of, 118; family life, 95, 98–100; freedmen, blocked integration of, 110, 111; Jefferson College, influence on, 102, 113, 119, *129*, 150, 160, 175; Louisiana College, purchase of, 4, 93, 101–2; plantation life, 94–98, 103, 108; property deed to college, 195–96; remains, relocation of, 173; secured free education for male relations, 116–17
Aime family tomb, 100, 173
Ainsworth, Harold R., 182
Algiers, Louisiana, 120, 124, 144
All Hallows College (Salt Lake City, Utah), 150
Alonzo de Castile (play), 143
alumni, 112; of Centenary College, 63; of St. Mary's Jefferson College, 140, 144, 149, 155, 165–66, 177; of St. Mary's Jefferson College pressed for support, 156, 159–60

Alumni Hall, 126, 144, 151, 155, 181
American Association for the Advancement of Science, 171
American District (New Orleans). *See* Faubourg St. Marie
American instructors (Anglo-American), 140
American Journal of Science and Arts, 62
American Revolutionary War, 34
Americanization, 5, 125, 176; of Jefferson College, 136, 139, 148–49, 167
ancienne population, 15, 53, 55, 100, 117, 118, 120, 136, 175; Creoles in Louisiana, history of, 14–22; decline of, 1, 140, 168, 177–78; educational ideology of, 6, 13, 31, 38, 45, 49–51; ethno-political struggles with Anglo-Americans, 6, 26–29, 37, 62, 67; Freemasonry, 45–46, 89–91; George Washington Cable controversy, 168–71; impact of Civil War on, 104–6; influence on Collège d'Orléans, 29–33, 37; influence on Jefferson College, 3, 39–40, 160; maintenance of French language, 32, 37; profiting from slavery, 24–25; social relations with *française étrangers*, 93; and state politics, 37–38; wealth through sugar, 22–26. *See also* Aime, Valcour; Burr conspiracy; Creoles, French; sugar barons
ancienne régime. *See* ancienne population
Andry and Bendernale (architectural firm), 154
Angers, France, 33, 111
Anglo-Americans, 3, 5, 13, 22, 31, 38, 41, 55, 57, 64, 70, 92, 104, 168, 176; education, 15, 32–33, 37; ethno-political struggles with French Creoles, 1, 6, 16–17, 18–20, 26–29, 32–33, 37, 40, 62, 67, 103, 175; Freemasonry, 45–46, 89–91; in New Orleans, 54, 62; at St. Mary's Jefferson College, 140–41; racism, 107–8, 171; as sugar planters, 25–26, 60, 136
Annals of Electricity, Magnetism, and Chemistry, 62
apostolic delegation to Cuba, 144
Appeal to the Colored Citizens of the World, 114
Archdiocese of New Orleans, 10
architecture: of antebellum structures, 2, 4, 7, 20, 44, 178; of Collège d'Orléans, 32; of French Quarter, 15–16; of Jefferson College, 43–44, 59–61, 175; of Manresa Retreat, 11–12; of St. Mary's Jefferson College, 151, 154
Ascension Parish, 2, 118
Ashland Plantation (Belle Helene Plantation), 61
Assumption Parish, 120
athletics, 4, 48, 141, 144; baseball at St. Mary's Jefferson College, 149; football games at St. Mary's Jefferson College, 148–49. *See also* recreation
Atlas Historique et Géographique de la France Dupuis les Temps les Plus Reculés Jusqu'a non Jours, 69
Audubon Park Sugar School, 136
Avegno, B., 102

Babin, E., 102
Bac (instructor at Louisiana College), 80–81
bachelor of arts degree, 30, 46, 77, 84, 122
bachelor of science degree, 122, 153
Bacot, Barbara Sorelle, 43
Baird, Absalom, 109
Banks, Nathaniel Prentice, 105, 109, 110
Barat, Mother General St. Madeline Sophie, 115
Barrouquere, Elizabeth, 177
Barry, John M., 162
bathing house, 47
Baton Rouge, Louisiana, 61, 65, 110, 138, 179; Louisiana State University relocated to, 140, 149; as northern border of the "gold coast," 1, 25–26, 60
Battle of New Orleans (1815), 32
Battle of Wagram, 52
Baude, Henry, 206
Baudier, Roger, 93, 166
Bauer, Craig A., 1
Bayonne, France, 64
Bayou Lafourche, 26
Bayou Teche, 26
Bay St. Louis, Mississippi, 149, 163
Beauvais, Armand Julie, 38
Bell, Caryn Cossé, 69
Bellanger, Father Henri (Marist president of Jefferson College), 10, 110–13, 125, *131*, 144, 172, 185, 197–99, 203; as educator of slaves, 114; as pastor of Holy Name of Mary Church, 124
Benedictines, 111, 154
Bienville, Jean-Baptiste Le Moyne de, 15
Biever, Albert, 156
Bigot, John Baptist, 138, 139, 185
Bladon Springs, Alabama, 99
Blanc, Antoine, 65, 81, 102
Blanc, Evariste, 36
Blenk, James Hubert, *132*, 158, 173; alumnus of Jefferson College, 5, 10; involvement with Loyola University, 155–56; president of St. Mary's Jefferson College, 143–45, 153; science hall named for, 154, 155
Blenk Science Hall, *134*, 154, 156, 181
Blessed Peter Faber Conference Room, 2

Blineau, Oliver, 102
"Blues," name for St. Mary's Jefferson College athletic teams, 148–49
Bocage Plantation, 61, 172
Bonaparte, Joseph, 34
Bonaparte, Louis-Napoléon (Napolean III), 72
Bonaparte, Napoleon, 27, 34, 52, 53, 96
bonfires on the levee, 138, 142
Bonny, Eugene, 194
Bon Séjour Plantation (Oak Alley Plantation), 43, 172
Bordeaux, France, 36, 69, 73, 98
Boudousquie, Adolphe, 191, 197
Boué, Father, 56
Bougère, Achille, D., 56
Bourbon Monarchy, 34, 36, 52
Bourgeois, A., 115
Bourgeois, Joseph, 199
Bourgeois, Lillian C., 4, 46, 105, 113, 140
Bourgoyne Street, 193
Boyd, David French, 140–41
Brackenridge, Henry Marie, 25
Bradshaw, Jim, 163
Brasseaux, Carl A., 19
Bridgewater, Massachusetts, 176
Bringier, Amedee, 61
Bringier, S. W., 102
Bringier family, 61
Brow, Seraphin, 102
Brunett, Louis J. B., 90
Bryant, Robert, 181
buildings: as archival sites, 6, 9; Collège d'Orléans, 36; Jefferson College, 43–44, 58, 60, 104, 112, *126*; Manresa House of Retreats, 2–3, 11–12, 172, 181, 183; St. Mary's Jefferson College, 150, 154. *See also* architecture
Buman, Nathan A., 19
Burnside, John, 106, 118–19
Burnside, Louisiana, 66
Burr, Aaron, 28
Burr conspiracy, 26–29, 35
Burthe, Dominique Francois, 41, 42, 52
Butin, Romain, 206
Butler, Benjamin Franklin, 105

Cabanocey: The History, Customs, and Folklore of St. James Parish, 140
Cabanocey Plantation, 41, 98
Cabinet de Lecture, 70
Cable, George Washington, 14, 17, 168–71
Caddo Parish, 161
Caedmus (author of "Ode to Jefferson College"), 178
Canby, Edward Richard Sprigg, 109
Canonge, Ernest, 46
Canova, Dominico, 55
Caprice Plantation, 14
Capuchin Friars, 31
Carondelet Street, 74
Carr (professor), 32
Carriere, Antoine, 102
Carriere, Oliver P. (judge), 10, 43, 182
Carroll, Reverend J., 205
Castellanos, Henry C., 32, 36, 93
Castellon, 73
catalogues (academic), 47, 87, 93, 120, 172; Louisiana College, 72, 74, 79, 93; St. Mary's Jefferson College, 120, 121–22, 146, 153–54, 156, 160, 163, 177
catechism, 46
Catholic Church: African Americans and, 36; French Creoles and, 6
Catholic Education Association, 161
Catholic Feast Days, 142
Catholicism: at Jefferson College, 46, 48, 102, 80–81; at Louisiana College, 79; revival of, 5; at St. Mary's Jefferson College, 4, 8, 137
Catholic Mass, 110; attendance of Creoles, 175; attendance of slaves, 114; at Jefferson College, 46, 48, 80; at Manresa Retreat, 184; at St. Mary's Jefferson College, 157, 159
Catholic Students: at Jefferson College, 46; at Louisiana College, 79–81; at St. Mary's Jefferson College, 8
Catholic University of Ireland, 143
Cemetery of Loyasse, 124–25
Centenary College of Louisiana (Louisiana Centenary College), 63–64
Centenary College of Mississippi, 63
Central Coal and Coke Company, Missouri, 180
Centre College, 40
Certificates of Mercantile Training, 46
Chamberlain, Jeremiah, 40
Chapelle, Placide-Louis, 142, 144
Chapel of the North American Jesuit Martyrs, 11
Charity Hospital (New Orleans), 84
Charters Street, 34
Charvet, Lucien, 36
Chateau d'If, 99
Chaurain, Father Stephen, 124
Chauvin, Reverend F., 205
Chevet, F., 62, 64
cholera, 70
Chopin Plantation, 14
Christmas, 118, 145; levee bonfires during, 138–39
church pew rental, 80

Citizens Bank of Louisiana, 66, 68, 79, 86, 191–93
Civil Digest of 1808, 29
Civil Rights Act of 1875, 107
Civil War: Jefferson College during time of, 3, 4, 110, 175; Louisiana and planters during time of, 91, 97, 103–6, 115, 152; Marists during time of, 110–12; and Reconstruction, 106–8. *See also* freedmen's schools
Claiborne, William Charles Cole: governor of the Territory of Orleans, 27–31, 33; promoted higher education, 29–32
Claire, E. J., 181
Clark, Daniel, Jr., 29
climate, Louisiana, 7, 22, 137, 138
Cline, Rodney, 3, 8, 14
Clouet, Alex de, 79, 84
Coca-Cola Company, 159
code noir, 114
Cohen, Michael David, 47
Colin, Father Jean-Claude, 110–11
Coliseum, Rome, 99
College Bank, 121
Collège d'Orléans, 7, 42, 47, 52–53, 93, 170, 175; administration of, 32–36; alumni of, 36–37;
College Infirmary, 140
College of Baton Rouge, 59
College of Belley, 111, 143
College of Louisiana: as an Anglo-American institution, 7, 37–38, 40–41, 92–93, 175; awarded federal monies, 54; Creoles critical of, 42; closure of, 63–64, 75, 87, 93; established by Claiborne, 29–31
College of St. Chamond, 110
College of Sts. Peter and Paul, 65
College of the Immaculate Conception, 65, 155
College of William and Mary, 30, 43
College Point, 7, 123, 143
Collins, Reverend J., 205
commercial curricula at Jefferson College, 121–22, 125, 144, 157
Confederacy, 103, 108, 113, 119
Confederate Army, 46, 104
Confiscation Acts of 1861 and 1862, 109, 113
Congregation of the Most Holy Redeemer (Redemptorists), 143
Conti Street, 70
Convent, Louisiana, 2, 10, 42, 68, 77, 112, 119, 134, 140, 172, 173, 180
Convention of Louisiana (1844), 15
Convent of the Sacred Heart, 42, 115, 116, 135, 150
Copes, Joseph S., 84, 87–88, 128
Coste, J., 111
Cottman, T., 102
cotton, 13, 42, 55, 66, 105, 137, 139
Cotton Centennial Exposition (1884 World's Industrial and Cotton Centennial Exposition), 169
Count of Monte Cristo, The, 99
Courier, 59
Courveille, Jean-Claude, 111
Covington, Louisiana, 154, 181
Covington Female Academy, 54
Crane, I., 115
Crane, J., 115
Creoles, French: Anglo-Americans and, 16–17, 26–27; constructed geneology of, 17–20; education of, 8, 31–32; *francaise etrangers* and, 20; political coalitions between French immigrants and, 27; social class and, 20–22; sugar and, 22–26. *See also* ancienne population; sugar barons
Creole, etymology of, 15–16, 169
Creoles of Color, 6, 16–17, 19, 168, 170–71
"Creoles of History and the Creoles of Romance," 169
Creoles of Louisiana, The, 17, 169
Crozet, Claudius, 52–56, 75, 185
Cuba, 57, 60, 96, 99, 139, 144
Cry, Eugene, 192
Cyr, Father Alcime, 161, 186

Dagneau, Father Phillip H., 10, 45, 137, 152, 157, 160, 162, 166, 177–78
Daily Delta, 72, 91
Daily Picayune, 72
Dakin, Charles Bingley, 61
Dakin, James Harrison, 61
Damaré, Dr., 79, 82, 84
Damaré, J., 115
Dargo, George, 29
Dartmouth College, 70
Dauphine Street, 59, 72, 75, 193–94
D'Avezac, Jules, 32
Davis, Robert W., 102
Dawdy, Shannon Lee, 20, 31
Dawson, John Charles, 27
De Boré, Jean Étienne, 22–23, 36, 171
Declaration of Independence, 92
Declouet, Alex, 84
De La Vergne, Hugues Jules, 54
Deléry, Simone de la Souchére, 33, 35, 53, 61
Delogny, Charles Robin, 46
Delogny Plantation, 14
De Marigny, Jean-Bernard Xavier Philippe, 39
Democratic Party, Anglo-American, 22

Dempster, Orton A., 73
Denegre, James D., 75, 79, 84, 191–94
Derbigny, Pierre Augustin Charles Bourguignon, 37, 40
De Sennegy, René (*Aldric Lettin* de *la* Peychardière), 39, 58–59, 109, 112, 115–16, 125
Destrehan Plantation, 172
discipline: at St. Mary's Jefferson College, 4, 120, 147, 158; at Jefferson College, 44, 48–49, 102
Dix, Dorothy, "Letter Box" advice column, 164
Dogué, Charles Oscar, 101, 102, 105, 185
Donaldsonville, Louisiana, 104, 120
Dooley, Edwin L., Jr., 43
Dominicans, 111
Doricourt, D., 193–94
Drake, Benjamin M., 63
Drexel, Katherine, 115
Druilhuet, Edmond, 192, 206
Ducatel, A., 194
Dufau, Camille, 69
Dufau, Cecile Street, 69
Dufau, Jules, 69
Dufau, Louis, 3–4, 9–10, 119, 173, 185; and conflicts with clergy, 80–83, 116, 172; death of, 91; as educator, 70; emphasized teaching of linguistics, 77; forged ties with University of Louisiana, 86–90; involvement with Masons, 89–91; negotiates Louisiana College financial problems, 78–80, 84–86; personal life of, 69–70; as president of Louisiana College, 68, 73–78, 175; relocated college to St. James Parish, 74. *See also* Street family tomb
Dufau, Marie, 69
Duffel, E., 115
Duffell, Albert, 102, 115
Dufont, N., 205
Dufossant, Joseph Soniat, 193–94
Dugue, Frs., Jr., 79, 84
Dugué, Oscar, 196
Dumaine Street, 95, 117
Dumas, Alexandre, 99
Duncan (First Municipal Council building doorkeeper), 70
Duncan, Patricia L., 44
Duplantier (legal witness), 194
Dupré, Jacques, 38, 42
Durel, Reverend U., 204
Dyer, John P., 37

Easley, John, 199
East Feliciana Parish, 7, 37, 38, 42, 63, 189
Ecole Impériale Polytechnique, 33, 52
Ecole Normale Supérieure, 33
Edgard, Louisiana, 181
education, higher, 13; history of in Louisiana, 8–11
Elba (Mediterranean island), 52
Elkins, Louis, 73
Emion, F., 73
Emory University, 43
English Common Law, 26
English garden, 95–96
Enrollments. *See individual colleges*
Ernest A. Carrere's Sons Real Estate Firm, 179
Esplanade Avenue, 70, 92
Eustis, George, 46
Evangeline: A Tale of Acadie (poem), 57
Everett, Alexander Hill, 56–59, 62, 185
Everett, Erastus, 70
Evergreen Plantation, 172
Exchange Alley, 70, 91

Faculty Disagreements, 172
Falgoust, Glenn, 99
Faubourg Marigny, 39, 54
Faubourg Tremé, 32, 36
Faubourg St. Marie (American District), 54, 62, 155, 176
Favre, Reverend Father Pére, 123, 137
Fay, Edwin Whitfield, 9, 40, 93
Felicité Plantation, 98, 106, 172
Ferchand, J. B., 194
Ferry, Alexis, 75, 79, 84, 98, 101, 113, 195, 197, 199
Fertel, Rien, 37
Ficklen, John R., 75
Field, Martha R. (Catherine Cole), 135
Fifteenth Amendment, 107
financial gifts and donations: to Jefferson College, 42, 62; to St. Mary's Jefferson College, 3, 115–16, 119, 160
Financial Panic of 1837, 54–55
Fire of 1842, 10, 175
First Municipal Council (lecture hall), 70
Fleming, Walter L., 141
Follett, Richard, 14, 24
Forstall, E. J., 103
Forstall, Victor, 102
Fortier, Alcée, 8–9, 76, 93, 119
Fortier, Estelle, 173
Fortier, Florent Louis, 98, 100–101, 112–13, 171, 195, 197–99
Fortier, Gabriel, 120
Fortier, James, 146
Fortier, Louis, 173

Fortier, Marie Felicité, 94
Fortier, Michel II, 95
Fortier, Michel Jean, 46, 94
Fortier, Septime, 75, 79, 84, 98, 101, 113, 195, 197–99
Fortier family tomb, 173
Fossier, Albert E., 18
Foulhouze, James, 90–91
Foyer Maconnique Lodge, 91
française étrangers, 18, 20, 27, 67, 68, 74–75, 90, 93, 137; hired by Louis Dufau, 72. *See also* French immigrants
Franklin, Benjamin, 70
Franklin College, 54, 64, 188, 189
Freedmen's Bureau (United States Bureau of Refugees, Freedmen, and Abandoned Lands), 108–10, 113
"Freedman's Case in Equity," 170–71
freedmen's schools, 106, 109
Freemasons, 45–46, 89–91
Free School Act of 1847, 73, 83
Freetown, Louisiana, 69, 91–92
French architecture. *See* architecture
French Creoles. *See* Creoles, French
French immigrants, 1, 6, 9, 18–20, 26–27, 29, 45, 54, 68–69, 72, 74, 85, 90. See also *française étrangers*
French legal custom, 26
French Quarter (Vieux Carré), 34, 78, 86, 117, 171; devastated by fires, 15–16; governed by French Creoles, 54; planters resided in, 25, 41; schools in the, 31, 36, 52, 72–74, 76, 92
French Revolutions, 27, 33, 35, 68
French Rite (Freemasonry), 46, 90
Frere, Adrien, 79

Gabriel Aime (riverboat/schooner), 96, 118
Gallier, James, Jr., (architect) 101
Gallier, James, Sr., (architect) 60–61
Garden District (New Orleans), 156
Gardere, Fergus, 102
Gardiner, Captain W. D., 147
Gast, Elinor Judith, 179
Gaud, Father Henry, 137, 185
Gaudet, J., 199
Gaudet, Martha, 138
Gaudet, T. R., 102
Gaunt (instructor in penmanship), 73
Gauthreaux, J., 115
Gayarré, Charles Étienne Arthur, 36–37, 93, 168–71
Gehman, Mary, 171
Generes, J. L., 102
Genois, Celestine, 193–94
Gens Du Couleur. *See* Creoles of Color
Genti, Mr. (creditor), 198
Gentil, Jean-Sylavain, 72
Georges, Father Henri, 153, 164, 186, 205–7
Georgetown University, 31
German immigrants, 119, 176
Gibbens, Durbin J., 182
Gipsey (Steamer), 77
Girod, Nicholas, 34
Girod Street Cemetery, 61
Godberry, J. W., 102
gold coast (Louisiana River Road region), 1, 26, 45, 78, 182; industrialization and change along the, 94, 140, 160; land seized by federal government, 109; plantations along the, 61, 135, 172; as site of Jefferson College, 64, 68, 75, 120, 125, 175; at time of Civil War, 103, 105, 106. *See also* River Road *and individual parishes*
gonorrhea, 98
Grand Coteau, Louisiana, 64–65, 154–55, 159, 181
Grandissimes: A Story of Creole Life, The, 169
Grand Lodge of Ancient Freemasons of South Carolina, 90
Grand Lodge of the State of Louisiana, 90
grand tour, 31, 62, 98, 99
Granet, Bernard, 52, 185
gratis education, 46
Green, Jennifer, 21
Green, F. (professor), 142
Grimes, Father John Joseph, 138, 185
Guerin, Widow, 193
Guillot, Joseph, 32
Guillot, Pierre, 35
Guizot's Law, 111
Gurlie, Claude, 32

Hamilton, Alexander, 28
Hamilton, William S., 38
Hardy, Emmett Louis, 179
Harington, Thomas, 194
Harris, Thomas H., 177–78, 179
Harris County Glenwood Cemetery, 92
Harvard College, 31, 43, 57
Harvard University Law School, 46
Hawaii, 139
Hebert, Ignace, 46
Hebert, Louis, 46
Hébert, Paul Octave (governor), 5, 75, 79
Hebert, R., 205
Henriette Delille Street, 36
Henry, Father Thomas, 139, 186

higher education. *See* education, higher
Hill, Mary Bernardine, 10, 40, 144
Holy Family, 11, 117
Holy Name of Mary Church, 124, 144
Holy Rosary Cemetery, 148
Hospital Street, 193
Hotel Monteleone, 150
Houmas House Plantation, 66, 118, 172
Howard-Tilton Memorial Library, Tulane University, 174
Hundred Days campaign, 52
Hunt, Alfred N., 35
Hunter, Robert E., 43
hurricane, 137

immigrant labor, 136
Imperial Artillery School, 52
indigo, 13, 22
Ingalls, Thomas R., 55–56
intermarriage: between cousins, 20; between Creole daughters and Anglo-American sons, 168; interfamily marriage, 95
interracial marriage, 171
Irish Chanel (New Orleans), 56
Irish immigrants, 19, 21, 26, 56, 60, 104, 106, 120, 136, 175; Creole disdain for, 119; and support for American politicians, 176
irritation of the brain. *See* meningitis
Italian immigrants, 19, 55, 136, 151–52, 160
Italy, students from, 120

Jackson, Louisiana, 63–64
Jacob, Emile, 198
Jacobs, Claude F., 114
Janssens, Jean-Baptiste, 182
Jeannine, Mary Tircuit, 140
Jefferson, Thomas, 3, 27, 30, 38–39, 43, 45, 48–49
Jefferson Athletic Club, 144
Jefferson College: Acadians and, 46–47; closure of, 64–67; construction of, 43–44; Creole values of, 45–47; curriculum, 49–51; daily schedule of students, 47–48; destroyed by fire, 58–59; discipline at, 48–49; English language at, 49; financial assistance of state legislature, 62–64; formation of, 3–4, 38–42; presidents of, 51–58; rebuilding of, 59–62; slavery and, 43, 47
Jefferson College Corporation, 113
Jefferson College High School, 156
Jefferson College Record, 146, 149–50
Jefferson Military College, 30, 43
Jefferson Military College Foundation, 30
Jefferson Parish, 69, 92, 196

Jesuits. *See* Society of Jesus
Jesuit High School (New Orleans), 180, 182
Jim Crow laws, 152, 160
John Burnside and Company, 118
Johns Hopkins University, 108
Johnson, Governor Henry, 37
Jones, Edward, 92
Jourdan, A., 115

Kansas City, Missouri, 180
Kantz, Joseph (captain), 146–47
Kappel, Vernon Elmer, 9, 76, 92
Kaslow, Andrew F., 114
Keith, Lee Boarman, 180
Keith, Richard Henry, 180
Kendall, John Smith, 93
Kenny, Michael, 180
Kerr, Donal, 111
King Bee Tobacco, 152–53

Labarre, L., 84
LaBarre, Valcourt, 75, 79
Labauve (creditor), 198
Labranch, Godegrande, 46
LaBranche, Similien, 46
Lachance, Paul F., 27
ladies' auxiliary, 161
Lafourche (steamer), 117
Lafourche Parish, 22, 105, 120
LaGarde, Lynn, 159
Lakanal, Joseph, 33–35, 52; linked to regicide, 35
Lambert, Pierre, 32
Landry, N., 115
language instruction, 122; ancient languages, 49–50, 154; and Dufau's French grammar text, 70–71; French versus English, 32, 56, 73, 77, 108, 152, 176; written language, 51
Lanusse, Paul, 32
Larose, Louisiana, 148
Laura Plantation, 172
Lapice, B., 103
La Tourrette's Reference Map of the State of Louisiana, 66, 93
Lavillebeuvre, E. F., 117
lead acetate, 99
League of the Sacred Heart, 143
Leaumont, Walter, 182
LeBourgeois, Joseph, 102
LeBreton, Dagmar Renshaw, 1, 6, 14, 99, 104, 117, 177
Leche, Richard Webster, 182
Leliévre, Jacques-Felix, 70–71
Le Maitre, Delphine, 170

Le Petite Versailles (Aime plantation), 4, 95, 96, 99–1, 104, 117–19; destroyed in fire, 173
Leroy, Gustave, 91, 191–94
Letchford, W. H., 102
letter censorship (censorship of the press), 72
levees, 21, 35, 96, 104, 173; bonfires on the, 138–39, 142; flood of 1927, 162–63; repairs of, 137–38
liberal arts curricula, 3, 5, 57, 76; Creole preference for, 6–7, 49, 136; at Jefferson College, 49, 51; at St. Mary's Jefferson College, 120–22, 124, 137, 144, 161, 172; and "The Fathers of the Society of Mary," 201
Lincoln, Abraham, 103, 105–6
Lisbony (freemason), 91
Lock, Colonel Samuel H., 120
Long, Huey Pierce, 179
Longfellow, Henry Wadsworth, 57
Louis-Charles, Duke of Normandy and Prince Royal of France, 34
Louisiana Agriculturalist and Mechanics Association, 7
Louisiana Artillery, 46
Louisiana College (New Orleans): curriculum, 73; *française étrangers* hired as instructors at, 72; relocated to Jefferson College campus, 75– 88, 92–93; yellow fever impacts, 74–75
Louisiana College (St. James Parish), 3–4, 9, 67–68, 93, 117, 120, 139, 185; Civil War, during, 101, 105, 116; closure of, 92; connection with University of Louisiana, 85–88; curriculum and language, 77–78; Dufau's legacy at, 174–75; financial troubles at, 83–85; relocated from New Orleans to, 75–76; returned as Jefferson College, 119; role of religion at, 79–83; tuition increase at, 79
Louisiana Constitution of 1812, 37
Louisiana Constitution of 1852, 83
Louisiana Constitution of 1864, 113
Louisiana Courier, 73
Louisiana Historical Society, 169
Louisiana legislature, 53, 56, 59, 64, 163, 176; against school punishment practice, 44; appropriation of funds towards education, 31, 38, 40, 46, 62–63; divisions between French Creole and Anglo-Americans in the, 33, 37–38, 40; and the "free school act," 73, 83; granted charter for Louisiana College, 75, 86; and St. Mary's Jefferson College, 121, 145, 179
Louisiana Office of State Parks, 64
Louisiana Ordinance of Secession, 103
Louisiana Purchase, 18, 30, 45
Louisiana Real Estate and Development Company, 161
Louisiana State Board of Education, 156
Louisiana Statehood, 1812, 29
Louisiana State House of Representatives, 46
Louisiana State Seminary of Learning, 64, 120
Louisiana State University, 64, 136, 140–41, 152, 157, 182; as athletic rival to St. Mary's Jefferson College, 148–49
Louisiana Sugar Planter, 136
Louisiana Sugar Planters Association, 136
Louis Philippe I, 97, 111
Louis XVI (Louis Auguste), 33–34
Lowry, R. P., 165–66
Loyola, St. Ignatius, 3, 12, 65; house of prayer named for, 11; and Manresa, 166, 181–83
Loyola Hall, 2
Loyola University New Orleans, 149, 155–57, 166, 172
L'Université et Ses Adversaries, Quelques Observations Sur le Libelle Intitulé "le Monopole Universitaire, Destructeur des Lois et de la Religion," 69
Lusher, Robert, 47, 55, 64
Lutcher High School, 165
Lycée Louis-le-Grand, 69
Lyman, John Baptiste, 92
Lyon, France, 65, 124, 143

Macacona, Costa Rica, 157–58
Madame John's Legacy (townhouse), 95
Madonna and Child (painting), 101
Maisounabe, Jean Baptiste Leon, 64–66, 166
Malarcher, J., 115
Manresa, Spain, 181
Manresa Board of Director's Building Committee, 183
Manresa House of Retreats, 2, 10, 172, 178; description of retreats at, 183–84; renovations and repairs to, 182–83
Manresa House of Retreats (Corporation), 182
Manuel des Verbes Français, Réguliers et Irréguliers, 69
Maquire, Father William, 161–62
Marc, Saint, 73
Mardi Gras, 145
Marie Antoinette, 34
Marion, Father Abbe, 112
Marist Act of Incorporation, 1875, 201–4; charter established, 124

Marist College (Atlanta, Georgia), 150
Marist Corporation, 164, 180
Marists. *See* Society of Mary
Marist tombs, 172
Marksville, Louisiana, 120
Marquis de LaFayette, Marie-Joseph Paul Yves Roch Gilbert du Motier, 34
Martin, P., 115
Martin and Himel Law Offices (Lutcher, Louisiana), 181
Masons. *See* Freemasons
Matrana, Marc M., 97
Mazureau, Etienne (attorney general), 41–42, 52, 187
McCulloch, Gary, 112
mechanization of sugar industry, 1
Medical College of Louisiana, 55, 87
Melancon, N., 102
Melson, Caye A., 182
meningitis, 92
Metairie, Louisiana, 161
Micelie, Sam A., 182
Miles, E. H., 160
military instruction: military cadet training, 141, 146; military training, 78–79, 146; Student Army Training Corps unit established, 147
Military Operations of General Beauregard, The, 46
Miller, Randal M., 114, 143
Millet, Widow Pierre, 205
mineral rights, 180, 182, 206
Mire, E. C. E., 103
Mississippi Conference of the Methodist-Episcopal Church, 63
Mississippi Department of Archives and History, 30
Mississippi Gulf Coast, 120
Mississippi River: bricks of clay out of, 12; College Point located along the, 7; Jefferson College and importance of, 42–43; Jefferson College students bathing in, 47
Mississippi River Flood of 1927, 4, 125, 164
Mississippi Territory, 29–30
Miss S. S. Hull's Young Ladies' Institution, 74
Mobile, Alabama, 36, 65, 120, 149
Moore, Governor Thomas Overton, 103
Moniteur du Sud, 70–71
Montegin, Sidney, 139
Morand Plantation, 32
Moreau-Liset, Louis, 32
Morel, Octave, 91
Morgan, Lee, 29, 63
Morning Star, 121,156
mosquitoes, 10, 73–74, 99, 137
Mossy, Jules, 102
Mount Vesuvius, 99
Murphy, Tim, 183
Murrillo, Bartolomé Esteban, 101

Napoleon House (restaurant and bar in New Orleans), 35
Natchez, Mississippi, 30
National Register of Historic Places, 64, 183
National Republican Party, 22, 37
Nelson, William Hamilton, 32
New England Educational System, 176
New Orleans, Louisiana, 5, 15, 23, 31, 34, 41–43, 52, 55, 59–60, 84–85, 112, 117, 121, 138, 143, 155, 169, 171, 191, 197; antisecessionists, site of, 91; Catholic education in, 155–56, 166; cemeteries in, 173; Claudius Crozet in, 54; Collège d'Orléans, 7, 31–32, 35–37; districts of, 54; Dufau in, 69–72, 77, 86–87, 89, 91; and the economic panic of 1973, 136; education system in, 108, 115, 176; Eliza Ripley in, 96; Faubourg Marigny, 39; flood of 1927, 162–63; freedmen's schools, as site of, 109; Jesuit High School of, 180–82; Louisiana College, as site of, 72–76, 78; Loyola University, as site of, 172; Masonic lodges, as site of, 45–46, 89–91; *placage,* as site for, 16–17; planters in, 8, 25, 47, 95; Rex Parade, 145; Territory of Orleans, 15, 25, 27–28; Union troops in, 104–5; 1884 World Exposition in, 169; yellow fever in, 99
New Orleans As It Was: Episodes of Louisiana Life, 36
New Orleans College. *See* Collège d'Orléans
New Orleans fires (1788 and 1794), 15
New Orleans Masonic Library, 41
New Orleans Medical Association, 84
New Orleans Province of the Society of Jesus, 180, 182
New York Charities Aid Association, 171
New York City: Masonic Lodges in, 90; New York industrial exposition of 1953, 94
New Year's Eve, 138
Nicholas, Robert Carter, 41
Nicholls, Francis Redding Tillou, 142
Niehaus, Earl Francis, 9
Norman, Benjamin Moore, 93
Norman's Chart of the Lower Mississippi River, 93
Northampton, Massachusetts, 171

northern teachers, 109, 176
Nottoway Plantation, 172

oak allées, 43, 116
Oak Alley Plantation, 43, 172
Oakland College (Alcorn State University), 84
O'Brien, Michael, 25, 45, 117
Oceania (South Pacific) Islands, 111
Ockman, Widow Antoine, 206
"Ode to Jefferson College," 178
Odin, Jean-Marie, 110, 137
oil, drilling for, 161, 180, 206
Oliver, John, 102
Old Creole Days, 169
Old State Capital Building (Baton Rouge), 61
orchestra, St. Mary's Jefferson College, 142, 146
Orleans High School, 70
Orleans Parish, 68, 75, 101, 143
Osiris (Masonic) Lodge, Number 8, 90
Ostini, Father Anthony, 183
Our Lady of Good Counsel Building, 2, 183

paddle-wheel ships, 42
Paigot, G. B., 204
Parent, Aimee Francoise, 41
Paris, Father Auguste Simone, 80
Paris, France: Crozet transferred to, 52; Dufau taught in, 69; Felicie Amie dies in, 100; Lakanal in, 34
Paris Public Education Committee, 33
Patin, Marie Louis, 40, 95
Peltier, Harvey, 12
Père Lachaise Cemetery, Paris, France, 36
Perilloux (creditor), 198, 199
Persac, Marie Adrien, 93
Peterkin, Darryl L., 6
Petit, Father Joseph, 161, 186
Petiville (creditor), 198
Peytavin, John L., 112, 177
Pfleger, E., 165
Phi Kappa Sigma Fraternity, 88
Philippines, 139
Pilié, Joseph, 43
Pinson, Félix, 36
Pitot, James, 32
placage, 16–17
Plaisance, Henry J., 147–48
Plaisance, Irvin P., 182
planters. *See* sugar barons
Plaquemines Parish, 105, 120, 162
pneumonia, 36, 118
Point Coupee, Louisiana, 120
political feud, 27
Pompeii, 99
Pope Clement XIV, 111
Pope Gregory XVI, 110
Pope Leo XIII, 144
Pope Pius VII, 111
Pope Pius IX, 81
porters' lodges, 2, 11, 44, 126, 142
Portraits Littéraires: Nouvelle Orléans, 71
Powell, Lawrence N., 16, 29, 169
Power, Edward John, 9
powerhouse: Manresa workmen's shed, 2; powered St. Mary's Jefferson College electricity, 153
Précis Historique des Régences, Précédé d'une Introduction et Suivi des Pièces Justificatives et Documents Historiques et Législatifs, 69
preparatory department, 5, 76–78, 122, 126, 156–57
president's house, 3, 11, 44, 78, 126; as College Hotel, 78; as St. Ignatius House of Prayer, 11
Price's Water Queen (riverboat), 143
Princeton University, 41
Procès ou Assassinat juridique de Louis XVI (trial or legal murder of), 34
Protestants, 26, 37, 40, 45, 63, 79
public high schools, 157
public lectures, 70–71, 169
Puerto Rico, 139, 144

Quinn, Father Peter F., 158–59, 161, 186

racial segregation, 110, 140
Randall, David, 38
Rapides Parish, 64
Rapier, Father George James Horton, 139, 162–64, 185–86
Reconstruction, 1, 5, 8 16, 21, 105–8; Creole planters at time of, 176; and Freedman's Bureau, 109; and racial segregation, 140, 152; St. Mary's Jefferson College at time of, 115–16, 120, 125
recreation: in college catalogues, 144; daily recreation periods, 47–48; football and baseball, 148; music, 141–42; recreation rooms, 44, 181. *See also* athletics
Reeves, William D., 10
religious education, 31; African Americans and, 114; at St. Mary's Jefferson College, 113, 198
religious retreats, 11, 159
Renaudier, O., 204
Rex Parade (School of Design), 145
Rich Bend, 14

Richbend Plantation, 98
Richardson, William, 12
Richmond Academy, 29
Riddell, John Leonard, 84
Riddell, William Pitt, 73, 84, 87–89
Rightor, Henry, 93
Rilleux, Norbert, 24
Ripley, Eliza, 96
riverboats, 42, 142, 151. *See also individual names of vessels*
River Road, 1–3, 154, 178
Robert, A., 115
Robert, Alexie, 197
Robertson, Samuel M., 46
Rochefort, M. M. Nicolas, 32–33, 36–37, 170
Rockaway Marine Pavilion, Long Island, 61
Rodrigue, John C., 24
Roland, Charles P., 45, 104, 178
Roman, Alfred, 46, 101, 113, 191–95, 197
Roman, André Bienvenu (governor), *127*; Creole education, supporter of, 38, 75; death of, 117; Jefferson College, involvement in, 41–44, 110–12; Louisiana College, inaccurately credited for opening, 93; tomb of, 117, 173
Roman, Jacques Etiènne, 40
Roman, Jacques Telesphore, 95
Roman, Joseph, 120, 173
Roman, Josephine, 95
Roman, L., 103
Roman, Onesime, 120
Roman, Sidney, 120
Roman family, 108
Roman plantation, 1, 66, 95, 111
Romeville High School, 165
Roselius, C., 102
Royal Street, 41, 52, 91
Rudd, Robert Mark, 42
"rule of silence," 183
Rummel, Archbishop Joseph Francis, 166
Rutgers University, 70

Sacher, John M., 38
Sacred Heart of Jesus Chapel, 2, *130*
Saint Domingue (Haiti), 22; immigrants from, 27, 29, 32, 36, 43, 69; slave revolt of 1791, 24
Salter, Father John M., 180
San Francisco Plantation (St. Frusquin), 56, 172
Saxon, Lyle, 162
Scarborough, William Kauffman, 106, 119
scarlet fever, 91
Schmandt, Raymond H., 8
"School of Design, The" (Rex Parade), 145
School of Fine Arts, Dublin, 60
scientific curricula, 153, 172
Scribner's Monthly, 169
Scottish Rite (Masonic Lodge), 89. *See also* Freemasons
Scottville, Texas, 120
Sears, Claudious Wistar, 87–89
Sears Roebuck catalogs, 152–53
Second District of Public Schools (New Orleans), 84
Second Plenary Council of Baltimore, Maryland, 1864, 115
secularism: antebellum era, 6; Creoles, French, and, 45, 48, 175; education shift from, 38, 184
segregation in schools, 108–9, 113–14, 152, 160. *See also* freedmen's schools
Seminary of Saint Magloire of the Doctrinaires, 33
Seré, V., 115
servants, 19, 47; quarters of, 140
Shannon, Reverend Mother Anna, 115
sharecropping, 1, 8, 105–8, 136
shareholders: of Jefferson College, 102, 108, 113, 115, 195–97; of Jefferson College opposed by Marists, 123–24; of Louisiana College, 79, 83
Sharon Female College, 84
Shaw, John Angier, 176
Shaw, John William (archbishop), 135, 158–59, 163–64
Sheppard, C. M., 102
Shreveport, Louisiana, 64
Shugg, Roger W., 21, 31
Simon, Edouard, 79, 84
Sisters of the Blessed Sacrament, 115
Sisters of the Sacred Heart, 110–11; city named in honor of, 42; and education, 31; established freedmen school, 115; missionary work, 110, 114
Sitterson, J. Carlyle, 26, 31, 108
slaves, 1, 6, 16; and Creoles, 15, 16–17; as free labor on sugar plantations, 14, 22, 24–25, 96, 170; as freedmen, 103–7, 115; and Freedmen's Bureau, 108–9; as integral to class system, 19–22, 39; at Jefferson College, 43–45, 47, 78; racial segregation of, 114, 140; ruled by the *Code Noir*, 114; sexual relations between masters and, 170; shared church pews with students, 80; uprisings, 24, 27, 29
Smith, Father Richard H., 147, 186
Smith, J. Frazer, 168
Smith, Thomas Ruys, 169
Social Life in New Orleans: Being Recollections of my Childhood, 96

Société Catholique d'Education Religieuse et Littéraire (Catholic Society for Religious and Literary Education), 180
Society of Jesus (Jesuits), 110–11; early interest in Jefferson College, 64–66; Jefferson College, purchase of, 166, 180; Jefferson College campus, renamed buildings on, 181; Loyola University New Orleans, founding of, 155; Manresa Retreat, 2, 4, 10, 181–82, 184; sale of St. Mary's Jefferson College, 205–7
Society of Mary (Marists): drilled for oil, 161; Jefferson College, acts of transfer, 197–99; Jefferson College to, transfer of, 10; Society of Jesus, akin to, 111; St. Mary's Jefferson College, administration of, 116, 120, 123, 125, 152, 156, 177; St. Mary's Jefferson College, established, 3, 4, 8, 111–13, 124; St. Mary's Jefferson College governing board, literary conflict with, 115, 123–24; St. Mary's Jefferson College, sold, 179–81sugar parishes, experiences in, 110, 137–38; and impact of 1927 Flood on the College, 162–64, 172–73
Solar, F., 69
South American students, 152, 160
Southern Art Union, 171
Southern Association of Colleges and Schools (SACS), 5, 161
Southwestern Louisiana Industrial Institute (University of Louisiana at Lafayette), 149, 157
Spanish architecture, 15–16
Spanish in New Orleans, 15–16
spiritual exercises, 3, 66, 181–84
Spring Creek Academy, 64
Spring Hill College, 65, 149
St. Alphonsus Church, 56
St. Andrew Masonic Lodge, Number 5 (New Orleans), 90
St. Ann Street, 31
State Item, 160
State School for the Deaf, Baton Rouge, 179
State Teachers Association, 84
Stations of the Cross, 12, 117, 184
statues, 11; Mary; 12, 137; St. Ignatius Loyola, 12; St. Joseph, 12, 137
St. Augustine's Church, 36
St. Bernard Parish, 105, 162
St. Charles College, 64–65, 155, 159
St. Charles Hotel, 60
St. Charles Parish, 94–95, 105
St. Claude Avenue, 36
St. Claude Street, 32
steamboats. *See* steamships
steam calliopes, 42, 143
steam engine power, 24, 49, 95
steamships, 1, 57, 77, 80, 96, 112, 117, 151. *See also* riverboats *and individual vessel names*
Stephens, Reverend A., 30
St. Francis Xavier, 181
St. Hélène, Island, 34, 53, 96
St. Ignatius House of Prayer, 11, 181
St. James Catholic Church, 100, 112, 117, 173
St. James Parish, 1, 2, 8; enslaved population in, 25; historic documents in, 9
St. James sugar refinery, 95, 100, 119
St. John the Baptist Parish, 103, 120
St. Joseph, 11, 137
St. Joseph Plantation, 98, 172
St. Joseph's Church: Paulina, Louisiana, 164; Thibodaux, Louisiana, 12
St. Joseph's College, 31, 40
St. Joseph's School for Freedmen, 115
St. Louis Cathedral (New Orleans), 173
St. Louis Cemetery: Number 1, 171, 173; Number 2, 173; Number 3, 173
St. Louis Exchange Hotel, 59
St. Louis Street, 70, 91
St. Martin Parish, 22, 105
St. Mary's Brass Band, 122, *133*, 141–42, 145–48; linked to cadet corps, 146
St. Mary's Cannoneers, 46
St. Mary's College: Baltimore, Maryland, 31, 40; Dunbalk, Ireland, 143; Ilchester, Maryland, 143
St. Mary's Hall, 2, 131
St. Mary's Jefferson College, 3–4, 8, 124, 135; alumni of, 165–66; athletics at, 148–49; brass band, 141–42; curriculum at, 121–22, 137, 143, 153; decline of, 159–65; dining hall constructed, 151; diocesan support wanes, 156; early years, 115–16; expansion of, 140; first Marist college in the US, 113; grounds of, 137; levee bonfire tradition initiated by, 138–39; memoirs regarding, 10; powerhouse fire at, 153; Student Army Training Corps at, 147–48; student newspaper, 149–51; teaching certificates granted by, 145; Valcour Aime instrumental in, 119; varied student body at, 136, 141, 152–53
St. Michael's Academy, 115
St. Michael's Church, 93, 110, 160, 172, 181; Creole members of, 151–52; Dufau controversy with, 80–83; Jefferson College stu-

dents attended, 46, 48, 80; Marist control of, 110–12; provided theological instruction to Jefferson College students, 56, 137
St. Michael's Parish, 151–52
St. Michael's Parish Cemetery, 172
St. Michael's Parochial School, 181
Story, Sydney, 5
Storyville (red-light district, New Orleans), 5, 171
St. Patrick's Church, 61
St. Peter Canisius Hall, 2
Straight University, 92
Street, Angelique Leroy, 69
Street, Timothy Benedict, 69
Street family tomb, 173–74
St. Stanislaus College, 149, 163
St. Stephens, Mississippi, 30
Student Army Training Corps (SATC), 147–48
student recreation (St. Mary's Jefferson College), 44, 47–48, 141, 144, 148, 181; bathing, 47, 183; choir, 143; debate team, 143, 148; orchestra, 142, 146; publication venues, 177
sugar barons, 1, 4, 7–8, 15, 20, 24–25, 39, 45, 49, 93–167
sugar cane, 1, 14, 22–24, 45, 96
sugar parishes, 2, 4, 8, 14, 20, 26, 41, 54–57, 62, 67–68, 78–86, 102, 115, 138, 160, 162, 163, 167, 172, 175; educational institutions in the, 109; Marists in the, 110; residents of, 103–5, 107
sugar planter debt, 21, 54–55, 106, 118, 139
sugar planter wealth, 1, 4, 14, 20, 22–26, 95
Supreme Council of 33ds for the Sovereign and Independent State of Louisiana (a division of Scottish Rite freemasons), 90

teaching curricula (toward teaching certificate), 145
Teddy (student), letter to his mother, 157–58
Teinturier (professor of mathematics), 33
Terrebonne Parish, 22, 105, 120
Testut, Charles, 71–72, 93
Tewksbury, Donald G., 7
Thelin, John R., 7–8, 174
theology, 33, 45, 80
Thibodaux, Louisiana, 12
Thibodeauxville, Louisiana, 120
Thirteenth Amendment, 107
Tholomier, Father Claude Anthony, 80–83, 172
Thorndike's Principles of Teaching, 145
Thouvenin, Matthew, 132, 146, 151
Thouvenin Dining Hall, 151, 181, 186, 126
Times Picayune, 9, 59, 162, 164
Tinker, Edward Larocque, 93
tobacco: agricultural crop, 13; smoking and chewing of, 39, 49
Toledano, Ben, 102
Toledano, Roulhac B., 96, 119
Torrey (mathematics instructor), 73
"To the Friends of Catholic Education," 160
Tourrette, John La, 66
Trahan Architects, 183
Transylvania College, 31
Trask, Benjamin H., 74
Treaty of Versailles, 148
Tregle, Joseph G., 15
Tremé, Claude, 32
Tremé (*faubourg*). *See* Faubourg Tremé
Troop of Horses (St. Charles Parish), 95
Trudeau, Chas F., 177–78
Trudeau, René (captain), 95
tuition, 5, 44, 46, 86, 115, 136; at Jefferson College, 56, 102; at Louisiana College, 78; at St. Mary's Jefferson College, 115–16, 150, 160–62, 177
Tulane, Paul, 119
Tunnel, Ted, 103
tutors, 5, 27, 176; Gabriel Amie toured Europe accompanied by a, 98; French as preferred language for, 176, 187–88; hired by French Creoles, 31, 83, 95; French immigrants as, 72
typhoid, 123

Une Paroisse Louisianaise, 109
uniforms, 146
Union Association of New Orleans, 91
Union troops: housed at college, 104; relocation of, 113; occupation of Jefferson College by, 3, 104
University of Angers, 33
University of Edinburg, 73
University of France, 69, 73
University of Louisiana (Tulane University), 7, 70, 73, 84, 92, 119; connection with Louisiana College, 85–87, 89; Dufau appointed to faculty of, 88
University of Virginia, 43, 48, 108
Ursuline Nuns, 31, 111
Ursuline Street, 36, 193
US War Department, Jefferson College Student Army Training Corps and, 147–48

Vacherie Grade School, 165
vacuum panning: St. James Sugar Refinery

and, 95; sugar industry and method of, 24, 49
Valcour Aime's Guards, 104
Varnado, Otto S., 9
Vatican, Rome, 99
Vavasseur, Jean, 42
Vavasseur Plantation, 42
Vella, Christina, 62
Vermillionville, Louisiana, 120
Versailles, France, 94
Vieux Carré. *See* French Quarter
Villa Gonzaga Retreat House, 181
Virgin Mary, 11, 173
Voltaire, 45, 50

Wakelyn, Jon L., 20
Walker, David, 114
Waguespack, Wilhelm J., 46
War of 1812, 32, 95
Washington, DC, 28, 31, 164–65
Washington, Henry Augustine, 19
Washington, Mississippi, 30, 43
Washington and Lee College, 9
weather. *See* climate
Webre, Antoine, 81
Webre, B. S., 102
Webre, Edgard, 206
Weber, Father N. A., 165
West Point (United States Military Academy), 52–53, 55, 87
West Point fatigues, 142, 146
Whig Party, 22, 37, 57
White Hall Grade School, 165
white supremacy, 16, 171
Whitney Plantation (Habitation Haydel), 56, 172
Wilkinson, James (general), 28–29
Wilson, Samuel, Jr., 10, 43, 60
Winchester, Mrs., 103
Winters, John D., 104
Wlute's Art of Teaching, 145
World's Industrial and Cotton Centennial Exposition of 1884, 169
World War I, 147, 157

Xavier, Louis Stanislas (Count of Provence, Louis XVIII), 34
Xavier Hall, 181, 183

Yale College, 31, 41, 73
Yarrut, Louis H., 179, 184
Yazoo and Mississippi Valley Railroad, 151
yellow fever, 42, 73–76, 78, 99
York Rite (Freemasonry), 46, 90